D1000010

BIOSTRATIGRAPHY OF FOSSIL PLANTS

James M. Schopf, 1911-1978

This symposium volume is dedicated to James M. Schopf who effectively bridged areas of Botany such as plant morphology, anatomy, systematics and nomenclature with areas of Geology such as stratigraphy, historical paleontology and the formation of coal. James M. Schopf was a steadfast friend, counselor and supporter of those who studied fossil plants.

QE
904
.A1
B56

BIOSTRATIGRAPHY OF FOSSIL PLANTS

Successional and Paleoecological Analyses

Edited by

David L. Dilcher
Indiana University

and

Thomas N. Taylor
Ohio State University

Dowden, Hutchinson & Ross, Inc.
Stroudsburg Pennsylvania

328564
Tennessee Tech. Library
Cookeville, Tenn.

The editors are indebted to William H. Gillespie for his help with typesetting and proofreading.

Copyright © 1980 by **Dowden, Hutchinson & Ross, Inc.**
Library of Congress Catalog Card Number: 79-27418
ISBN: 0-87933-373-1

All rights reserved. No part of this book may be reproduced or transmitted in any form or by any means—graphic, electronic, or mechanical, including photocopying, recording, taping, or information storage and retrieval systems—without written permission of the publisher.

82 81 80 1 2 3 4 5
Manufactured in the United States of America.

Library of Congress Cataloging in Publication Data

Main entry under title:
Biostratigraphy of fossil plants.
Includes index.
CONTENTS: Banks, H. P. Floral assemblages in the Siluro-Devonian.—Phillips, T. L. Stratigraphic and geographic occurrences of permineralized coal-swamp plants-upper Carboniferous of North America and Europe.—Pfefferkorn, H. W. and Gillespie, W. H. Biostratigraphy and biogeography of plant compression fossils in the Pennsylvanian of North America. [etc.]
1. Paleobotany. 2. Paleontology, Stratigraphic. I. Dilcher, David L. II. Taylor, Thomas N.
QE904.A1B56 560'.1'7 79-27418
ISBN 0-87933-373-1

Distributed world wide by Academic Press, a subsidiary of Harcourt Brace Jovanovich, Publishers.

CONTENTS

PREFACE

Paleoecology may be defined as the discipline that attempts to reconstruct ecosystems of the geologic past and their changes through time. Most of the methods utilized in ecology are theoretically applicable in paleoecology. However, the paleoecologist works with a volume of rock that contains components (fossils) that are preserved only in a spatial relationship; and it is these components that are utilized in reconstructing the paleoecosystem. "Paleosuccession," that is, progressive changes in the paleoecosystem, is indicated by the fossil assemblages found in a succession of facies along a vertical axis. It is in the concept of vertical succession of taxa that biostratigraphy has its foundations. Historically, paleoecology developed concomitantly with systematic paleobotany, and today is more dependent upon taxonomy and systematics than ever before. One of the first to employ fossil plants as indicators of paleoecology was Lyell, who recognized plant-bearing beds as evidence of a terrestrial environment. Another significant contribution was made by Heer, who recognized subtropical floral elements in Tertiary sediments of the Arctic, and commented upon the vertical and geographic variability of a vegetation being closely associated with climatic change. From that point onward, paleoecology rapidly developed, with important contributions being made by Reid and Chandler who were among the first to demonstrate the importance of community analysis in paleoecology. Because of the nature of the fossils they worked with (fruits and seeds), these authors also had to consider features associated with transport phenomenon prior to deposition. In the United States, paleoecology flourished just after the turn of the century—in part because of the discovery of numerous excellent fossil floras; improved techniques of working with fossil plants; and an emerging awareness by researchers of the important interface between ecology, evolutionary biology, and geology. Today paleoecologists must not only consider the systematics of an organism under investigation, but also must take into account such factors as transport, geochemistry, biological adaptation, diagenesis, site of burial, rate of sedimentation, characteristics of the source vegetation, and regional geology, to new a few. These variables, all of which must be evaluated in paleoecological research, become compounded when biostratigraphic correlations are attempted. In the case of plant megafossils, the problems are even more difficult because of the general fragmentary nature of the "species" being investigated, and the numerous problems associated with determining whether a paleosuccessional species truly reflects genetic change or constitutes environmentally induced modification.

The papers selected for this volume not only encompass different periods of geologic time and plant megafossils preserved in differing modes, but also represent different approaches used in the analysis of fossil plants relative to their paleoecology and biostratigraphy. Each contribution demonstrates how a basic set of variables may be analyzed to provide understanding of the paleoecology and paleosuccession of an assemblage of plant megafossils.

The paper by Banks, "*Floral assemblages in the Siluro-Devonian,*" is an excellent presentation of the "big picture" needed in considerations of early land plant evolution and biostratigraphy. The author has established seven generic assemblage-zones, beginning in the Late Silurian (Pridolian) and extending to the Famennian (Upper Devonian). These have been designated I—*Cooksonia*, II—Zosterophyllum, III—*Psilophyton*, IV—*Hyenia*, V—*Svalbardia*, *VI–Archaeopteris*, and VII—*Rhacophyton;* and, like other biostratigraphic indexes, these designations transgress time rock units in some instances. Banks believes that, to be of value biostratigraphically, each fossil assemblage must be analyzed relative to the level of morphological evolution (e.g., terminal versus adaxial sporangial position, leafless stems versus microphyllous or megaphyllous stems). He further underscores the necessity of extensive collecting and the application of modern techniques to the study of plant megafossils. The potential liabilities of using plant megafossils in Devonian biostratigraphy are also discussed. Principal among these is the importance of precise taxonomy based upon complete character suites. The biostratigraphic application of Devonian plant megafossils is only now becoming fully recognized for its potential. There can be little doubt that as paleobotanists continue to piece together an understanding of the earliest land plants, their utility as stratigraphic tools will be increased. This may be most important for the earliest forms, which are often small and possess a low degree of organ differentiation and tissue complexity.

This volume contains two papers that consider Upper Carboniferous (Pennsylvanian) vegetation, with the contributions differing in the types of fossils analyzed and the biostratigraphic techniques employed. In the paper by Pfefferkorn and Gillespie, "*Biostratigraphy and biogeography with plant compression fossils in the Pennsylvanian of North America,*" the importance of impression-compression fossils as biostratigraphic markers is discussed. These authors indicate that because the sequence of first occurrences of new species is quite similar throughout North America and Europe, the fossils provide an excellent method of correlating Pennsylvanian sediments of the Amerosinian floral realm. Small differences are evident in impression-compression floras during the Lower and Middle Pennsylvanian, with greater differences occurring during the Upper Pennsylvanian. However, these differences are not great enough to mask similarities in the floras from different geographic regions. They suggest that there were four floral provinces in North America during the Pennsylvanian (Acadian Province, Inter-Appalachian Province, Cordilleran Province, Oregonian Province), each distinguished by a small number of endemic taxa. Because Pennsylvanian plant megafossils are not restricted to coal-bearing sequences, they are common enough to be stratigraphically useful over large geographic areas. In addition to a discussion of the biostratigraphic application of impression-compression floras, Pfefferkorn and Gillespie present an extensive bibliography of Pennsylvanian floras for twenty-eight states, northern South America, and eastern Canada.

Structurally preserved Pennsylvanian plants preserved in coal balls are the megafossils discussed by Phillips in "*Stratigraphic and geographic occurrences of permineralized coal-swamp plants—Upper Carboniferous of North America.*" These fossil plants have provided an enormous amount of basic information about the anatomy, morphology, and evolution of Pennsylvanian vascular plants because of the exquisite preservation of their tissue systems. In this contribution, the geographic and stratigraphic positions of coal balls have been plotted in five major regions, including the Donets Basin of the U.S.S.R.; European countries; and the American Appalachian, Eastern Interior, and Western Interior regions. Moreover, the author has indicated the position of various coal-swamp taxa within several selected coals that may be utilized as ecological indicators for the establishment of general patterns of Pennsylvanian swamp vegetation. The genera and species of five major groups of vascular plants (lycopods, ferns, arthrophytes, seed ferns, and cordaites) are recorded for each stratigraphic level (coal) in which they are known. The biostratigraphic value of coal-ball plants will become increasingly useful as ecological parameters are more accurately defined. It is important to note that these Pennsylvanian vegetation-swamp complexes may provide a method whereby some of the smaller, less well-defined coals can be more accurately positioned stratigraphically.

Possibly one of the more neglected aspects of plant megafossil biostratigraphy involves the complex nature of depositional phenomenon, and the sorting of plant fossils prior to, and during, fossilization. Because of load sorting, plant fossil assemblages may reflect a highly biased sample source for the vegetation, which may greatly alter concepts about the ecological and evolutionary history of a particular assemblage. For example, lacustrine sediments deposited in a small basin will contain a rather uniform population of leaves, while in a larger basin, the increased distance from shore will result in a greater number of sun leaves being present in the assemblage. In addition, small plants will not be able to transport parts by wind as far as taller plants; understory plants in a closed canopy may also be poorly represented as fossils for the same reason. Such physical and biological factors, which may ultimately influence a fossil sequence, are discussed by Spicer in his contribution titled, "*The importance of depositional sorting to the biostratigraphy of plant megafossils.*" The author makes special note of the necessity of comparing suites of fossils that have undergone similar depositional histories. This most important, and generally overlooked factor, is especially critical in dealing with fine divisions of geologic time.

The contribution by Schopf and Askin, "*Permian and Triassic biostratigraphic zones in southern land masses,*" considers the value of plant fossils as biostratigraphic indexes in correlating Permian and Triassic sediments in the southern continents. Because zonation has been based upon tetrapod fossils, which in many areas are lacking or poorly represented, plant fossils, if properly interpreted, provide possible stratigraphic markers for use in correlation with other indexes. After defining the problem, the authors discuss the various types of biozones as defined by the *International Stratigraphic Guide* (Hollis D. Hedberg, ed.). The remainder of the paper presents a review of both plant megafossils and microfossils for the southern continents, and points out those that may be most significant in biostratigraphy. For now, palynomorph assemblage zones and megafossil range zones provide the best method of correlation for the southern continents.

Plant megafossils of Triassic age are also discussed by Ash, "*Upper Triassic floral zones of North America.*" Upper Triassic sediments in North America are known to contain at least 140 well-characterized plant species. Despite the abundance of these species, there has been little previous attempt to differentiate these rocks into zones based upon plant megafossils. In this contribution, Ash defines three floral zones for the Upper Triassic. The *Eoginkgoites* Zone is present at the base of the Chinle Formation in the southwest and slightly lower in the Newark Group of the eastern United States. The *Dinophyton* Zone is slightly younger and includes an extensive flora in numbers of species. It is delimited by the loss of *Eoginkgoites* and the first appearance of *Dinophyton*. The flora of the third zone is poorly known and dominated by ferns and cycadophytes. These three Upper Triassic floral zones have been correlated with the Triassic stages in Germany using pollen grains and spores, and provide important biostratigraphic information that may expand the usefulness of plant megafossils in other geographic areas.

The Succor Creek Flora is a composite assemblage of at least a dozen Miocene florules from Oregon and Idaho. In the paper by Taggart and Cross, "*Vegetation change in the Miocene Succor Creek Flora of Oregon and Idaho: A case study in paleosuccession,*" the authors have recorded variations in the vegetation and plotted these data against the stratigraphy of the individual florule sites. Stratigraphic control has been aided by pollen types arranged in paleoassociations. These include: montane, conifer, bottomland-slope forest, swamp, pine, and xeric. A sixth category includes undetermined pollen types, or those that could not be associated with the other five types. Megafossil taxa were assigned to these paleoassociations and a vegetational model developed for the area. These data suggest that the climax forest communities of Succor Creek involved mesic deciduous components in the valley bottoms and along low slopes. Montane conifer elements persisted on higher slopes. Taggart and Cross suggest that volcanic ash falls and gas venting were responsible for the disruption of the climax forest vegetation. The approach presented in this paper provides an appropriate framework in which to consider paleosuccession sequences of Tertiary vegetation, and in a very real sense, more accurately interpret complex community types like those existing today. Such studies provide an opportunity for the paleoecologist to determine whether changes in the vegetation resulted from biological succession or were the result of physical processes.

Some of the most exquisitely preserved angiosperm remains have been recovered from middle Eocene (Claiborne Formation) sediments in the southeastern United States. The fossils typically occur in disjunct clay and lignite lenses that are believed to represent abandoned channel fillings on an ancient flood plain. One of the principal problems in working with these fossils has been the inability to accurately locate the fossil beds stratigraphically. In their contribution, "*Biostratigraphic analysis of Eocene clay deposits of Henry County, Tennessee,*" Potter and Dilcher have found some angiosperm leaf types that are specific for a particular clay lens or pit. In other instances, certain cuticular features, such as trichomes, may ultimately provide an important fine-stratigraphic marker. Studies of this type can be useful in formulating model systems that will allow phylogenetic trends to be distinguished from paleosuccessional stages and local ecological differences.

Chemical fossils have not been generally perceived as biostratigraphic markers, even though they fulfill all of the requirements necessary to be useful. Recent technical advances have made it possible to examine complex organic compounds that are present in rocks of varying geologic age. Although all organic compounds produced by organisms are not uniformly preserved as chemical fossils, the analysis of certain compounds, or groups of compounds, provides an opportunity to biostratigraphically zone the rocks in which the organic residues are found. Brooks and Niklas, "*The chemistry of fossils: Biochemical stratigraphy of fossil plants*," discuss compound extraction techniques and provide several examples of the potential use for such techniques. In addition to providing base-level data that compliment the results obtained from other techniques (e.g., transmission and scanning electron microscopy and elemental analysis), chemical fossils may also provide valuable information about biochemical relationships among certain fossils, which may assist in more completely understanding their paleoecology and evolution.

There can be little doubt that plant fossils will play an increasingly important role in biostratigraphy in the years to come. There are still numerous problems that must be solved by each investigator during the analysis of a fossil assemblage if the geographic and stratigraphic correlations are to be valuable. Nevertheless, as the papers in this volume so graphically demonstrate, paleobotanists are now evaluating fossil plants as once dynamic biological components that were affected by all of the factors influencing their environment and their deposition. The willingness of investigators to apply multidimensional approaches in dealing with plant fossils promises that, in the years ahead, a fuller understanding of fossil plant assemblages will lead to more accurate discussion of community interactions and a better understanding of paleosuccession.

David L. Dilcher
Thomas N. Taylor

LIST OF CONTRIBUTORS

Sidney R. Ash
Department of Geology-Geography
Weber State College
Ogden, Utah 84408

Rosemary A. Askin
Institute of Polar Studies
The Ohio State University
Columbus, Ohio 43210

Harlan P. Banks
Division of Biological Sciences
Cornell University
Ithaca, New York 14853

Jim Brooks
Geochemistry Section
The British National Oil Corp.
150 St. Vincent Street
Glasgow G2 5LJ
Scotland

Aureal T. Cross
Department of Botany and Plant Pathology
Michigan State University
East Lansing, Michigan 48824

David L. Dilcher
Department of Biology
Indiana University
Bloomington, Indiana 47405

William H. Gillespie
West Virginia Department of Agriculture
Charleston, West Virginia 25305

Karl J. Niklas
Division of Biological Sciences
Cornell University
Ithaca, New York 14853

Hermann W. Pfefferkorn
Department of Geology
University of Pennsylvania
Philadelphia, Pennsylvania 19104

Tom L. Phillips
Botany Department
University of Illinois
Urbana, Illinois 61801

Frank W. Potter, Jr.
Department of Biological Sciences
Fort Hays State University
Hays, Kansas 67601

James M. Schopf (deceased)
Department of Geology and Mineralogy
The Ohio State University
Columbus, Ohio 43210

Robert A. Spicer
Department of Botany
Imperial College of Science and Technology
Prince Consort Road
London S.W.7
Great Britain

Ralph E. Taggart
Department of Botany and Plant Pathology
Michigan State University
East Lansing, Michigan 48824

1

FLORAL ASSEMBLAGES IN THE SILURO-DEVONIAN

Harlan P. Banks

SUMMARY

This paper is a preliminary attempt at a biostratigraphic classification of Late Silurian and Devonian strata on the basis of plant megafossils. Seven biozones (generic assemblage-zones) are proposed tentatively: I, *Cooksonia* Zone—late Ludlow through Pridoli; II, *Zosterophyllum* Zone—Gedinnian through middle Siegenian; III, *Psilophyton* Zone—middle Siegenian through Emsian; IV, *Hyenia* Zone—Eifelian into early Givetian; V, *Svalbardia* Zone—early Givetian through late Givetian; VI, *Archaeopteris* Zone—Frasnian through middle Famennian; VII, *Rhacophyton* Zone—middle Famennian through Tn la (of Tournaisian). For each zone, both the characteristic genera and numerous newly evolved morphological characteristics are given. The paper emphasizes the sources of error in such a classification and the work that needs to be done in order to increase the precision of the zonation. The zones proposed are compared with those suggested by some palynologists.

INTRODUCTION

Devonian Floras by E. A. Newell Arber was published posthumously in 1921. In this brief book Arber made the first attempt to record the stratigraphic occurrence of Devonian megafossils. From his lists of plants and the strata in which they occurred he concluded that one could recognize an older, *Psilophyton*, and a younger, *Archaeopteris*, flora. Arber and later Halle (1916, p. 4) recognized that these two floras differed more from one another than did the *Archaeopteris* flora from those of the Lower Carboniferous. Arber's analysis supported the then developing concept that Devonian land plants were the most primitive of vascular plants and that evolutionary change characterized the Devonian Period.

Kräusel (1937) refined Arber's work by distinguishing three successive floras during Devonian time: An Early—*Psilophyton*, a Middle—*Hyenia* and a Later—*Archaeopteris* flora.

Leclercq (1940) supported Kräusel's tripartite division but suggested that *Protopteridium* replace *Hyenia* as a name for Middle Devonian assemblages because *Protopteridium* included a larger number of species with a wider geographic distribution. Ironically, Leclercq and Bonamo (1971) reduced several species of *Protopteridium* to one and (1973) proposed a new generic name, *Rellimia*, for that species.

My own experience with Devonian vascular-plant megafossils parallels that of many others (e.g., Seward, 1931, p. 334) who have recognized that fossil plants are not confined by stratigraphic boundaries (Figure 1.1). Nevertheless, papers in which plant fossils are used in biostratigraphic analyses are increasing in

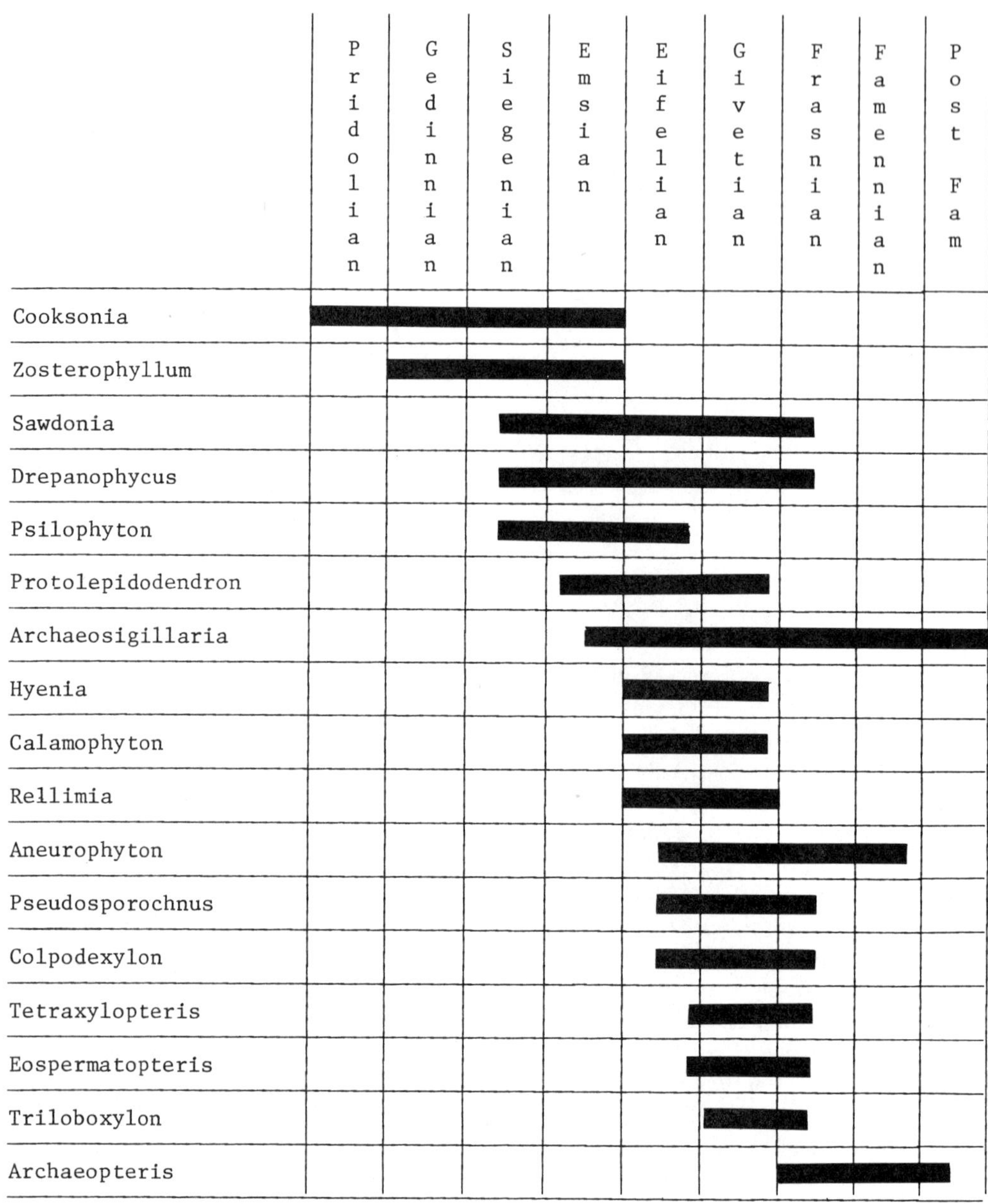

FIGURE 1.1 *Selected genera of Devonian plants whose range extends across Stage boundaries. (In many cases a single species is involved. All genera are referred to in the text and the citations that accompany them provide the data for this figure. See also the Note for Table 1.1).*

both number and sophistication (e.g., Remy and Havlena, 1962; Phillips, Peppers, et. al., 1974; McGregor, 1977; Retallack, 1977). The following account is a preliminary attempt to find some biostratigraphic order in the rapidly accumulating data on Devonian plants. Similar attempts have been made by Petrosian (1968) and Senkewich (1968). Perhaps the inadequacy of these efforts will stimulate a search for the kind of precise straitigraphic data that are needed for a more satisfactory zonation.

A survey of reasons why the present effort is so tentative will serve a useful purpose if it suggests directions for future research.

1. Paleobotanists have been concerned more with morphology, anatomy, and taxonomy than with stratigraphy. As a result, some collections of Devonian megafossils are recorded as coming from Lower, Middle, or Upper Devonian strata or from one or another Stage of Devonian time (Table 1.1), intervals too extended for precise biostratigraphy. Other collections are recorded from certain formations but rarely does one know precisely the level within the formation.
2. Even when the formation yielding megafossils is known, its age in terms of international standard stages is poorly understood.
3. The major cause of difficulty in the correlation of fossiliferous horizons arises from the terrestrial (or continental) nature of the most productive deposits. These were inland and perhaps intermontane deposits (e.g., the Rhynie chert), or were former lakes, ponds, or streams on a coastal plain (e.g., the Catskill area of eastern New York State). They are now preserved chiefly as lenses of varying size. Lacking critical invertebrate fossils, the marine equivalents of these terrestrial deposits are difficult to decipher. Hope for the future rests heavily on palynology (see, for example, McGregor, 1977; Richardson, 1974). The dearth of data on the position of fossiliferous lenses within a given formation results in large measure from the random scattering over the Devonian coastal plain of the ponds and lakes whence the lenses originated. The lenses are uncovered sufficiently to permit mass collecting only when quarrying, road-building, dam-building, or similar large-scale activities coincide with the lens. Even then, the paleobotanists collecting actively in the area often fail to arrange the coordination with stratigraphers that could lead to detailed mapping of the formation and the precise position of the fossiliferous lenses. This will be corrected only when its significance becomes apparent to all workers.
4. Megafossils of plants, unlike those of many invertebrates, are usually preserved as *fragments* of varying size.

TABLE 1.1 *Subdivisions of Silurian and Devonian periods with approximations of time of beginning and duration. Right column shows seven tentative floral generic assemblage zones.*

System	Series	Stage	Zone	Began (m.y)	Dur. (m.y)	Plant megafossil generic assemblage-zones
Mississippian ?		Tournaisian	Tn lb	?		
			Tn lb (part)			?
	Upper		Tn la	345		*VII Rhacophyton*
		Famennian		353	8	
		Frasnian		359	6	*VI Archaeopteris*
Devonian	Middle	Givetian		365	6	*V Svalbardia*
		Eifelian		370	5	*IV Hyenia*
	Lower	Emsian		374	4	*III Psilophyton*
		Siegenian		390	16	
		Gedinnian		395	5	*II Zosterophyllum*
Silurian	unnamed	Pridolian		405	10	*I Cooksonia*
	Ludlow			415	10	
	Wenlock			425	10	
	Llandovery			435	10	

Modified from Harland et al., 1964; Martinsson, 1969, Cocks et al., 1971; House et al., 1977.

NOTE: Detailed studies of numerous groups of organisms, particularly conodonts, cephalopods, and plant spores, are demonstrating that some strata usually assigned to the Tournaisian are probably of Devonian age. The Belgian Tn 1a and lower Tn 1b appear to be of post-Famennian Devonian age. This likelihood is shown in the table by a query at the Devonian-Mississippian boundary (see Streel, 1972; Sandberg et al., 1972; Richardson, 1974; House et al, 1977, pp. 5, 9; Streel and Traverse, 1978; McGregor, 1979).

The importance of this point is treated thoroughly by Beck (1970).

a. Some fragments have been described solely from an examination of immediately visible features that vary with preservation. These features may represent a true outer surface or simply an exposed inner surface such as the cortex. The observed differences may result in two generic names for a single plant. Application of paleobotanical techniques may reveal features such as epidermal cell patterns and again result in the description of a new taxon because the pattern is unknown for any preexisting taxon. Differences in the preservation of vascular tissue (coaly compressions vs. cellular permineralizations) may also lead to the erection of two taxa when, if the two had been preserved in the same way, one taxon would suffice. For example, Edwards and Richardson (1974) described a new genus, *Salopella*, that has many of the characteristics of *Rhynia*. But because *Salopella*, a compression fossil, lacks all the tissues that are preserved in the permineralized *Rhynia*, they erected a new taxon.

b. Fragments are found occasionally in organic connection. Others share specific morphologic features that provide convincing evidence that they were parts of one plant. Still others have been recorded as parts of one plant solely on the basis of a belief that they must have been together. The result is poor interpretation. One example is the description of *Eospermatopteris* (Goldring, 1924) as an arborescent dawn seed fern. The stumps, to which nothing has ever been found attached, today retain the name *Eospermatopteris*. Goldring's unattached branching systems are now regarded as *Aneurophyton* (Krausel and Weyland, 1923). The unattached "seeds" were shown by Kräusel and Weyland (1935b) to contain spores and are thus sporangia of unknown affinity. The "microsporangia" are probably inorganic markings.

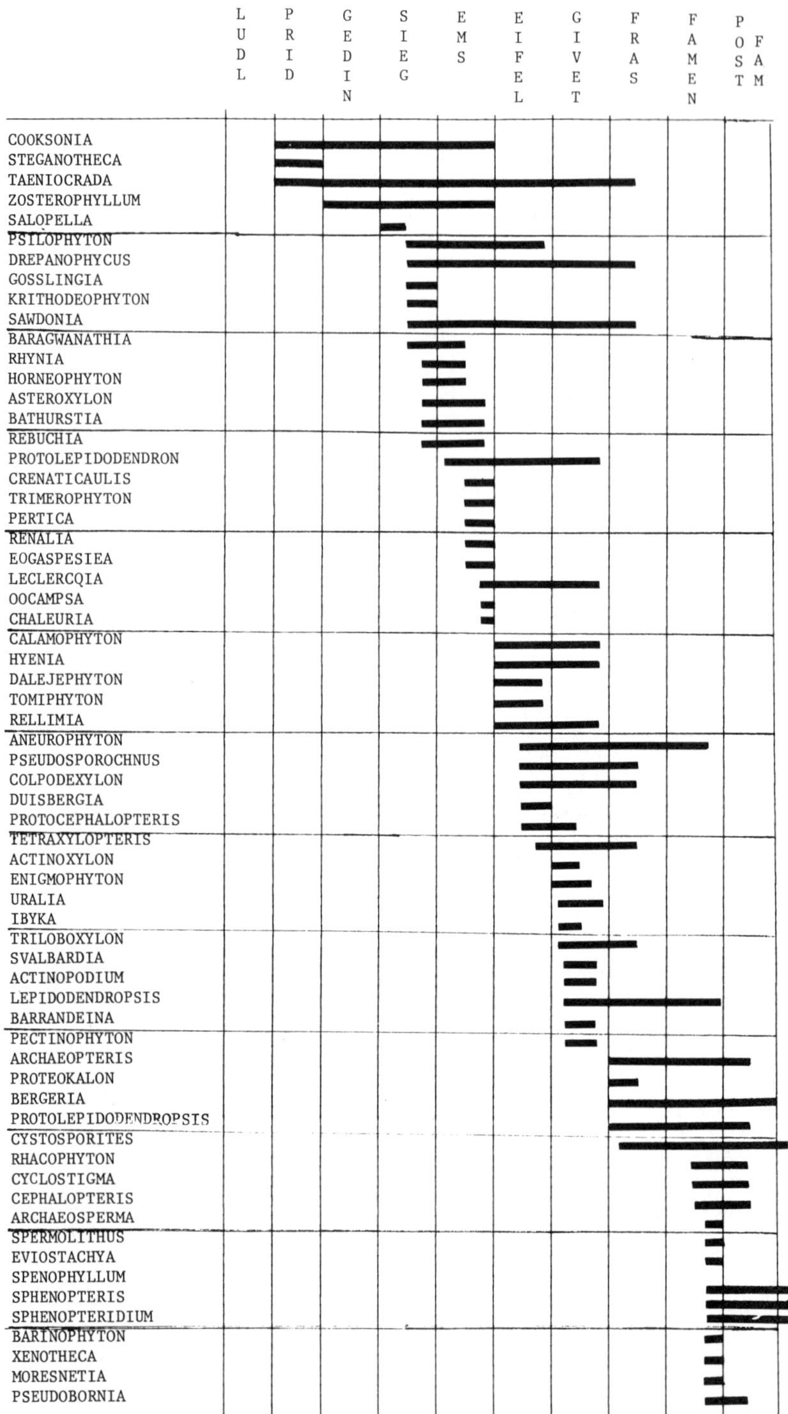

FIGURE 1.2 *Stratigraphic ranges of some genera of Devonian megafossils. (Data are from the sources cited under the various assemblage-zones. See Note for Table 1.1).*

The above explanations demonstrate why the listing of all supposed species that are recorded at a given level in time can yield a completely misleading interpretation. Edwards and Richardson (1974) also discuss these difficulties in the study of Devonian plants. Richardson (1974) emphasizes the problems that confront a stratigraphic palynologist. They apply equally well to those concerned with megafossils. Therefore, in this preliminary survey I propose to select taxa that are fairly well documented, taxa whose stratigraphic position is well or moderately well known, taxa whose geographic distribution may have some breadth, and taxa with which I have some degree of familiarity. Obviously this is a biased survey. I propose further to limit the discussion to generic assemblages because (1) many Devonian genera are monotypic, at least in the literature if not in real life, and (2) many species of Devonian plants are founded on minor characteristics that would be insignificant in living organisms. Omissions from my assemblages are mainly intentional at this stage in the development of a biostratigraphy of Devonian plants. Approximate stratigraphic ranges of most of the plants mentioned are given in Figure 1.2.

The fragmentary nature of Devonian plants suggests that supplemental evidence might be useful in the recognition of assemblage-zones. For example, seeds have been found in Devonian strata but only in the Famennian Stage. Thus, isolated seeds can provide just as good an indication of Famennian strata *(Assemblage-zone VII, Table 1.1)* as do the sphenopsids *Eviostachya* and *Pseudobornia*. The same holds for other morphological features that can suggest at least the oldest assemblage zone to which they can be assigned. For example, fragments of secondary xylem as currently understood are no older than Eifelian, lateral sporangia no older than early Gedinnian, adaxial sporangia no older than middle Siegenian. That is the rationale for including isolated morphological or anatomical characters along with generic names when characterizing an assemblage-zone.

In order to avoid the implication that all Devonian plants are poorly understood as whole plants, let me cite two examples to dispell that myth. The Givetian plant *Leclercqia complexa* Banks, Bonamo, and Grierson, 1972, yielded almost all the morphological, anatomical, and histological data one could desire in an extant plant. Subsequent papers (Bonamo and Grierson, 1973; Grierson, 1976) further expanded the understanding of ontogenetic development of spores and the detailed histological structure of tracheids. We now have a nearly complete picture of *Leclercqia complexa* as a trailing, herbaceous lycopod. *Archaeopteris* was thought for many years to be an Upper Devonian fern known widely around the world by its morphology, and *Callixylon* was thought to be a Late Devonian gymnosperm known by its anatomical structure. Beck (1960) found the two in organic connection. Carluccio, Hueber, and Banks (1966) reported the anatomical structure of *Archaeopteris* and Beck (1971, 1977) has continued to add significant details. The plant was a tall tree with an extensive root system. It is now called *Archaeopteris*, a highly evolved, well understood progymnosperm whose distribution was world-wide in Frasnian and Famennian time.

ASSEMBLAGE-ZONE I

Cooksonia Zone
Late Ludlovian to end of Pridolian
(ca. 406-395 mya)

The genus *Cooksonia* occurs at several horizons through the Pridolian (Downtonian) of Wales (Lang, 1937; Edwards, 1979). Two other genera of

TABLE 1.2 *Comparison of tentative megafossil generic assemblage-zones in Devonian strata with palynological zones suggested by McGregor (1977) and Richardson (1974). See Note for Table 1.1*

Series or Stage	Present Paper	McGregor, 1977	Richardson, 1974
Post Famennian Tn 1b Tn 1a	?		 *V. nitidus*
FAMENNIAN	*Rhacophyton* Assemblage—zone VII		*V. pusillites* *S. lepidophytus* *L. cristifer*
FRASNIAN	*Archaeopteris* Assemblage—zone VI		*optivus—bullatus*
GIVETIAN	*Svalbardia* Assemblage—zone V	*devonicus—orcadensis*	*Triangulatus* *Densosporites devonicus*
EIFELIAN	*Hyenia* Assemblage—zone IV	*velata—langii*	*Rhabdosporites langii* *Acinosporites acanthomammillatus*
EMSIAN — UPPER	*Psilophyton*	*annulatus—lindlarensis* / *Grandispora*	*Calyptosporites biornatus—proteus*
EMSIAN — LOWER	Assemblage—zone III	*annulatus—lindlarensis* / *sextantii*	*Emphanisporites annulatus*
SIEGENIAN		*caperatus—emsiensis*	*Dibolisporites cf. gibberosus*
GEDINNIAN	*Zosterophyllum* Assemblage—zone II	*micrornatus—proteus*	*Emphanisporites micrornatus* *Streelispora newportensis*
PRIDOLIAN	*Cooksonia* Assemblage—zone I	chulus—?vermiculata	Synorisporites tripapillatus

vascular plants occur in Pridolian strata, *Steganotheca* (Edwards, 1970a) and *Taeniocrada* (Obrhel, 1962). Edwards and Davies (1976) have recently reported *Cooksonia*-like fossils from late Ludlovian strata in Wales thus marking the base of Assemblage-zone I (Table 1.2). At present there are no acceptable reports of older megafossils of vascular plants. The genera named belong to the subdivision Rhyniophytina (-opsida) Banks, 1968 and Edwards, 1970a. Zone I corresponds closely to the tentative spore zones of McGregor and of Richardson (see Table 1.2).

Morphological characteristics of plants in the *Cooksonia* Assemblage-zone are:

small, simple, little differentiated vascular plants

dichotomous branching

no leaves

no roots

terminal sporangia (sparse)

slender xylem strand (perhaps centrarch based on analogy with younger fossils)

homospory

Localities where the *Cooksonia* assemblage has been collected include Czechoslovakia (Obrhel, 1962), New York State (Banks, 1973), Podolia (Ishchenko, 1969), and possibly Libya (Daber, 1971). In each of these areas there appears to be one occurrence of the assemblage only. In Wales (Lang, 1937) the assemblage has been found at several geographic localities and several horizons throughout the vertical extent of Pridolian strata. Welsh specimens are the only ones from which histological and anatomical data and spores have been obtained.

ASSEMBLAGE—ZONE II
Zosterophyllum Zone
Gedinnian to middle Siegenian
(ca. 395-380 mya)

Zosterophyllum (see Edwards, 1972), a new type of vascular plant, Zosterophyllophytina (-opsida) Banks, 1968, appears near the base of the Gedinnian. The rhyniophytes *Cooksonia* and *Taeniocrada* are still present and a new representative, *Salopella* (Edwards and Richardson, 1974;) is found in upper Gedinnian or lower Siegenian strata. The same authors point out that the late Gedinnian-early Siegenian flora has more in common with older (early Gedinnian) than with younger (middle Siegenian) floras. Zone II corresponds closely to McGregor's *micrornatus-proteus* Zone but less well to Richardson's zones (see Table 1.2).

Morphological characteristics of plants in Zone II include those listed for Zone I as well as lateral sporangia; H branching; and probably, exarch primary xylem (although this character has not been demonstrated earlier than middle Siegenian time).

Localities yielding floras of the Zone II type include Scotland (Edwards, 1972), Spitsbergen (Høeg, 1942), and Czechoslovakia (Obrhel, 1968), where they appear to be Gedinnian. In Belgium and Wales, the age is between late Gedinnian and early Siegenian (see Edwards and Richardson, 1974, and papers cited therein).

ASSEMBLAGE-ZONE III
Psilophyton Zone
Middle Siegenian to end of Emsian
(ca. 380-370 mya)

The middle Siegenian saw the advent of many more and varied vascular plants. Two new subdivisions appeared, Lycophytina (opsida) Banks, 1968, and

Trimerophytina (opsida) Banks, 1968. Plants of uncertain taxonomic position were represented by *Krithodeophyton* (Edwards, 1968).

Rhyniophytes were represented by *Cooksonia* and *Taeniocrada*, zosterophylls by *Zosterophyllum* and two new variants, *Sawdonia* (see Hueber and Banks, 1967; Hueber, 1971a) and *Gosslingia* (see Edwards, 1970b). *Psilophyton* is the first trimerophyte (see Banks, Leclercq, and Hueber, 1975), *Drepanophycus* the first lycopod (see Stubblefield and Banks, 1978), and *Krithodeophyton* represents the enigmatic barinophytes.

Later in the Siegenian, or perhaps at the start of the Emsian, the *Rhynia* types, *Rhynia* and *Horneophyton*, are found in Scotland, *Yarravia* and *Hedeia* of the Australian Siegenian-Emsian may prove to be more advanced than typical rhyniophytes. At the same time the lycopods *Asteroxylon* and *Baragwanathia* appear in Scotland and Australia respectively.

By Emsian time the new rhyniophytes *Eogaspesiea* (Daber, 1960) and *Renalia* (Gensel, 1976) appear on the Gaspé Peninsula; new zosterophylls are *Rebuchia* (see Hueber, 1970), *Crenaticaulis* (Banks and Davis, 1969), and *Bathurstia* (Hueber, 1971b); new trimerophytes are *Trimerophyton* (Hopping, 1956) and *Pertica* (Kasper and Andrews, 1972); new lycopods are *Protolepidodendron* (see Kräusel and Weyland, 1932), *Leclercqia* (see Kasper, 1977, abstr.) and *Kaulangiophyton* (Gensel, Kasper, and Andrews, 1969). Near the close of Emsian time two possible precursors of progymnosperms appear. They are *Chaleuria* (Andrews, Gensel, and Forbes, 1974), which appears to demonstrate incipient heterospory, and *Oocampsa* (Andrews, Gensel, and Kasper, 1975). Both genera are found in New Brunswick, Canada. Their age is based on studies of associated spores (McGregor, pers. comm., 1978). Zone III corresponds less well to spore zones which are more precisely defined (Table 1.2). Several new morphological characteristics appear in Zone III:

branching more abundant and more complex

pseudomonopodial branching a result of overtopping; (Some branches appear to be lateral branches, determinate in their growth so far as can be determined.)

enations present on some zosterophylls

microphylls present on lycopods

rhizomes (plants in Rhynie chert)

roots still absent unless structures in the zosterophylls *Crenaticaulis* and *Gosslingia* represent adventitious roots (see Banks and Davis, 1969)

adaxial sporangia (in lycopods, Kräusel and Weyland, 1935a)

first paracytic stomata (*Drepanophycus* — Stubblefield and Banks, 1978)

fertile branching systems terminated by masses of sporangia (Banks, Leclercq, and Hueber, 1975)

only homospory observed on spores preserved in situ

complex dehiscence mechanism observed in *Zosterophyllum* (see Edwards, 1969a)

centrarch primary xylem (Banks, Leclercq, and Hueber, 1975; *Krithodeophyton* — Edwards, 1968; Hueber, 1968)

exarchy in lycopods (Fairon-Demaret,

1971) and in zosterophylls (Edwards, 1969a, b; 1970b)

incipient heterospory (*Chaleuria* — Andrews, Gensel, and Forbes, 1974)

In the present state of our knowledge of late Siegenian through Emsian time there seems to have been a steady progression of evolutionary innovations. It is possible that subsequent, more detailed, stratigraphic studies may suggest subdividing this *Psilophyton* Zone somewhere near the Siegenian-Emsian boundary and possibly also higher in the Emsian (see Richardson's zonation in Table 4).

Localities that have yielded Assemblage III are world-wide in distribution (see Banks, 1975). They include many localities in North America (see Andrews et. al., 1977), Spitsbergen (Høeg, 1942), Norway (Halle, 1916), Wales (Edwards, 1968, 1969a), Scotland (Lang, 1932; Kidston and Lang, 1921), western Europe (Stockmans, 1940), Poland (Zdebska, 1972), Siberia (Petrosian, 1968), Australia (Lang and Cookson, 1935), and China (Lee and Tsai; 1977).

ASSEMBLAGE-ZONE IV
Hyenia Zone
Eifelian
(ca. 370-365 mya)

Calamophyton Kräusel and Weyland 1926 (see Schweitzer, 1973), and *Hyenia* Nathorst, 1915 (see Schweitzer, 1972), mark the base of Zone IV. They are accompanied by continuing members of Zone III such as *Psilophyton.* The lycopod *Protolepidodendron* Krejci (see Obrhel, 1968), the progymnosperm *Rellimia* Leclercq and Bonamo, 1973 (see Mustafa, 1975; Leclercq, 1940), and the possible progymnosperm *Dalejephyton* (see Obrhel, 1968) are also early members of the zone. By mid-Zone IV, further evolutionary ramification is apparent with the appearance of the progymnosperm *Aneurophyton* Kräusel and Weyland, 1923, the cladoxylalean *Pseudosporochnus* (see Obrhel, 1962; 1968), and the lycopod *Colpodexylon* Banks, 1944. *Duisbergia* Kräusel and Weyland, 1929 (see Schweitzer, 1966), and *Protocephalopteris* Ananiev, 1960 (see Cornet, Phillips, and Andrews, 1976), are two other additions.*Tomiphyton* (see Stepanov, 1967) is regarded by Russian workers as a fern, but I am inclined to think that when its anatomy is found it will prove to be another progymnosperm. Toward the close of Zone IV, the progymnosperm *Tetraxylopteris* and the large stump *Eospermatopteris* appear (see Mustafa, 1975). Zone IV covers a longer time span than do spore zones that are initiated at about the same time (Table 1.2).

New morphological characteristics introduced during Zone IV are:

an approach to the whorled position of leaves and sporangiophores (Kräusel and Weyland, 1926)

an approach to the sporangiophore of articulatae (Leclercq and Andrews, 1960)

secondary xylem (cambium and secondary phloem not yet demonstrated but probably present) in progymnosperms (*Rellimia (Protopteridium)* Mustafa, 1975)

precursors of megaphyllous leaves demonstrated by ultimate appendages of progymnosperms (*Aneurophyton* — Serlin and Banks, 1978)

complex fructifications in progymnosperms (*Rellimia* Leclercq and Bonamo, 1971)

plants with pteridophytic reproduction but with gymnospermous secondary xylem (*Tetraxylopteris* — Beck, 1957).

Localities at which this assemblage is found include eastern North America (Andrews, Gensel and Forbes, 1974), Belgium (Leclercq, 1940), Spitsbergen (Høeg, 1942), Czechoslovakia (Obrhel, 1968), West Germany (Schweitzer, 1972, 1973), and Siberia (Petrosian, 1968).

ASSEMBLAGE-ZONE V
Svalbardia Zone
much of the Givetian
(ca. 365-359 mya)

The most notable feature of the *Svalbardia* Zone is the abundance and the wide geographic range of genera that had already appeared in Zone IV, the *Hyenia* Zone. For this reason, one could argue that the designation of a specific zone is unnecessary. Also, the boundary between the Eifelian and Givetian stages is not well defined and the exact position of some of the plant localities with respect to the boundary is imprecise, often depending on the interpretation of the stratigrapher cited. Thus, my designation of a Zone V starting somewhat above the supposed Eifelian-Givetian boundary is to be regarded as an approximation. The argument favoring a Zone V is the appearance, as an index, of a group of advanced progymnosperms. This group consisting of *Actinoxylon* Matten, 1968; *Svalbardia* Høeg, 1942; and *Actinopodium* Hoeg, 1942, is certainly closely related to *Archaeopteris* (see Carluccio, Hueber, and Banks, 1966; Beck, 1970). The three genera might even be interpreted as members of the genus *Archaeopteris*, as Beck (1970) explains in detail. At present we have insufficient data to reach a decision. I prefer to keep them distinct, as possible precursors of *Archaeopteris* and as an index of the start of a new zone that precedes the striking and widespread appearance of the *Archaeopteris* flora in Zone VI. *Actinoxylon* is the first genus of the group to appear, near the base of Givetian. *Svalbardia* and *Actinopodium* are a little younger. Another progymnosperm, *Triloboxylon*, also appears early in Givetian time (see Matten, 1974; Figure 1.1).

Actinoxylon, used here as an index of the start of Zone V, is repesented only by axes permineralized by iron pyrite. This suggests the strong possibility of finding *Actinoxylon* at many other localities if enough investigators spend the time to seek out and prepare pyritized specimens. Givetian and especially Frasnian fossils collected on the Devonian coastal plain in New York State have consistently yielded large numbers of well-preserved pyritized plants. Two excellent examples are *Leclercqia* Banks, Bonamo, and Grierson, 1972, and *Tetraxylopteris* Beck, 1957 (Bonamo and Banks, 1967); Scheckler and Banks, 1971a). Recently, Mustafa (1975) has found similarly preserved plants at Lindlar, Germany. His polished slices are excellent and provide strong support for my belief that the *Archaeopteris* "precursors" can be found to have been widespread. Zone V is probably less precise than corresponding zones based on spores (Table 1.2).

Plants of Zone IV that become common and widespread in Zone V are *Protolepidodendron*, *Pseudosporochnus* (see Leclercq and Banks, 1962), *Hyenia*, *Calamophyton*, *Rellimia (Protopteridium)*, *Aneurophyton*, and *Protocephalopteris*. Other genera in the assemblage are *Eospermatopteris* Goldring, 1924; *Uralia* and *Pseudouralia* (see Petrosian, 1968); *Barrandeina* Stur, 1882; *Pectinophyton* Høeg, 1935; *Archaeosigillaria* Kidston, 1901; *Drepanophycus* Gőpp, 1852; *Colpodexylon* Banks, 1944; *Leclercqia* Banks, Bonamo, and Grierson, 1972; *Lepidodendropsis* (see Iurina, 1969); *Enigmophyton* (Høeg, 1942); and *Ibyka* (Skog and Banks, 1973).

New morphological characteristics that appear during this interval are:

arborescent habit (*Eospermatopteris; Lepidodendropsis* — see Iurina, 1969)

ligule (*Lerlercqia* —see Grierson and Bonamo, 1979)

abundant secondary xylem (from progymnosperms)

primary xylem with numerous protoxylem strands (Scheckler and Banks, 1971a — *Triloboxylon*)

secondary phloem (*Triloboxylon* — see Scheckler and Banks, 1971a)

periderm (*Triloboxylon* — see Scheckler and Banks, 1972)

Localities at which the Svalbardia Assemblage occurs in some abundance include Bohemia (see Obrhel 1962, 1968); Germany (see Schweitzer 1966, 1972, 1973); Belgium (see Leclercq, 1940; Stockmans, 1948); eastern New York State (see for example Bonamo, 1977; Grierson, 1976; Grierson and Banks, 1963; Bonamo and Banks, 1966; Bonamo and Grierson, 1972); U.S.S.R., Siberia (see Iurina, 1969; Petrosian, 1968), Spitsbergen (see Høeg, 1942).

ASSEMBLAGE-ZONE VI
Archaeopteris Zone
Frasnian to middle-Famennian;
(ca. 359-349 mya)

Zone VI, like Zone V, is heavily populated by genera continuing from older strata; *Aneurophyton* (see Serlin and Banks, 1978); *Tetraxylopteris* Beck, 1957; *Pseudosporochnus* (abundant unpublished material in Cornell University Paleobotanical Collection); *Archaeosigillaria* (see Grierson and Banks, 1963); *Drepanophycus* (see Stubblefield and Banks, 1978); *Colpodexylon* Banks, 1944; *Triloboxylon* Matten and Banks, 1966; *Eospermatopteris* (see Serlin and Banks, 1978); *Sawdonia* (see Hueber and Grierson, 1961).

The newer plants in this assemblage include *Protolepidodendropsis (Bergeria)* from Spitzbergen (see Schweitzer, 1965), which may occur at the end of Zone V as well; *Proteokalon* Scheckler and Banks, 1971b; and *Cystosporites* (see Chaloner and Pettitt, 1964).

The plant that serves as an index for the start of Zone VI is *Archaeopteris*. Its planated, webbed, ultimate appendages fit most criteria of leaves; and both vegetative and fertile specimens are readily referable to the genus. It is quite possible that genera of Zone V, such as *Svalbardia*, may be demonstrated to be early species of the genus *Archaeopteris*. It could also be true that species of *Svalbardia* formed a series from those with finely dissected to those with undissected leaves. Nevertheless, until a demonstration of these possibilities has been achieved, I prefer to consider *Archaeopteris* as an index of Zone VI and *Svalbardia*-like plants as precursors, in Zone V. *Callixylon*, a gymnospermous secondary wood, is found associated with *Archaeopteris* around the world. Beck (1960) showed the two genera in organic connection and *Archaeopteris* now refers to a whole plant.

Nonetheless, isolated bits of secondary wood can still be assigned to *Callixylon* and can help serve as an index of Zone VI. The abundance and wide distribution of *Archaeopteris* account for its value as an index fossil.

Morphological characteristics added during Zone VI are:

megaphyllous leaves (planted, webbed ultimate appendages — *Archaeopteris*)

heterospory (*Archaeopteris; Barinophyton* — see Pettitt, 1965)

root system (*Archaeopteris* — as opposed to adventitious roots)

mixed pith (*Tetraxylopteris* — see Matten and Banks, 1967)

pith (*Archaeopteris* — see Arnold, 1930 for *Callixylon;* Carluccio, Hueber, and Banks, 1966 for *Archaeopteris)*

complex fructification of progymnosperm (*Tetraxylopteris* — see Bonamo and Banks, 1967)

seed megaspores (one functional and 3 aborted spores — see Chaloner and Pettitt, 1964)

Zone VI is best developed, most widespread, and most studied in North America (see Arnold, 1930; Banks, 1944; Beck, 1960; Bonamo and Banks, 1967; Carluccio, Hueber, and Banks, 1966; Grierson and Banks, 1963; Matten and Banks, 1966; Scheckler and Banks, 1971a,b; Serlin and Banks, 1978; Stubblefield and Banks, 1978) and southwestern Siberia (see review by Petrosian, 1968, and papers cited therein).

ASSEMBLAGE-ZONE VII
Rhacophyton Zone
Middle Famennian
(ca. 349-? mya)

The assemblage characteristic of Zone VI continues into Famennian strata without any well-marked change. *Rhacophyton* appears at about middle Famennian time (see Cornet, Phillips, and Andrews, 1976) and serves as an index for Assemblage-zone VII. This assemblage is best represented in strata of late and post Famennian age (e.g., Belgium, Stockmans, 1968; Bear Island, Schweitzer, 1967, 1969; and Siberia, Ananiev and Graizer, 1957; Ananiev and Eganov, 1957; Ananiev and Michailova, 1958). On Bear Island, *Protolepidodendropsis, Archaeopteris, Pseudobornia,* and *Cyclostigma* are all found in the post-Famennian Tn 1a (Schweitzer, 1967, 1969; Owens and Streel, 1970). This age assignment is based on Owen and Streel's report (1970) of Kaiser's discovery of spores characteristic of Tn 1a in the Tunheim Series in which these megafossils occur. Comparable studies, may ultimately clarify the age of all Famennian and post-Famennian megafossil-bearing localities (see Streel and Traverse, 1978; McGregor 1979).

Plants accompanying *Rhacophyton* include *Pseudobornia, Cyclostigma, Archaeopteris, Sphenophyllum, Sphenopteridium, Protolepidodendropsis, Sublepidodendron, Cephalopteris* (see Schweitzer, 1967, 1969), *Eviostachya* (Leclercq, 1957), *Barinophyton* (Arnold, 1939), *Archaeosperma* and *Cystosporites* (Pettitt and Beck, 1968), *Leptophloeum* (White, 1905), *Aneurophyton, Sphenopteris, Xenotheca, Moresnetia* (Stockmans, 1948), and *Spermolithus* (Chaloner, Hill and Lacey, 1977). Plants endemic to the U.S.S.R. are listed in Petrosian (1968). Zone VII corresponds well with Richardson's upper Famennian spore zone (Table 1.2).

New morphological characteristics are:

probably the first true gymnosperm (because of the presence of unattached seeds)

complex cones definitely assignable to Sphenopsida (*Eviostachya*—see Leclercq, 1957; and *Pseudobornia*—see Leclercq, 1964)

precursors of zygopterid coenopterid ferns *(Rhacophyton*—see Leclercq, 1951).

seeds, both radially (*Archaeosperma* Pettitt and Beck, 1968) and bilaterally symmetrical (*Spermolithus*—see Chaloner, Hill, and Lacey, 1977).

Zone VII is well developed on Bear Island (Schweitzer, 1967, 1969) and in Ireland (see Chaloner, 1968); Belgium (see Stockmans, 1948); and the U.S.S.R. (see Petrosian, 1968). Other well-dated occurrences are northwestern Pennsylvania (Pettitt and Beck, 1968); southwestern New York (Arnold, 1930); and West Virginia (Cornet, Phillips and Andrews, 1976; Phillips, Andrews and Gensel, 1972).

DISCUSSION

Evidence from the time of appearance of Devonian plant megafossils permits the recognition of successive generic assemblage-zones. Seven such zones can be defined. Each zone is named for an index fossil: I, *Cooksonia;* II, *Zosterophyllum;* III, *Psilophyton;* IV, *Hyenia;* V, *Svalbardia;* VI, *Archaeopteris;* VII, *Rhacophyton.* Assurance that a collection represents a given zone demands a representative collection of fossils and an analysis of the relative state of evolution of their morphological features. The latter refers to such characteristics as (1) leafless vs. microphylls or megaphylls; (2) rootless vs. adventitious roots or a root system; (3) herbaceous habit vs. arborescent; (4) sporangia terminal vs. lateral or adaxial; (5) sporangia borne on unmodified structures vs. specialized branch systems;(6) homospory vs. heterospory or seed bearing; (7) protostele vs. mixed pith, or siphonostele or dissected siphonostele; (8) exarchy vs. centrarchy, mesarchy, or endarchy; (9) primary growth only vs. secondary growth added.

Essential to the foregoing is mass collecting from active quarries where the chance of uncovering a rich fossiliferous lens (former pond, lake, or similar site of deposition) is good. Equally important is the application to the material of modern laboratory techniques. It is no longer adequate to describe the few characteristics that appear on the surface of a fossil. Some anatomical and histological details are required to establish genera that may approach a natural classification. Many of the references cited in this paper have been chosen because the fossils have been subjected to these techniques. Many genera endemic to one or another region are omitted here simply because, descriptively, they do not meet these criteria.

One test of the validity of the seven assemblage-zones for biostratigraphic analysis is the ability to use them to achieve chronostratigraphic correlation. Although the units so distinguishable may be overly extensive, rocks containing fossils of, for example, Assemblage-zone II are readily separable from those containing fossils of Assemblage-zone III on the basis of a good collection of fossils from each. The same is true of the other zones. Furthermore the units can be identified at several localities, another test of validity. For example, in Wales, Podolia, Czechoslovakia, and New York State one finds exclusively fossils of Assemblage-Zone I—*Cooksonia* in rocks that have been assigned to the Pridolian Stage. The distribution of the zosterophyll assemblage is worldwide (Banks, 1975) and clearly indicates an approximate equivalence between the rocks at the many localities. Future studies will undoubtedly allow the identification of rocks representing shorter intervals of time and their correlation over a broader geographic range. The present analysis is simply a start in that direction.

There are potential flaws in the assemblages presented here. Klitzsch, Lejal-Nicol, and Massa (1973) have reported a flora of psilophytes and lycopods from the Acacus Formation in the Mourzouk Basin in Libya. The Acacus is described as extending from middle to upper Llandoverian to overlying Siegenian strata. The age of the fossils, found near the summit of the Acacus, is therefore said to be between late Llandovery and early Siegenian. If, indeed, the plant remains are Silurian or even early Devonian in age, they represent a considerably earlier development of both trimerophytes (*Psilophyton* of Klitzsch et al.) and lycophytes than I have indicated. However the Libyan fossils are poorly preserved and have yielded neither anatomical nor histological details. In my opinion the specimen referred to *Psilophyton* shows no characteristic of generic value. The lycopods are decorticated axes indicating only the subdivision to which they belong, Lycopsida. A true outer surface and some leaves are a minimal requirement for assigning them to a genus. They resemble decorticated lycopods of middle to lower Upper Devonian strata in other parts of the world. If they alone were to be considered, the strata would be assigned to Zone IV, V, or VI. Yet if they are Silurian to early Devonian in age, my assemblage-zones must be revised. Unfortunately I have been unable to obtain satisfactory evidence of their age. As a result I have excluded them until such time as their age can be agreed upon by independent observers.

A second example is the presumed Silurian *Baragwanathia* in Australia (Garratt, 1978). If it did appear in Silurian rocks, this lycopod antedates considerably the middle-Siegenian time I have assigned to the oldest lycopod. I have endeavored unsuccessfully to obtain convincing evidence of the Silurian age of this new find of *Baragwanathia*. Again I must conclude that there is not yet sufficient reason to change my succession of zones, although I am well aware that they may have to be modified subsequently. I urge only that the change be based on evidence that is acceptable to a number of critical observers.

Another example is to be found in Petrosian (1968) and Senkewich (1968). A careful reader will note that certain genera are reported there from strata older than those I have reported. There are several reasons for the apparent discrepancy. First there are different interpretations in the Russian literature. For example, floras that Ananiev refers to as Early Devonian are considered by Petrosian to be Eifelian (Petrosian, 1968, p. 582). However early Eifelian in many Russian papers is equivalent to late Emsian of western workers (Rzhonsnitskaya, 1968). Obviously one needs precise stratigraphic positions and detailed morphological data before generalizing about the apparently greater age of some of the Russian material. Secondly, I am uncertain that some of the genera placed in the Givetian by Pterosian are the same organism that I cite, using the same name, in Frasnian strata. Only long and detailed first-hand study can clarify the many problems of this type. Until these discrepancies have been resolved, I propose to continue to refer only to identifications that I consider justified by the evidence. Thirdly, I suspect that if many of the "Primofilices" reported by the Russian workers were studied anatomically and histologically, they would prove to be progymnosperms.

Collectively, however, the assemblage-zones proposed here, by Petrosian (1968), and by Senkewich (1968) are remarkably similar. Both Russian colleagues recognize Eifelian, Givetian, Frasnian, and Famennian assemblages. When techniques designed to yield anatomical, histological, and morphological data are applied to the vast Russian paleoflora, it is likely that the global similarity among Devonian floras will be even more striking than it is today.

An example of the taxonomic problems raised is the genus *Lepidodendropsis*, which is shown in Figure 1.2 as originating in the Givetian. This also means that arborescent lycopods lived in Givetian time. *Lepidodendropsis* was considered originally to be restricted to Mississippian strata. It was later reported from late Famennian rocks. There is no question that Iurina (1969) has splendid, large, anatomically preserved lycopods. The problem is whether they should be referred to the genus *Lepidodendropsis*. If they are, then the genus appears early.If they are not, then the genus continues to characterize late Famennian or early Mississippian strata. My inclination would be to erect a new genus for the Givetian fossils from Kazakhstan, on the basis of the absence of a leaf abscission scar and the small size of the leaves. This early appearance of arborescent lycopods with an exarch protostele and a wide zone of bark suggsts that certain Russian Devonian floras may antedate those in other parts of the world.

A fourth example of problems associated with assemblage-zones is the frequent revision of the age of a Formation by continuing research. In Figure 1.2, *Barinophyton* is listed as appearing late in the Famennian. An occurrence of this genus in southwestern New York (Arnold, 1939) is upper Famennian according to the geologic map of the state. However, the Geological Survey of Pennsylvania is considering a change that will lower the position of the formation bearing *Barinophyton*, perhaps even as far as the Frasnian (J.D. Grierson, pers. comm.).

Finally, variation in the use of names always presents a challenge. David White (1902) referred to *Dimeripteris* in the flora of Perry, Maine, U.S.A. Kräusel and Weyland (1941) placed the same material in *Rhacophyton*. Current research is not exempt. The *Psilophyton princeps* of Hueber and Grierson (1961) is now *Sawdonia ornata* Hueber (1971a). These changes are routine but they serve to caution the reader to expect some discrepancies between the text of this paper and certain of those cited.

In conclusion, it is to be hoped that the crude state of biostratigraphic studies of Devonian plant megafossils will encourage paleobotanists to obtain better stratigraphic data and to extract more anatomical and morphological data from their specimens.

ACKNOWLEDGMENTS

The writer wishes to acknowledge with thanks the valuable and pertinent suggestions on the manuscript that were made by Dr. D. C. McGregor.

REFERENCES

Ananiev, A. R. 1960. Study of the Middle Devonian flora of the Saian — Altai Mountain Region. *Bot. Zhur.* 45:649-666. (In Russian: transl. in Ithaca.)

Ananiev, A. R., and E. A. Eganov. 1957. The age of the Bistriansk series in the southeast of western Siberia correlated with the discovery made of *Cyclostigma Kiltorkense* Haughton in the vicinity of Oujour. *Dokl. Akad. Nauk SSSR*. 113: 402-406. (In Russian: transl. by S. I. G. du B. R. G. M. Paris.)

Ananiev, A. R., and M. I. Graizer. 1957. On the flora of contiguous Devonian and Carboniferous layers of the Minusinsk trough. *Dokl. Akad. Nauk SSSR.* 116:997-1000. (In Russian: transl. in Ithaca.)

Ananiev, A. R., and J. V. Mikhailova. 1958. On the age of the deposits of the lower part of the Minusinsk series in connection with the discovery of *Lepidodendropsis hirmeri* Lutz in the Samokhvalsk suite. *Dokl. Akad. Nauk SSSR.* 123:1081-1084. (In Russian: transl. in Ithaca.)

Andrews, H. N.; P. G. Gensel; and W. H. Forbes. 1974. An apparently heterosporous plant from the Middle Devonian of New Brunswick. *Palaeontology* 17:387-408.

Andrews, H. N.; P.G., Gensel; and A. E. Kasper. 1975. A new fossil plant of probable intermediate affinities (Trimerophyte-Progymnosperm). *Canadian Jour. Bot.* 53:1719-1728.

Andrews, H. N.; A. E. Kasper; W. H. Forbes; P. G. Gensel; and W. G. Chaloner. 1977. Early Devonian flora of the Trout Valley Formation of northern Maine. *Rev. Palaeobot. Palynol.* 23:255-285.

Arber, E. A. N. 1921 *Devonian floras. A study of the origin of Cormophyta.* Cambridge: Cambridge Univ. Press.

Arnold, C. A. 1930. The genus *Callixylon* from the Upper Devonian of central and western New York. *Pap. Michigan Acad. Sci., Arts, Letters* 11:1-50.

Arnold, C. A. 1939. Observations on fossil plants from the Devonian of eastern North America. IV. Plant remains from the Catskill Delta Deposits of northern Pennsylvania and southern New York. *Contrib. Mus. Paleontol. Univ. Michigan* 5:271-314.

Banks, H. P. 1944. A new Devonian lycopod genus from southeastern New York. *Am. Jour. Bot.* 31:650-659.

Banks, H. P. 1968. The early history of land plants. *In Evolution and Environment: a symposium presented on the occasion of the 100th anniversary of the foundation of Peabody Museum of Natural History at Yale University.* E. T. Drake, ed. New Haven: Yale Univ. Press, pp. 73-107.

Banks, H. P. 1973. Occurrence of *Cooksonia,* the oldest vascular land plant macrofossil, in the Upper Silurian of New York State. *Jour. Indian Bot.* 50A: 227-235.

Banks, H. P. 1975. Palaeogeographic implications of some Siluro-Early Devonian floras. In *Gondwana Geology,* K. S. W. Campbell, ed. Canberra: Australian National Univ. Press, pp. 75-97.

Banks, H. P.; P. M. Bonamo; and J. D. Grierson. 1972. *Leclercqia complexa* gen. et sp. nov., a new lycopod from the Late Middle Devonian of eastern New York. *Rev. Palaobot. Palynol.* 14:19-40.

Banks, H. P. and M. R. Davis. 1969 *Crenaticaulis,* a new genus of Devonian plants allied to *Zosterophyllum,* and its bearing on the classification of early land plants. *Am. Jour. Bot.* 56:436-449.

Banks, H. P.; S. Leclercq; and F. M. Hueber. 1975. Anatomy and morphology of *Psilophyton dawsonii,* sp. n. from the Late Lower Devonian of Quebec (Gaspé) and Ontario, Canada. *Palaeontographica Americana* 8:75-127.

Beck, C. B. 1957. *Tetraxylopteris schmidtii* gen. et sp. nov., a probable pteridosperm precursor from the Devonian of New York. *Am. Jour. Bot.* 44:350-367.

Beck, C. B. 1960. The identity of *Archaeopteris* and *Callixylon. Brittonia* 12:351-368.

Beck, C. B. 1970. Problems of generic delimitation in paleobotany. *Proc. North Am. Paleontol. Conv., 1969,* (Chicago), C, pp. 173-193.

Beck. C. B. 1971. On the anatomy and-morphology of lateral branch systems of *Archaeopteris. Am. Jour. Bot.* 58:758-784.

Beck, C. B. 1977. Preliminary report on the architecture of the primary vascular system of *Callixylon. Bot. Soc. Am. Misc. Ser. Pub.* 154:33-34.

Bonamo, P.M. 1977 *Rellimia thomsonii* (Progymnospermopsida) from the Middle Devonian of New York State. *Am. Jour. Bot. 64:* 1272-1285.

Bonamo, P. M. and H. P Banks. 1966. *Calamophyton* in the Middle Devonian of New York State. *Am. Jour. Bot.* 53:778-791.

Bonamo, P. M., and H. P. Banks 1967. *Tetraxylopteris schmidtii:* Its fertile parts and its relationships within the Aneurophytales. *Am. Jour. Bot.* 54:755-768.

Bonamo, P. M., and J. D. Grierson, 1973. Sporophylls, sporangia and spores of *Leclerqia complexa. Am. Jour. Bot.* 60(4 supp.): 16.

Carluccio, L. M.; F. M. Hueber; and H. P. Banks. 1966. *Archaeopteris macilenta,* anatomy and morphology of its frond. *Am. Jour. Bot.* 53:719-730.

Chaloner, W. G. 1968. The cone of *Cyclostigma kiltorkense* Haughton, from the Upper Devonian of Ireland. *Jour. Linn. Soc. London* (Bot.) 61:25-36.

Chaloner, W. G.; A. J. Hill; and W. S. Lacey. 1977. First Devonian platyspermic seed and its implications in gymnosperm evolution. *Nature* 265:233-235.

Chaloner, W. G., and J. M. Pettitt. 1964. A seed megaspore from the Devonian of Canada. *Palaeontology* 7:29-36.

Chaloner, W. G. and A. Sheerin. 1979. *Devonian macrofloras.* In "The Devonian System,"*M. R. House, C. T. Scrutton, and M. G. Bassett, eds. Special Papers in Palaeontology No. 23,* pp. 145-161.

Cocks, L. R. M.; C. H. Holland; R. B. Rickards; and I. Strachan. 1971. A correlation of Silurian rocks in the British Isles. *Jour. Geol. Soc., London* 127:103-136.

Cornet, B.; T. L. Phillips; and H. N. Andrews. 1976. The morphology and variation in *Rhacophyton ceratangium* from the Upper Devonian and its bearing on frond evolution. *Palaeontographica* 158B:105-129.

Daber, 1960. *Eogaspesiea gracilis* n. g. n. sp. *Geologie* 9:418-425.

Daber, R. 1971. *Cooksonia* — One of the most ancient psilophytes— widely distributed, but rare. *Botanique* 2:35-40.

Edwards, D. 1968. A new plant from the Lower Old Red Sandstone of South Wales. *Palaeontology* 11:683-690.

Edwards, D. 1969a. Further observations on *Zosterophyllum llanoveranum* from the Lower Devonian of South Wales. *Am. Jour. Bot.* 56:201-210.

Edwards, D. 1969b. *Zosterophyllum* from the Lower Old Red Sandstone of South Wales. *New Phytologist* 68:923-931.

Edwards, D. 1970b. Further observations on the Lower Devonian plant, *Gosslingia breconensis* Heard. *Phil. Trans. Roy. Soc. London* 258B:225-243.

Edwards, D. 1970a. Fertile Rhyniophytina from the Lower Devonian of Britian. *Palaeontology* 13:451-461.

Edwards, D. 1972. A *Zosterophyllum* fructification from the Lower Old Red Sandstone of Scotland. *Rev. Palaeobot. Palynol.* 14:77-83.

Edwards, D. 1979. A late Silurian flora from the Lower Old Red Sandstone of south-west Dyfed. *Palaeontology* 22:23-52.

Edwards, D., and E. C. W. Davies. 1976. Oldest recorded *in situ* tracheids. *Nature* 263:494-495.

Edwards, D., and J. B. Richardson. 1974. Lower Devonian plants from the Welsh Borderland. *Palaeontology* 17:314-324.

Fairon—Demaret, M. 1971. Quelques caractères anatomiques du *Drepanophycus spinaeformis* Göppert. *C. R. Acad. Sci.*, Paris, 273D:933-935.

Garratt, M. J. 1978. New evidence for Silurian (Ludlow) age for the earliest *Baragwanathia* flora. *Alcheringa* 2:217-224.

Gensel, P. G. 1976. *Renalia hueberi*, a new plant from the lower Devonian of Gaspé. *Rev. Palaeobot. Palynol.* 22:19-37.

Gensel, P. G.; A Kasper; and H. N. Andrews. 1969. *Kaulangiophyton*, a new genus of plants from the Devonian of Maine. *Bull. Torrey Bot. Club* 96:265-276.

Goldring, W. 1924. The Upper Devonian Forest of seed ferns in Eastern New York. *New York State Mus. Bull.* 251:50-92.

Grierson, J. D. 1976. *Leclercqia complexa* (Lycopsida, Middle Devonian): its anatomy, and the interpretation of pyrite petrifactions. *Am. Jour. Bot.* 63:1184-1202.

Grierson, J. D., and H. P. Banks. 1963. Lycopods of the Devonian of New York State. *Palaeontographica Americana* 4(31):220-295.

Grierson, J. D., and P. M. Bonamo. 1979. *Leclercqia complexa:* Earliest ligulate lycopod (Middle Devonian). *Am. Jour. Bot.* 66:474-476.

Halle, T. G. 1916. Lower Devonian plants from Röragen in Norway. *Kungl. Svenska Vetenskapsakad, Handl.* 57:1-46.

Harland, W. B.; A. B. Smith; and B. Wilcock, eds. 1964. The Phanerozoic Time-Scale. *Quart. Jour. Geol. Soc. London* 120S:1-458.

Høeg, O. A. 1935. Further contributions to the Middle Devonian flora of Western Norway. *Norsk Geol. Tidsskr.* 15:1-18.

Høeg, O. A. 1942. The Downtonian and Devonian flora of Spitsbergen. *Norges Svalbard-Og Ishavs-Undersokelser* 83:1-228.

Hopping, C. A. 1956. On a specimen of *"Psilophyton robustius"* Dawson from the Lower Devonian of Canada. *Proc. Roy. Soc. Edinburgh* 66B:10-28.

House, M. R.; J. B. Richardson; W. G. Chaloner; J. R. L. Allen; C. H. Holland; and T. S. Westoll. 1977. A correlation of Devonian rocks of the British Isles. *Geol. Soc. London, Spec. Report No.* 7, 110pp.

Hueber, F. M. 1968. *Psilophyton:* the genus and the concept. In D. H. Oswald. ed., 1968, vol. 1, pp. 815-822.

Hueber, F. M. 1970 *Rebuchia:* a new name for *Bucheria* Dorf. *Taxon* 19:822.

Hueber, F. M. 1971a. *Sawdonia ornata:* a new name for *Psilophyton princeps* var. *ornatum. Taxon* 20:641-642.

Hueber, F. M. 1971b. Early Devonian land plants from Bathurst Island, District of Franklin. *Geol. Surv. Canada. Paper 71-28* pp. 1-17.

Hueber, F. M., and H. P. Banks. 1967. *Psilophyton princeps:* the search for organic connection. *Taxon* 16:81-85.

Hueber, F. M., and J. D. Grierson. 1961. On the occurrence of *Psilophyton princeps* in the early Upper Devonian of New York. *Am. Jour. Bot.* 48:473-479.

Ishchenko, T. A. 1969. The Cooksonian paleoflora in the Skala horizon of Podolia and its stratigraphical significance. *Geol. Zhur* (Kiev.) 29:101-109. (In Russian: transl. by Geol. Surv. Canada.)

Iurina, A. L. 1969. Devonian flora of central Kazakhstan. Central Kazakhstan Dept. of Geology. *Geol. of Central Kazakhstan* 8:1-143.

Kasper, A. E., Jr. 1977. A new species of the Devonian lycopod genus *Leclercqia* from New Brunswick, Canada. *Bot. Soc. Am. Misc. Ser. Pub. 154* p. 39.

Kasper, A. E., Jr., and H. N. Andrews, Jr. 1972 *Pertica*, a new genus of Devonian plants from northern Maine. *Am. Jour. Bot.* 59:897-911.

Kidston, R., and W. H. Lang. 1921. On Old Red Sandstone plants showing structure, from the Rhynie Chert Bed, Aberdeenshire. Part IV. Restorations of the vascular cryptogams, and discussion of their bearing on the general morphology of the pteridophyta and the organization of land plants. *Trans. Roy. Soc. Edinburgh* 52:831-854.

Klitzsch, E.; A. Lejal-Nicol; and D. Massa. 1973. Le Siluro-Dévonien a Psilophytes et Lycophytes du bassin de Mourzouk (Libye). *C. R. Acad. Sc.* (Paris) 277D:2465-2467.

Kräusel, R. 1937. Die Verbreitung der Devonfloren. 2nd *Congr. Internl, Stratigr. Geol. Carb. Compte Rendu* (Heerlen 1935), pp. 527-537.

Kräusel, R., and H. Weyland. 1923. Beiträge zur Kenntnis der Devonflora. I. Die Fundorte der beschriebenen Pflanzen. *Senckenbergiana* 5:154-184.

Kräusel, R., and H. Weyland. 1926. Beiträge zur Kenntnis de Devonflora. II. *Abh. Senckenberg. Naturfor. Gesell.* 40:115-155.

Kräusel, R., and H. Weyland. 1929. Beiträge zur Kenntnis der Devonflora. III. *Abh. Senckenberg. Naturfor. Gesell.* 41:315-360.

Kräusel, R., and H. Weyland. 1932. Pflanzenreste aus dem Devon. IV. *Protolepidodendron* Krejci. *Senckenbergiana* 14:391-403.

Kräusel, R., and H. Weyland. 1935a. Neue pflanzenfunde in rheinischen Unterdevon. *Palaeontographica* 80B:171-190.

Kräusel, R., and H. Weyland. 1935b. Pflanzenreste aus dem Devon. IX, Ein stamm von *Eospermatopteris* — bau aus dem Mitteldevon, des Kirberges, Elberfeld. *Senckenbergiana* 17:9-20.

Kräusel, R., and H. Weyland. 1941. Pflanzenreste aus dem Devon von Nord—Amerika. *Palaeontographica* 86B:1-78.

Lang, W. H. 1932. Contributions to the study of the Old Red Sandstone Flora of Scotland. VIII. On *Arthrostigma*, *Psilophyton*, and some associated plant remains from the Strathmore beds of The Caledonian Lower Old Red Sandstone. *Trans. Roy. Soc. Edinburgh.* 57:491-521.

Lang, W. H. 1937. On the plant-remains from the Downtonian of England and Wales. *Phil. Trans. Roy. Soc. London.* 227B:245-291.

Lang, W. H., and I. Cookson. 1935. On a flora, including vascular land plants, associated with *Monograptus*, in rocks of Silurian age, from Victoria, Australia. *Phil. Trans. Roy. Soc. London.* 224B: 421-449.

Leclercq, S. 1940. Contribution à 'l'etude de la flore du Dévonien de Belgique. *Acad. Roy. Belgique, Mém., Cl. Sci.* 12:1-65.

Leclercq, S. 1951. Étude morphologique et anatomique d'une fougère du Dévonien supérieur. Le *Rhacophyton zygopteroides* nov. sp. *Ann. Soc. Géol. Belgique, Mém.* 9:1-58.

Leclercq, S. 1957. Étude d'une fructification de Sphenopside à structure conservée du dévonien supérieur. *Acad. Roy Belgique Mém:* 14:1-40.

Leclercq, S. 1964. Recent studies on Devonian sphenopsids. 10th *Intern. Bot. Congr., Abstr.* (Edinburgh) 041:18.

Leclercq, S., and H. N. Andrews, Jr. 1960. *Calamophyton bicephalum*, a new species from the Middle Devonian of Belgium. *Missouri Bot. Garden Annals*, 47:1-23.

Leclercq, S., and H. P. Banks. 1962 *Pseudosporochnus nodosus* sp. nov., a Middle Devonian plant with cladoxylalean affinities. *Palaeontographica* 110B:1-34.

Leclerq, S., and P. M. Bonamo. 1971. A study of the fructification of *Milleria (Protopteridium) thomsonii* Lang from the Middle Devonian of Belgium. *Palaeontographica* 136B:83-114.

Leclercq, S., and P. M. Bonamo, 1973. *Rellimia thomsonii,* a new name for *Milleria (Protopteridium) thomsonii* Lang 1926. Emend. Leclercq and Bonamo 1971. *Taxon* 22:435-437.

Lee, H.H., and C. Tsia. 1977. Early Devonian *Zosterophyllum* - remains from southwest China. (Abstr. in English) *Acta Palaeontologica Sinica* 16:12-34.

Martinsson, A. 1969. The series of the redefined Silurian System. *Lethaia* 2:153-161.

Matten, L. C. 1968. *Actinoxylon Banksii* gen. et sp. nov.: a progymnosperm from the Middle Devonian of New York. *Am. Jour. Bot.* 55: 773-782.

Matten, L. C. 1974. The Givetian flora from Cairo, New York: *Rhacophyton, Triloboxylon* and *Cladoxylon.* Bot. Jour. Linn. Soc. 68:303-318.

Matten, L. C., and H. P. Banks. 1966. *Triloboxylon ashlandicum* gen. and sp. n. from the Upper Devonian of New York. *Am. Jour. Bot.* 53:1020-1028.

Matten, L. C., and H. P. Banks. 1967. Relationship between the Devonian progymnosperm genera *Sphenoxylon* and *Tetraxylopteris. Bull. Torrey Bot. Club.* 94:321-333.

McGregor, D. C. 1977. Lower and Middle Devonian spores of eastern Gaspé, Canada. II. Biostratigraphy. *Palaeontographica.* 163B:111-142.

McGregor, D. C. 1979. *Spores in Devonian stratigraphical correlation.* In "The Devonian System," eds. M. R. House, C. T. Scrutton, and M. G. Bassett, *Special Papers in Palaeontology No. 23* pp. 163-184.

Mustafa, H. 1975. Beiträge zur Devonflora. I. *Argumenta Palaeobotanica* 4:101-133.

Nathorst, A. G. 1915. Zur Devonflora des westlichen Norwegens. *Bergens Mus. Aarbok, 1914-1915* 9:12-34.

Obrhel, J. 1962. *Die Silur-und Devonflora Bohmens.* Symposium vol. of 2nd Internationalen Arbeitstagung über die Silur/Devon-Grenze und die stratigraphie von Silur and Devon. Bonn-Bruxelles 1960 pp. 180-185.

Obrhel, J. 1968. Die Silur-und Devonflora des Barrandiums. *Palaont.* 2B:635-793.

Oswald, D. H., ed. *International Symposium on the Devonian System,* Calgary, Canada: Alberta Soc. Petroleum Geol.

Owens, B., and M. Streel. 1970. Palynology of the Devonian-Carboniferous boundary. *Colloq. Stratig. Carbonifère, Congr. Colloq. Univ. Liège* 55:113-120.

Petrosian, N. M. 1968. Stratigraphic importance of the Devonian flora of the USSR. In D. H. Oswald, ed., 1968, vol. 2, pp. 579-586.

Pettitt, J. M. 1965. Two heterosporous plants from the Upper Devonian of North America. *Bull. British Mus. (Nat. Hist.) Geol.* 10:83-92.

Pettitt, J. M. and C. B. Beck. 1968. *Archaeosperma arnoldii* — a cupulate seed from the Upper Devonian of North America. *Contrib. Mus. Paleont. Univ. Michigan* 22:139-154.

Phillips, T. L.; H. N. Andrews; and P. G. Gensel. 1972. Two heterosporous species of *Archaeopteris* from the Upper Devonian of West Virginia. *Palaeontographica* 139B:47-71.

Phillips, T. L.; R. A. Peppers; M. J. Avcin; and P. F. Laughnan. 1974. Fossil Plants and Coal: Patterns of change in Pennsylvanian Coal Swamps of the Illinois Basin. *Science* 187:1367-1369.

Remy, W., and V. Havlena. 1962. The floral subdivision of Devonian, Carboniferous and Permian in the terrestrial-limnic Euroamerican flora of Europe. *Fortschr. Geol. Rheinland Westfalen* 3:735-752.

Retallack, G. J. 1977. Reconstructing Triassic vegetataion of eastern Australia: a new approach for the biostratigraphy of Gondwanaland. *Alcheringa* 1:247-277.

Richardson, J. B. 1974. The stratigraphic utilization of some Silurian and Devonian miospore species in the northern hemisphere: an attempt at a synthesis. *Intern. Symp. Belgian Micropaleont. Limits, Publ. No. 9* (Namur 1974) pp. 1-13.

Rzhonsnitskaya, M. A. 1968. Devonian of the USSR. In D. H. Oswald, ed., 1968, Vol. 1, pp. 331-348.

Sandberg, C. A.; M. Streel; and R. A. Scott. 1972. Comparison between conodont zonation and spore assemblages at the Devonian-Carboniferous boundary in the western and central United States and in Europe. *7th Congr. Intern. Stratigr. Géol. Carbonifère, Compte Rendu* (Krefeld, 1971), 1:179-203.

Scheckler, S. E., and H. P. Banks. 1971a. Anatomy and relationships of some Devonian progymnosperms from New York. *Am. Jour. Bot.* 58:737-751.

Scheckler, S. E., and Banks, H. P. 1971b. *Proteokalon,* a new genus of progymnosperms from the Devonian of New York State and its bearing on phylogenetic trends in the group. *Am. Jour. Bot.* 58:874-884.

Scheckler, S. E., and Banks, H. P. 1972. Periderm in some Devonian plants. In *Advances in plant morphology,* Prof. V., *Puri Commen. vol.,*Y. S. Murty; B. M. Johri; H. Y. Mohan Ram; and T. M. Varghese, eds., Meerut City, India: Praksahan, Sarita, pp. 58-64.

Schweitzer, H. J. 1965. Über *Bergeria mimerensis* and *Protolepidodendropsis pulchra* aus dem Devon Westspitzbergens. *Palaeontographica* 115B: 117-138.

Schweitzer, H. J. 1966. Die mitteldevonflora von Lindlar (Rheinland). I. Lycopodiinae. *Palaeontographica* 118B:93-112.

Schweitzer, H. J. 1967. Die Oberdevonflora der Bäreninsel. 1. *Pseudobornia ursina* Nath. *Palaeontographica* 120B: 116-137.

Schweitzer, H. J. 1969. Die Oberdevon-Flora der Bäreninsel. 2. Lycopodiinae.*Palaeontographica* 126B:101-137.

Schweitzer, H. J. 1972. Die Mitteldevon-Flora von Lindlar (Rheinland). 3. Filicinae—*Hyenia elegans* Kräusel and Weyland. *Palaeontographica* 137B:154-175.

Schweitzer, J. H. 1973. Die Mitteldevon-Flora von Lindlar (Rheinland). 4. Filicinae—*Calamophyton primaevum* Kräusel and Weyland. *Palaeontographica* 140B:117-150.

Senkewich, M. A. 1968. Devonian continental deposits of Central Kazakhstan. In D. H. Oswald, ed., 1968, vol. 2, pp. 1117-1127.

Serlin, B. S., and H. P. Banks. 1978. Morphology and anatomy of *Aneurophyton,* a progymnosperm from the Late Devonian of New York. *Palaeontographica Americana* 8(51):343-359.

Seward, A. C. 1931. *Plant Life Through the Ages.* Cambridge: Cambridge U. Press.

Skog, J. E., and H. P. Banks. 1973. *Ibyka amphikoma,* gen. et sp. n., a new protoarticulate precursor from the late Middle Devonian of New York State. *Am. Jour. Bot.* 60:366-380.

Stepanov, S. A. 1967. On the Middle Devonian Flora from the borders of the Kuzbas. *V. V. Kuibishev Tomsk State Univ. of the Order of the Red Banner of Labor* 63:102-117. (In Russian:

transl. in Ithaca.)

Stockmans, F. 1940. Végétaux éodévoniens de la Belgique. *Mem. Mus. Royal Hist. Nat. Belgique.* 93:1-90.

Stockmans, F. 1948. Végétaux du Dévonien supérieur de la Belgique. *Mem. Mus. Roy. Hist. Nat. Belgique* 110:1-85.

Streel, M. 1972. Biostratigraphie des couches de transition Dévono-Carbonifère et limite entre les deux Systèmes. 7th *Congr. Intern. Stratigr. Géol. Carbonifère, Compte Rendu* (Krefeld 1971), 1:167-178.

Streel, M. and A. Traverse. 1978 Spores from the Devonian/Mississippian transition near the Horseshoe Curve section. Altoona, Pennsylvania, U.S.A. *Rev. Palaeobot. Palynol* 26:21-39.

Stubblefield, S., and H. P. Banks. 1978. The cuticle of *Drepanophycus spinaeformis*, a long-ranging Devonian lycopod from New York and eastern Canada. *Am. Jour. Bot.* 65:100-118.

White, D. 1905 *Paleontology.* In "The Geology of the Perry Basin in southeastern Maine," G. O. Smith and D. White, *U.S. Geol. Survey, Prof. Paper 35,* pp. 35-103.

Zdebska, D. 1972. *Sawdonia ornata* (= *Psilophyton princeps* var. *ornatum)* from Poland. *Acta Palaeobot.* 13:77-98.

ADDENDUM

While this paper was in press, Chaloner and Sheerin (1979) published a detailed discussion of Devonian Macrofloras. They discuss and illustrate evolutionary innovations (here called "new morphological characters introduced in each assemblage-zone"). They include some that are not mentioned here and they discuss phylogenetic considerations as well. It will be useful to read this paper and their paper in concert. Edwards (1979) added two form genera of probable vascular plants to the flora of the Pridolian (Downtonian) strata of Wales—*Hostinella* and *Psilophytites.* The latter genus bears enations (spines) and forces us to add enation to the list of morphological characteristics in Assemblage—zone I. Edwards, Bassett and Rogerson (1979) (Edwards, D.; M. G. Bassett; and E.C.W. Rogerson 1979. The earliest vascular land plants: continuing the search for proof. Lethaia 12: 313-324.) reported probable *Cooksonia* from Bringewoodian strata (middle Ludlow) of the Ludlow Series in Wales that is older than the *Cooksonia*-like fossils that are cited herein as the oldest occurrence of a vascular plant.

2

STRATIGRAPHIC AND GEOGRAPHIC OCCURRENCES OF PERMINERALIZED COAL-SWAMP PLANTS — UPPER CARBONIFEROUS OF NORTH AMERICA AND EUROPE

Tom L. Phillips

SUMMARY

The geographic areas in which coal balls occur in the Pennsylvanian of the United States and equivalent systems of Europe are indicated on maps, and the stratigraphic positions of coals yielding coal balls are summarized in a chart of five major coal regions: Donets Basin of the U.S.S.R.; the European countries; the Appalachian, Eastern Interior, and Western Interior coal regions of the United States. Stratigraphic positions of coal balls from the regions are largely complementary with some interregional correlations of coals. Selected coals from the more than sixty-five coals reported to have yielded coal balls are stratigraphically integrated and occurrences of coal-swamp taxa are compiled from the literature and original sources. The indexing of major stratigraphic zones of swamp vegetation, their geographic distribution, and the patterns of plant evolution and swamp paleoecology have biostratigraphic significance. These zones may also have economic implications as to the kinds of coal reserves available. To the extent that the precise stratigraphic position of a coal is unknown or that specific correlations between coal regions are uncertain, recognized zones of swamp plants from coal balls are subject to change. The precise stratigraphic determination of coal sources has not relied on their in situ coal-ball contents.

INTRODUCTION

Coal balls from bituminous coal seams in the Pennsylvanian of North America and equivalent systems of Europe are the most abundant sources of anatomically preserved vascular plants during their 400 million year history. In more than sixty-five coals across the Upper Carboniferous of Euramerica, from the Ukraine to Texas, in situ peat layers were permineralized by the concretionary formation of coal balls. These scattered in situ deposits in Euramerica extend across about one-tenth of vascular plant history from the Namurian of Czechoslovakia and Germany to near the Virgilian-Wolfcampian boundary in Texas.

It is not surprising that our most detailed fossil records of land plants should be those of swamp plants. The in situ occurrences of anatomically preserved plants from various swamp environments in so many coals in a geologic period will allow exceptional opportunities to study evolution and swamp paleoecology of in situ plant communities. In order to understand evolution in swamp genera, their reproduction, and their vegetative morphology, it is necessary to reconstruct the patterns of swamp paleoecology and interrelate plant communities and their peculiar swamp environments. This is obviously a very long term goal, but serious evolutionary studies of the coal-swamp plants are also long-term projects.

In order to determine patterns of vegetational composition and change in Pennsylvanian coal swamps, data are assembled on plant occurrences (as in this paper), swamp communities, and quantitative measures of peat profiles, which collectively relate also to bituminous coal reserves. Many of the properties of coals are related to their original botanical composition at the peat stage and to the swamp environments in which the plants grew and were buried. Hence, the biostratigraphic zones of swamp plants preserved in coal balls and their evolutionary and environmental changes also depict patterns with potential economic consequences. As we seek insight into evolutionary events in coal swamps, these same data can help in understanding many differences among coal seams and provide a basis for their selective exploration and beneficiation.

During the late Carboniferous the continental plates of Europe and North America were apparently juxtaposed, allowing the distribution of similar kinds of swamp plants and floras in similar kinds of environments as distant as those of Iowa and the Ukraine. The climate, while quite variable during the late Carboniferous, was generally semitropical to tropical with the paleoequator coinciding, at least in part, with the great Appalachian coalfield and perhaps extending across some northern European coal basins (Schopf, 1975). Two major vegetational events occurred during the late Carboniferous of Euramerica. Differentiation of world-wide floral provinces took place, hence the term Euramerican; and floras underwent drastic changes during the Westphalian-Stephanian transition. No coal-ball occurrences are known within the transition; it is the greatest single stratigraphic hiatus in coal-ball occurrences.

Less apparent have been the many changes in swamp vegetation and community structure because of the sequential introductions and subsequent evolutionary paleoecological diversification of major plant genera during the Pennsylvanian. The resultant patterns in changing swamp environments, during the Westphalian in particular, from one coal region to another are to be sought, in part, from data on coal-swamp plant occurrences recorded in the literature. While those occurrence data are still quite unevenly scattered stratigraphically and geographically, cooperative efforts can help delineate important regional patterns. The coal swamps of the Illinois Basin area have served as a guide thus far (Phillips et al., 1974) despite stratigraphic limitations in coal-ball sequences. These, however, can be complemented by and compared to swamp patterns of the other coal regions of Euramerica. Very important to interpretations of plant occurrences are quantitative data such as those given by Phillips, Kunz, and Mickish (1977).

It is an acceptable working hypothesis that Europe and North America shared many of the same species of swamp plants and swamp environments during the late Carboniferous and that changes observed in the coal basins of one continent are probably paralleled by similar kinds of changes on the other. We are particularly seeking to establish how similar the changes are and how they differ. While there are many variables that should not be over simplified, we can gain much insight about Euramerican swamp plant evolution and paleoecology by combining the coal-ball data from these once-contiguous continental plates. We have not emphasized collective similarities by combining the occurrence records of all the vascular plants reported from Euramerican coal balls. Indeed, when we combine them, we must continue to take into account the rigor of the various contributions, the utility of the taxonomic concepts involved, and the inherent limitations of the available data. This compilation is only a first attempt at union of the scattered data.

There are about equal numbers of coals with coal balls in North America and in Europe but their stratigraphic occurrences are largely complementary; most of the European occurrences are below the Westphalian C and all but a few American occurrences are above the Westphalian B. Near the Westphalian B-C boundary coal balls occur in all five major coal-ball regions of Euramerica: Donets Basin of the U.S.S.R., coal basins of western Europe, the Appalachian region, and the Eastern and Western Interior coal regions.

During the past 124 years of coal-ball plant studies there was a general trend of activities that followed roughly stratigraphic succession from Europe to North America. Today, studies are interregional and much less restricted stratigraphically in the United States. While there is very little hesitation about using the same generic names for the coal-swamp plants in North America as in Europe, the application of the same specific epithets has been quite cautious in most cases. It is only in the past few decades that studies, based on coal balls from near the Westphalian B-C boundary and/or on long-ranging Westphalian species, have begun to document occurrences of the same species on both continents.

The contributions of the past are already so numerous and from so many geographic locations and stratigraphic intervals as to be unwieldy in assessing what patterns of plant distribution, plant evolution, and vegetational changes have already been established or appear likely. Actually, the numbers of coal-ball localities, upwards of 200, and described taxa, more than 130 genera and 350 species, from coal-balls and other permineralized late Carboniferous peats of Euramerica are small compared to those of compression floras or spore floras, and minuscule compared to studies of modern floras in the same areas of Euramerica. In turn, practically all the available coal-ball data are derived from about thirty coals spanning about thirty-five million years. The unique aspect of the Euramerican coal swamps in the late Carboniferous, however, allows us to piece together at least an overview of the swamp plant patterns because of the limited kinds of important swamp plants and environments. The distribution of coal-ball bearing coals across the entire period provides the stratigraphic and geographic information uniquely suited to evolutionary and paleoecological studies. Other patterns evident in the plots of occurrences are mentioned later.

One aspect of the task of compilation is that histological preservation of stems, leaves, fructifications and root systems allows rather precise identifications, and such studies often require considerable detail for identifications at the species level. Thus, there are relatively few detailed lists of "floras"; one can hardly pursue any major group of swamp plants without encountering undescribed taxa, doubtful species, and other problems requiring monographic study. For the majority of common genera,their kinds of organs are known at least in part, but usually they can not be grouped into a natural assemblage fully representative of a species. The dismemberment of aerial portions of the swamp plants, as they became litter, or as their fructifications were dispersed, has necessitated the use of generic names for many plants based on isolated organs.

In turn, species of these plants, represented by the organs, have been described; thus, there are many more names for known swamp plants of a given age than there are actual known species in the conventional sense. Because of the evolutionary changes occurring during the Pennsylvanian, some species names have no doubt been applied to plants representing evolutionary segments or stages of natural phylogenies which appear distinct because of gaps in the stratigraphic record. Developmental stages or anatomical levels of some plants have also received different names, but much progress has been made with some genera in this area of research.

GENERAL OCCURRENCES OF COAL BALLS

Geographic Regions and Stratigraphic Distribution

Reported stratigraphic occurrences of coal balls vary considerably among coals in the five major coal-ball regions of Euramerica. While the stratigraphic positions of the coals with coal balls are largely complementary in the Upper Carboniferous, there is an important gap in plant occurrences adjacent to the Westphalian-Stephanian boundary. Interregional correlations between some of the coals have been established, but there are many uncertain interregional relationships between coals or permineralized plant deposits of about the same age.

In the Donets Basin of the Ukraine, U.S.S.R., at least twenty-two coals (see Figure 2.1) have yielded coal balls (Zaritsky, 1959; Snigirevskaya, 1972) from Suites C2/2 through C7/2 of the Middle Carboniferous. Coal balls have not been reported from the Russian Upper Carboniferous, and the text usage of Upper Carboniferous refers to western Europe. Zaritsky's (1971; Zaritsky, Makedonov, and Salnikova, 1971) broad geological studies of concretions, including coal balls, indicated that carbonate concretions occur in twenty-nine of the middle Carboniferous coals of the Donets. Coal balls have been obtained for current paleobotanical studies from eighteen mines in the Donets (Figure 2.2) (N.S. Snigirevskaya pers. comm. 1976). While the number of plant occurrences reported by M. D. Zalessky and by N. S. Snigirevskaya (see References) is relatively small compared to the many coals, it is important to note that the plant taxa are the same as or very similar to those of other European (Zalessky, 1910b) and American coal swamps (Snigirevskaya, 1972) of about the same age. It is anticipated that more detailed data will be available from the U.S.S.R. in the near future.

Elsewhere in Europe at least six stratigraphic occurrences of permineralized peats occur (Hirmer, 1928; Schopf, 1941a). The countries include England, The Netherlands, Belgium, France, Germany (B.R.D.), Poland, Czechoslovakia, and Yugoslavia or adjacent areas in Hungary with a questionable occurrence in Spain (Figure 2.2). These include the earliest coal-swamp floras known in detail especially those from the Union or Halifax Hard seam of England (lower Westphalian A) and continental equivalents, the Bouxharmont of Belgium and the Finefrau-Nebenbank of the Netherlands and Germany; the Katharina seam at the Westphalian A-B boundary has also yielded many coal balls in the Ruhr and Aachen coal districts.

While the number of seams in the lower Westphalian that contain coal balls are few, the localities are numerous and the detail of data provided by the summaries of Stopes and Watson (1909),

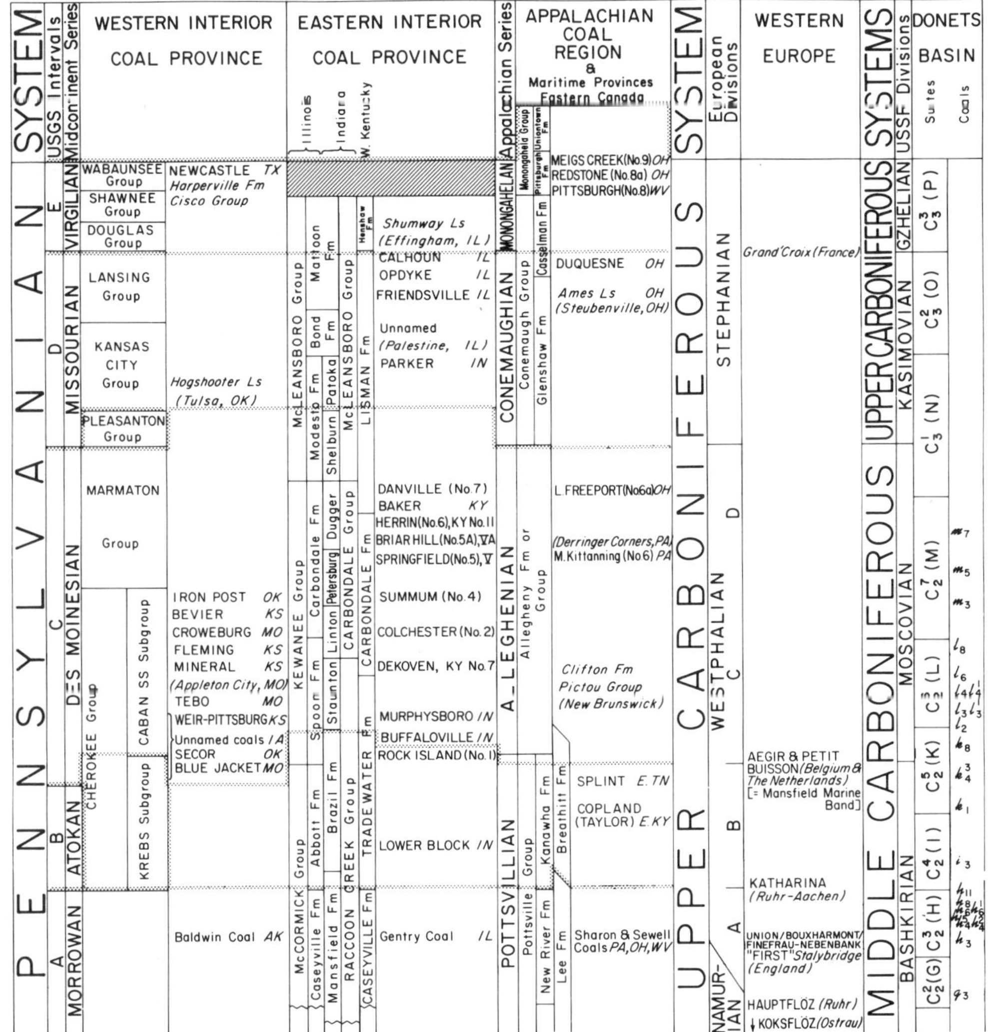

FIGURE 2.1 *Stratigraphic occurrences of coal balls in the Pennsylvanian of Euramerica.*

FIGURE 2.2 *Permineralized Upper Carboniferous plant deposits of Europe.*

Leclercq (1925), Koopmans (1928), and Hirmer (1928) are unsurpassed.

Above and below (Figure 2.1)these major western European sources of information are some important but less well-known occurrences of coal balls. These occur in the Namurian of Czechoslovakia (Koksflöz) and the Ruhr (Hauptflöz) (Stur, 1885; Kubart, 1914; Teichmuller and Schonefeld, 1955), and near the Westphalian B-C boundary in The Netherlands (Aegir) and in Belgium (Petit Buisson) (Koopmans, 1934; Leclercq, 1952). While there are no other bona fide occurrences of coal balls of established late Carboniferous age in western Europe, anatomically preserved plants from the Westphalian have been described from collections at coal mines in the Radnitz basin of Czechoslovakia (Corda, 1845; Kubart, 1911), and there is a report of coal balls in the province of Asturias in northwest Spain (Renier, 1926b), The silicified plants of the Stephanian A, Grand'Croix flora from the Loire (St. Etienne) basin of France are included in this compilation because of their importance in comparisons with coal-ball plant occurrences of the same approximate age. About 10 percent of the genera and almost 15 percent of the species of the Euramerican "coal-ball" floras have been described from them.

The Appalachian region of North America is the most recent area of coal-ball discoveries. The nine coals in which they have been found extend from at least as low as the Copland (Taylor) coal bed, Breathitt Fm, Pottsvillian Series of eastern Kentucky to coals in the Pittsburgh Fm, Monongahelan Series of West Virginia and Ohio. Coal balls also occur in Tennessee and Pennsylvania (Figure 2.3). There are few reports of coal balls from the Alleghenian whereas most of the coal balls in the Interior coal regions are of equivalent age (Des Moinesian). Early reports of Appalachian coal balls were given by Foster and Feicht (1946), Cross (1952, 1967), and Schopf (1961). The report by Baxter (1960) of coal balls in New Brunswick, Canada, is included with this region (Figure 2.1). Cross (1969) first reported a number of American coal-ball occurrences in his summary of Euramerican coal-ball stratigraphy. Since that time, coal balls have been reported by Good and Taylor(1974), McLaughlin and Reaugh (1976), Rothwell (1976b), and McCullough (1977). Of the Appalachian occurrences, those that are particularly important in the stratigraphic overlap between European and American coal balls are the ones in eastern Kentucky and eastern Tennessee. The Aegir horizon of Belgium and The Netherlands has similar significance as do the k_8, l_2, and l_3 coals from the Donets Basin.

The stratigraphic reference section of coal balls in the United States is that of the Eastern Interior coal region or the Illinois Basin (Illinois and adjacent portions of western Indiana and western Kentucky) which has seventeen coals containing coal balls. These range from possibly as low as the Lower Block Coal Member in the Brazil Fm, Pottsvillian of Indiana, up to the unnamed coal below the Shumway Limestone in the Mattoon Fm (Figure 2.1). In midcontinent terminology, the coal balls occur from the Atokan to the Virgilian. The occurrences of coal balls in this region are summarized by Phillips, Pfefferkorn, and Peppers (1973); and the broad stratigraphic patterns of changes in coal-swamp vegetation in the Illinois Basin are based, in part, on data from coal balls (Phillips et al., 1974). Coal balls are known from more than sixty mines and a number of stream-bank localities; most of these are shown in Figure 2.4. More than one-half of the localities are for the Springfield and Herrin Coal Members, which form the upper part of a successive group of Carbondale coals with coal balls, beginning with the Colchester Coal

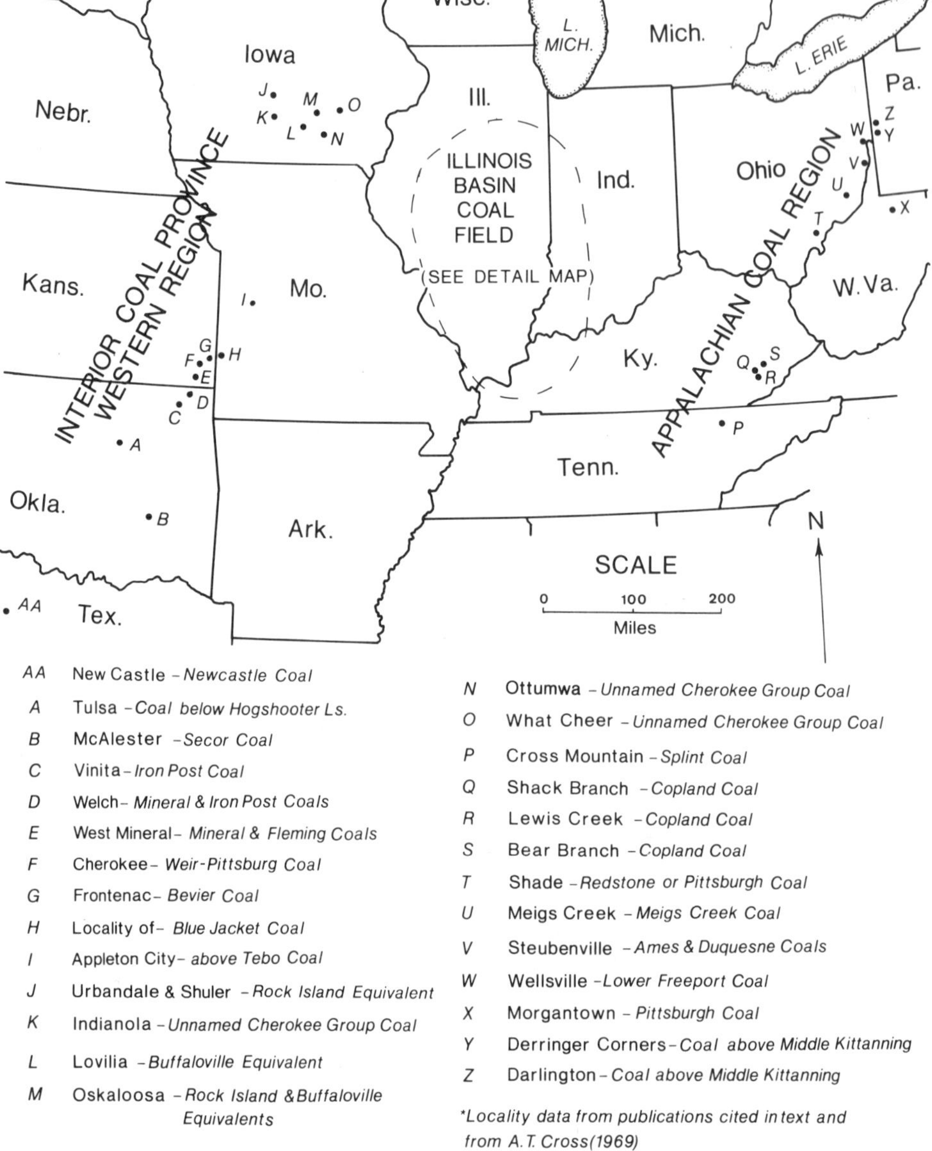

FIGURE 2.3 *Major U.S. coal-ball sites.*

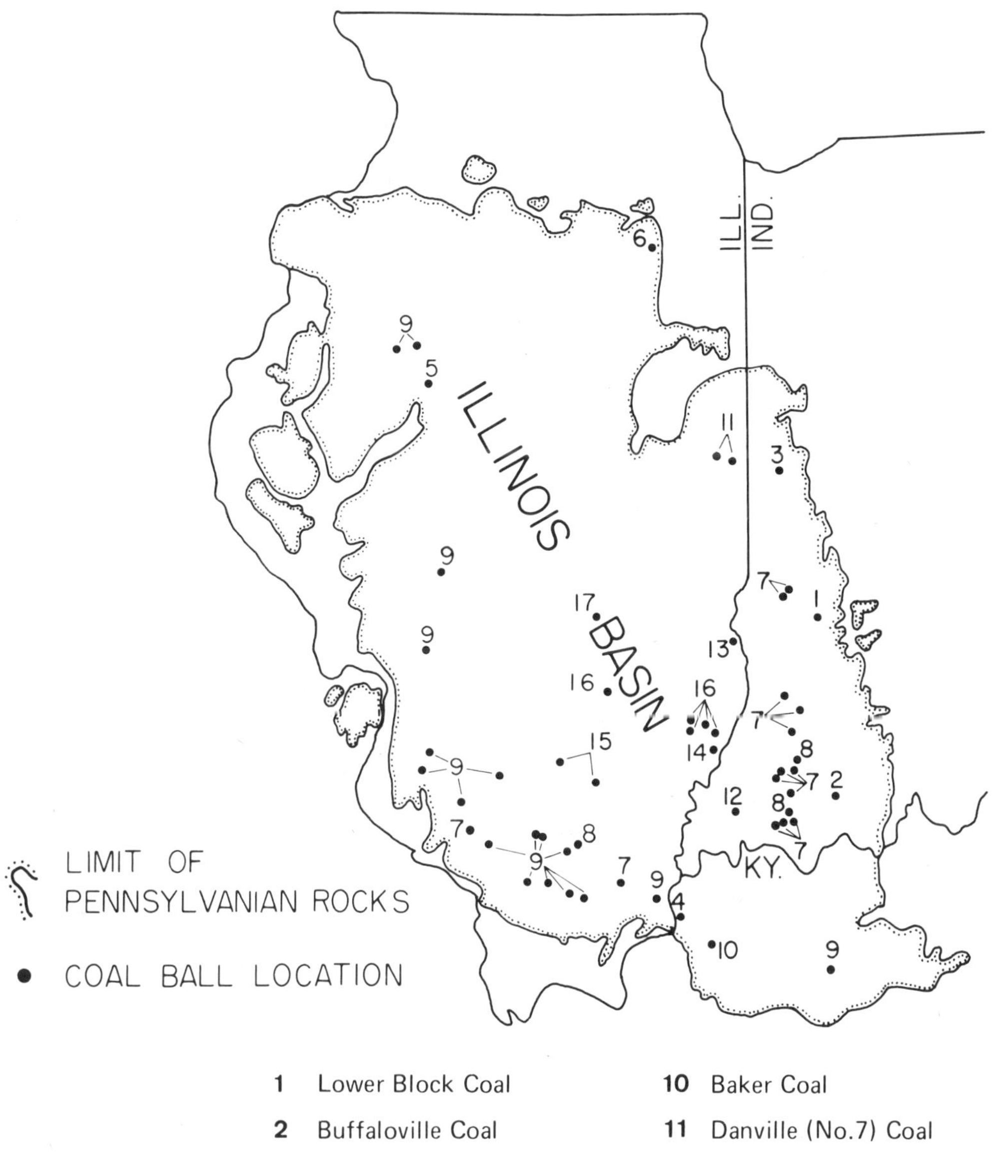

1 Lower Block Coal
2 Buffaloville Coal
3 Murphysboro Equivalent
4 DeKoven Coal
5 Colchester (No.2) Coal
6 Summum (No.4) Coal
7 Springfield (No.5) Coal
8 Briar Hill (No.5A) Coal
9 Herrin (No.6) Coal
10 Baker Coal
11 Danville (No.7) Coal
12 Parker Coal
13 Unnamed Coal
14 Friendsville Coal
15 Opdyke Coal
16 Calhoun Coal
17 Unnamed coal member of the Shumway Cyclothem

FIGURE 2.4 *Occurrence of coal balls in the Illinois Basin.*

Member. These permit some precise stratigraphic determinations of swamp changes and related plant occurrences; some of the coals in the Illinois Basin can be correlated with those of the Appalachian and Western Interior coal regions.

At least twelve coals in the Western Interior coal region of the United States have coal balls. Most of them are in Iowa, Kansas, Oklahoma, and Missouri (Figure 2.3). The coals range from at least as low in the Cherokee Group as the Secor (Oklahoma), Blue Jacket (Missouri), and a coal in Iowa, all of which are approximately equivalent to the Rock Island (No.1) Coal Member of northwestern Illinois, up to at least the Newcastle Coal in the Harperville Fm of the Cisco Group in northcentral Texas. The youngest of coal balls in North America, some of which are Permian in age, occur in the Young County area around Newcastle, Texas. Precise stratigraphic determinations for many Iowa coal balls are lacking; some correlate with the Rock Island (No. 1) Coal Member of Illinois, some are about at the position of the Buffaloville and others may occur within the approximate interval indicated by the bracket in Figure 2.1.

Stratigraphic Chart for Coal-Ball Occurrences

The stratigraphic chart (Figure 2.1) contains a minimum of formations and groups for each region and the series or division names that are used in those regions. An oversimplified arrangement has been used to consolidate stratigraphic occurrences of coal balls whereby a single listing is used for an entire region and, with the exception of the Eastern Interior Coal Province, only one stratigraphic column is given for American ones. This has necessitated adopting a generalized column for the Western Interior Coal Province, which is largely applicable to Iowa, Kansas, Missouri, and Oklahoma but not to Texas; the Newcastle Coal is simply inserted with formation and group names. The same has been done with the New Brunswick occurrence. The appropriate central Appalachian units have been added to the northern Appalachian section. In the Illinois Basin, both Midcontinent and Appalachian Series are recognized in Indiana and western Kentucky.

The use of Lower, Middle, and Upper Pennsylvanian in the United States for stratigraphic occurrences of coal balls has been quite irregular in the literature. The U.S. Geological Survey policy (Bradley, 1956, p. 2285) for the Appalachian region is: "The division between Lower and Middle Pennsylvanian is at the top of the New River and equivalent rocks, and the division between Middle and Upper Pennsylvanian is at the approximate boundary between the Allegheny and the Conemaugh." For the Eastern and Western Interior coal fields it states: "Lower is equivalent to the Morrow, Middle to the Atoka and Des Moines, and Upper to the Missouri and Virgil of the Mid-continent region." According to such usage, there are no known coal balls from the Lower Pennsylvanian of the United States.

In an attempt to relate the stratigraphic rock units of the series or divisions of the Pennsylvanian of the United States, the U.S. Geological Survey has instituted Intervals A through E (see plate 13 and articles in McKee and Crosby, 1975) which are indicated on the stratigraphic chart (dotted pattern) according to the more detailed treatments provided by Wanless (1975 a,b,c) on the Appalachian region, Illinois basin region, and Missouri and Iowa; by Stewart (1975) in Kansas, and by Frezon and Dixon (1975) for Oklahoma. Approximate stratigraphic relationships among American regions are derived, in part, from the above studies with the use of detailed state-based charts; but it should be noted that the above studies did not deal specifically with the interregional problems of coal correlations. Stratigraphic positions of coals in relationship to members used in the above articles were derived from Searight and Howe (1961), for Missouri; Jewett et al. (1968), for Kansas; Kosanke et al. (1960) and Hopkins and Simon (1975), for Illinois and equivalent coals; Shaver et al. (1970), for Indiana; Mullins, Lounsbury, and Hodgson (1965), for western Kentucky; Brant and De Long (1960), for Ohio; Dutcher et al. (1959), for western Pennsylvania; and the major correlation work by Wanless (1939). The reader is also referred to correlation charts by Moore et al. (1944), Havlena (1967), Aizenverg et al. (1960, 1975), and others in the publications of the International Congresses on Carboniferous Stratigraphy and Geology. The attempt to stratigraphically relate interregional and intercontinental coal-ball occurrences in the Upper Carboniferous obviously has many inherent uncertainties that could not be resolved.

STRATIGRAPHIC OCCURRENCES AND LOCALITIES OF COAL BALLS

Background and literature citations on important coal-ball sources, with comments on aspects of their stratigraphy, are given in the following summaries. They are discussed in stratigraphic sequence, first the European ones, then the American.

EUROPEAN COAL BALLS

There are some Lower Carboniferous occurrences of coal balls that should be mentioned although they are not included in this compilation. Coal-ball plants have been reported from the Visean (Lower Carboniferous) of Haltwhistle, Northumberland, England, in the Little Limestone coal, Carboniferous Limestone series (Absalom, 1929); from the Czechoslovakian part of the basin in the Jaklovec Series (Namurian A), by Dopita and Králik (1971) and from the "Upper Jankowice Seam" in the Poruba beds of the Polish part of the Cracow-Silesia basin, near Rybnik (Brzyski, 1965).

Namurian

The earliest report of coal balls in the Namurian was by Stur (1885); later Kubart (1908, 1910, 1914, 1931) described some of the plants. The coal balls occur in the Koksflöz Seam of the Ostrau-Karwin (Ostrava-Karviná) basin near Orlau, Czechoslovakia. The Koksflöz Seam is the highest in the Rand Group (Kubart, 1914, p. 15) which is below the Sattelflöz Group; the upper part of the latter is correlative with the Magerkohlen (Sprockhöveler) Schichten of the Ruhr (Namurian C).

Coal balls have been described from the Hauptflöz Seam, which is the upper coal in the Sprockhöveler Fm and near the top of the Namurian C in the Ruhr (Leggewie and Schonefeld, 1957, 1961). The silicified coal balls were obtained from the St. Hubertus Mine southwest of Essen-Kupferdreh (Teichmüller and Schonefeld, 1955; Leggewie and Schonefeld, 1959) and near Essen-Werden (Haas, 1975).

Westphalian A

The triangular-shaped Lancashire coalfield in England, where coal balls were first discovered (Hooker and Binney, 1855) in the Lower Coal Measures, is about 500 square miles, with the northern apex just N. of Burnley. The eastern boundary extends southward through Littleborough to Stalybridge, Cheshire. The coals yielding coal balls are the "First" coal at a road cut near Stalybridge and at the Hough Hill Colliery, Stalybridge, Cheshire (Stopes and Watson, 1909) and at localities extending northward from Oldham and around Littleborough in the Upper Foot Seam (Rochdale coal basin), which, in turn, joins the Lower Mountain Coal to form the Union Seam of the

Burnley coal basin. The latter contains most of the coal-ball localities of Lancashire. The Lancashire Lower Mountain, Upper Foot, and Union coals, and the Yorkshire equivalent, Halifax Hard coal, are those recognized by the Geological Survey of Great Britain in the "Seam Coding System for Coalfields of the Central Province of Great Britain," published in 1964. The "First" coal and the next higher Union or Halifax Hard are in the lower part of the Lower Coal Measures and are early Westphalian A in age.

The Lower Mountain and Upper Foot coals join to form the Union with the "line of union" occurring along a NW-SE line passing a few km SW of Burnley, roughly from Padiham to between Bacup and Todmorden, swinging eastward toward Yorkshire. Coal-ball sites near the "line of union" include Hapton Valley, Bacup, Old Meadows Pit, Cloughfoot, Dulesgate, Hill Top, Todmorden, and others. There are differences in the known distribution of some plants, particularly the cordaites, in the Lancashire coal field. The vegetation from localities along the "line of union" does not appear to be different from that known northward in the Burnley coal basin with the exception of Dulesgate. It is to the south of the "line of union" at Shore-Littleborough and near Oldham in the Upper Foot Seam where the greatest diversity in the swamp vegetation is encountered. The information is derived mostly from the coal-balls and some roof-nodules obtained at the Shore Mine of W. H. Sutcliffe. For this reason, the Upper Foot of Shore and Oldham is listed separately in occurrence charts; part of its contrasting diversity to the Union Seam is obscured by combining plant occurrences from Belgium, The Netherlands, and the Ruhr. There are separate occurrence lists for many of these western European localities (Stopes and Watson, 1909; Leclercq, 1925; Koopmans, 1928; Hirmer, 1928).

In Belgium, the Bouxharmont coal bed and equivalent coals (Beaujardin, Sainte Barbe, and Saurue) have yielded coal balls at Jupille, in the Hasard-Cheratte Mine at Micheroux near Liège and at the Wérister Mine near Fléron (also near Liège) (Leclercq, 1925, p. 73; Humblet, 1941; M. Streel, pers. comm., 1975). Coal balls in The Netherlands occurred in the Finefrau-Nebenbank equivalent at the Domaniale Mine at Kerkrade, Dutch Limburg (Koopmans, 1928) and at the Rheinpreussen Mine in the Ruhr.

The earliest recognition of coal balls in the Donets Basin was by Zalessky (1910b) from Coal h2/4 (first coal of Kalmious) of Suite C3/2, stratigraphically between the "Great Coal-ball Horizon" and the Westphalian A-B boundary (Hirmer, 1928; Zalessky, 1928).

Westphalian A-B Boundary

Coal balls have been found in the Katharina Seam of the Ruhr district of Germany in mines including the Dahlbursch, Dorstfeld, Hansa, Kölner, Vollmond, and Werne, and from areas around Altenessen, Duisberg, Hattingen, Langendreer, and Würm (Mentzel, 1904; Kukuk, 1909; slide collections at the Müseum für Paläeontologie, Berlin). Plants described by Felix (1886) were obtained from the Vollmond Mine at Langendreer, near Bochum, North Rhine-Westphalia, B.R.D. Felix (1886) referred to the seam as the Isabella. At Aachen (Aix-la-Chapelle) in the No. 1 Seam (equivalent to the Katharina), coal balls occurred at the Maria Mine.

Westphalian B-C Boundary

The second source of coal balls in Belgium and The Netherlands has yielded relatively few coal balls but it is of considerable importance in the stratigraphic comparisons of occurrences of European and American swamp plants. In Belgium the coal balls occur in the Petit Buisson at Bray (Bassin du Centre) (Leclercq, 1925; Renier, 1926a) and in the Maurage Horizon in the coal basin of Campine (Leclercq, 1952). These correspond to the Aegir Horizon at the top of the Westphalian B of The Netherlands and Germany. Coal balls were obtained from the Aegir in the Netherlands at the Emma Mine near Heerlen (Koopmans, 1934).

Westphalian Equivalents-U.S.S.R.

In 1910 Zalessky summarized the then rare occurrences of anatomically preserved plants in the Carboniferous of the Donets (Zalessky, 1910a), and in a subsequent paper he announced the discovery of coal balls in the Donets (Zalessky, 1910b). In the first paper he referred in part to two such occurrences below limestone I_3 in Suite C4/2, from which he illustrated *Calamites* and *Sphenophyllum*, and below limestone K_5 in Suite C5/2, from which his *Lepidostrobus bertrandii* was described (Zalessky, 1908). These species were not collected by Zalessky and at the time were not considered to be from coal balls. When he summarized the five stratigraphic occurrences of coal balls in the Donets, no coal from Suites C4/2 or C5/2 was listed (Zalessky, 1928).

Most of the recent studies of coal balls in the Donets basin have been based on specimens from coal k_8, Suite C5/2, Moscovian, near Pervomaisk in the Lugansk Region (Snigirevskaya, 1958, 1961, 1962b, 1964, 1967). The k_8 is in the lower Westphalian C and is about the same (or slightly older) age as the Rock Island Coal of the Illinois Basin. Other coals of Moscovian age extend across the Westphalian C into the D. Those coals with coal balls in the Donets Basin that are listed by Snigirevskaya (1972) are included on the stratigraphic chart.

Stephanian A

Plants from the Grand'Croix area of France are in silicified limnic peats not found in coal. The redeposited permineralized specimens occur in two layers of a massive conglomerate, the Poudinges de Grand'Croix, several hundred m above the coal measures of and near the top of the Rive de Gier Fm, which is late Stephanian A (de Maistre, 1963a,b; Vetter, 1971).

AMERICAN COAL BALLS

The stratigraphic positions and nomenclature of American coals with coal balls (Figure 2.1) largely follow the previous charts by Schopf (1941a) and by Cross (1969); most of the important localities are well established in recent literature cited herein and comments are restricted to certain of the coals. There are some changes and explanations necessary for certain localities and for some of the stratigraphic positions of coals on the chart (Figure 2.1). These are dealt with in ascending stratigraphic order.

Middle Pennsylvanian-Interval B

Lower Block Coal Member of Indiana

Coal balls from the Lower Block Coal, Brazil Fm, upper Pottsville, of Greene Co., Indiana were stratigraphically identified as to source by means of Indiana Geological Survey collection numbers and location. A search of the abandoned Long and Price Strip Mine area and exposed profiles of the Lower Block could not verify the occurrence of coal balls in the coal at that site. The plant assemblages in the coal balls are unusual because of the abundance of calamites and fern foliage but there is no resolution as to the age or locality of the coal balls. Because of such reservations, no plots of plant occurrences are given.

Copland (Taylor) Coal Bed of Eastern Kentucky

The Copland coal bed below the Magoffin Member of the Breathitt Fm of the upper Pottsville in the Central Appalachian basin is considered the oldest known coal-ball source in North America (Schopf, 1961). Wanless (1975a) correlates the Magoffin Limestone Member of the Breathitt Fm of Kentucky with the lower Mercer Limestone Member of the Pottsville of Ohio and Pennsylvania and with the Winifrede Limestone Member of the Kanawha Fm of West Virginia. These are placed above the middle of the Kanawha, and the Copland coal bed below what would be about middle Westphalian B in western European terminology. In the literature (Good and Taylor, 1974) they are referred to as Lower Pennsylvanian; if the Lower-Middle Pennsylvanian boundary is between the New River Fm and Kanawha Fm (Appalachian), then the coal balls are from the lower part of the Middle Pennsylvanian.

"Splint Coal" of Cross Mountain-Tennessee

The southern most occurrence of coal balls in the Appalachian region is from the Cross Mountain area near Careyville, Anderson County, eastern Tennessee (McLaughlin and Reaugh, 1976). The seam is referred to as the Splint coal and its stratigraphic position, as shown on the chart (Figure 2.1) is uncertain. According to McLaughlin and Reaugh (1976), it could be post-Pottsvillian (lower Allegheny), but Pennsylvanian strata younger than Kanawhan have not been found in Tennessee (Wanless, 1975a). The coal balls could possibly be as old as those from eastern Kentucky or at least as old as the Kanawha, but the stratigraphic details of the occurrence have not yet been published. Studies of the flora are at a preliminary stage (R. E. McLaughlin and Edward Masuoka, pers. comm., 1978).

Middle Pennsylvanian-Interval B-C

Rock Island (No. 1) and Buffaloville Coal Members of Illinois Basin and Iowa Coal Balls

The Rock Island (No. 1) Coal of Illinois is an important reference coal, hence it has been placed on the chart (Figure 2.1) even though no coal balls are known from it in Illinois. According to R. A. Peppers (pers. comm., 1978), at least one coal-ball source in Iowa correlates with the Rock Island. The coal was mined at the Angus Coal Company strip mine about 1 km NW of Oskaloosa, Iowa. Other midcontinent coal-ball sources stratigraphically placed at about the same position with the Rock Island are the Secor Coal from near McAlester, Oklahoma, and the Blue Jacket Coal of Missouri. Since examined collections from the Secor Coal of Oklahoma are small, plant occurrences are given separately from those of Iowa of about the same age. The Secor Coal occurs in the Boggy Fm near McAlester (Hendricks, 1937; see also Mamay and Yochelson, 1962.)

At three other Iowa mines, one long-since abandoned and two active, coal balls occur in a coal that correlates approximately with the Buffaloville of Indiana, just above the Rock Island (Peppers, pers. comm., 1978): the Atlas (= Lost Creek) Mine, 5 km NW of Eddyville near Oskaloosa; the Star Mine SW of Oskaloosa; and the Weldon Mine just N of Lovilia. In an attempt to stratigraphically bracket the unnamed coal near Lovilia with coal-ball plant occurrences, Schabilion, Brotzman, and Phillips (1974) placed it higher. Plant occurrences from these three mines are plotted as the Buffaloville equivalent. Coal balls do occur in the Buffaloville Coal Member of Indiana, but the collections are too small to merit plots with those from Iowa.

There are many coal-ball localities in unidentified coals in Iowa (see map in Darrah, 1941b, p. 36) from which very informative coal balls have been studied. Some of these include the following mines: Banner (near Indianola), Carbon Hill, Givin, Ford, Mich, Patik, Shuler, Shute and Lewis (near Ottumwa), Tillotson (near Ottumwa), Urbandale, What Cheer, and Williamson (3, 4, 5). Many of these Iowa localities were actually the tipples of mines that were abandoned at the time and some were probably from common gob areas for several pit operations, mostly in the Oskaloosa-Eddyville area. The sites shown in Figure 2.3 roughly encircle the area of known sources of Iowa coal balls. Iowan coals have been, in part, named (Landis and Van Eck, 1965); the problem is the lack of correlation and identification of the coals. The Bevier Coal Member occurs in Iowa, but inferences that any coal-ball locality occurs in that coal could not be verified (M. J. Avcin, pers. comm. 1978). Consequently, the plant occurrences plotted for the coal equivalents to the Rock Island represent the collective data for the vast majority of coal-ball studies in Iowa. Most of them are undoubtedly from the lower, middle part of the Cherokee Group, but as the type section of the Des Moinesian is in Iowa and its lower limit is as yet undetermined, the stratigraphic occurrences of the plants should be considered accordingly.

MIDDLE PENNSYLVANIAN-INTERVAL C

Weir-Pittsburgh Coal Member of Kansas

Coal balls were first discovered in the Weir-Pittsburgh by Darrah (1941b) and most of the subsequent studies of the plants has been by Leisman (see citations). The Weir-Pittsburgh is economically the most important coal in Kansas. All plant occurrences given are from the literature. Their paucity is closely related to the highly pyritic nature of most of the coal balls.

Tebo Coal Member of Missouri

Coal balls have been collected from *above* the Tebo Coal at the Pioneer Mining Company near Appleton City, Missouri. Little information on the plants is available.

Murphysboro Coal Member of Illinois and Reports in the Minshall and "Silverwood" of Indiana

Coal balls were reported from the Indiana Coal II ("Silverwood") at the Silver Island Strip Mine near Cayuga, Indiana by Feliciano (1942); Noé (1925) collected some of these and the collections have been recently reexamined. Schopf (1941a) indicated that according to Wanless (1939) the coal was actually the Minshall (correlative with the Rock Island of Illinois; Willman et al., 1975). During the past five years, abundant coal-ball collections have been studied from the same coal at the Maple Grove Strip Mine, also near Cayuga, Indiana; and according to Peppers (pers. comm., 1978) the unnamed Indiana coal is equivalent to the Murphysboro of Illinois, which is above the Buffaloville.

Coal in the Clifton Fm, New Brunswick, Canada

Baxter (1960) reported the Canadian coal balls from an unidentified coal in the Clifton Fm, Pictou Group. According to Hacquebard (1972), the Clifton Fm occurs in spore *zones a* and *b* of Barss and Hacquebard (1967) which correlate with the Westphalian C.

Mineral-Fleming Coal Members of Kansas

Where two or more seams are locally mined, the specific seam sources for discarded coal balls are not known. This is typically the case at the Pittsburg and Midway Coal Company mines in southeastern Kansas, where both the Mineral and Fleming Coals are mined. Coal balls occur much more abundantly in the Fleming (Mamay and Yochelson, 1962). The major locality is often given as West Mineral (or Mineral or Hallowell), Kansas. Stratigraphic occurrences from the literature and from collections are plotted for the Mineral and Fleming jointly because of such uncertainties. The coal-ball plants occurring in Kansas have been listed by Baxter and Hornbaker (1965) in a comprehensive treatment of such paleobotanical studies with referenced occurrences of taxa. Coal balls also occur in the DeKoven or Kentucky No. 7 Coal Member of the Illinois Basin, which is correlated with the Mineral Coal.

Colchester (No. 2) and Croweburg Coals

The Colchester Coal and its regional equivalents (including the Lower Kittanning Coal of the Appalachian region) have yielded relatively few coal balls in recent times although descriptions of coal-ball masses exist in the literature (Cady, 1915, p. 76) Small collections have been made from the Banner Mine just west of Peoria, Illinois (Damberger, 1970) and from the Croweburg from Vernon County, in extreme western Missouri. While none of these plants have been previously reported for the Colchester and equivalents, some coal-ball plants have been attributed erroneously to this coal from northeastern Illinois. The major coal-ball seam in the same area near Pit 11 (Mazon Creek biota) is the Summum (No. 4) Coal, which was mined in Pit 14. The coal balls were discarded in the spoils of Pit 11.

Bevier Coal Member of Kansas

Coal balls were first reported from the Bevier Coal in southeastern Kansas by Baxter (1951; see also Leisman, 1961; Mamay and Yochelson, 1962, pp. 201-202).

Iron Post Coal Member of Oklahoma

Coal balls have been reported from coal mines near Vinita and Welch, Craig County, in northeastern Oklahoma by Perkins (1976), who has shown coal-ball locations for the Iron Post as well as in the Mineral, Fleming, and Bevier Coals. The Iron Post is the upper-most coal in the Senora Fm of the Cabaniss Group (Subgroup on chart in Figure 2.1) in Oklahoma (Branson and Huffman, 1965). The plant occurrences for the Iron Post were derived from Perkins (1976) with the aid of some peels loaned by R. W. Baxter for species identifications of the lycopod trees. Cross (1969) first reported coal balls from the Iron Post.

Springfield-Harrisburg (No. 5) and Correlative Coals

More than one-half of the coal-ball occurrences in the Springfield Coal (formerly called Coal V) of Indiana are given by Canright (1959); they now number more than twenty in the Illinois Basin. While the coal balls from the Springfield Coal have not been studied as intensively as those of the Herrin, it has become evident that the vegetation of these two enormous deltaic swamps are quite similar where they can be compared from the same kinds of habitats. Consequently, plant occurrences are quite similar to the extent known. With few exceptions, however, those from the Springfield Coal are derived from upper delta regions and usually near stream channels, while those from the Herrin are all from the lower delta area.

The Middle Kittanning Coal of the Appalachian region is considered to be correlative with one of the coals in the Springfield, Briar Hill, and Herrin sequence. Wanless (1939, p.72) proposed correlation with the Herrin; Peppers (1970, p. 60) indicated correlation with the Springfield as more likely; and Kosanke (1973, p. 19) stated "that the Middle Kittanning-Princess No. 7 coal more than likely is not older than the Springfield, and it might be as young as the Briar Hill (No. 5a) coal of Illinois." The correlation of the Middle Kittanning is pertinent to the report of coal balls by Cross (1967) from a thin coal below the Washingtonville Marine Shale and above the Middle Kittanning in Lawrence County, western Pennsylvania (see Good, 1973). The locality is known as Derringer Corners or Darlington. In a field guide to the same area the coal balls were attributed to the Upper Kittanning (Ferm and Cavaroc, 1969).

Briar Hill (No. 5a) Coal Member of Illinois Basin

The Briar Hill Coal occurs in southeastern Illinois and adjacent portions of Indiana and Kentucky (the No. 10 Coal). Small coal-ball collections have been obtained from drill cores in southern Illinois and from the unnamed equivalent coal (formerly Coal Va) in Pike and Warrick Counties in western Indiana.

Herrin (No. 6) Coal and Correlative Coals

The Herrin Coal and its equivalent in western Kentucky (No. 11) have been the source of more coal balls than has any other American coal; most of the published literature on plant occurrences in the Herrin is based on plants from the vast Sahara Coal Company Mine No. 6 in southern Illinois between Harrisburg and Marion. Coal balls are known to occur in more than twenty-five mines across the lower delta area of the Herrin. It is the most economically important coal in Illinois.

Baker Coal Member of Western Kentucky

The coal balls from the abandoned Hart and Hart Coal Company Mine near Providence, Webster County, Kentucky, have been attributed to the Kentucky No. 11, 12, or 14 Coal. All of these are apparently from the same coal source known as the Baker Coal Member (R. A. Peppers, pers. comm., 1978) which is called the Allenby Coal in Illinois (Hopkins and Simon, 1975).

Danville (No. 7) Coal Member of Illinois

The Danville Coal has been a source of coal balls. However, early studies of the plants failed to distinguish the varied stratigraphic sources of specimens described "from the McLeansboro." Since 1960 (see Hopkins and Simon, 1975, p. 171), the Danville Coal has been the uppermost bed of the Carbondale Fm rather than at the bottom of the McLeansboro. While there is no basis for correlating the Lower Freeport Coal of Ohio with the Danville, the Lower Freeport apparently occurs at a stratigraphic position near the Danville.

Upper Pennsylvanian-Interval D

Until fairly recently, the known Upper Pennsylvanian coal-ball sources were largely restricted to the Illinois Basin. These include the Parker Coal of Indiana and those coals (Figure 2.1) from the unnamed coal near Palestine to the best known--the Calhoun Coal--all in the center of the basin. Cross (1969) reported another coal, below the Hogshooter Limestone near Tulsa, Oklahoma; and Rothwell (1976b) has reported two for which he has given lists of plant occurrences--the coal below the Ames Limestone and the Duquesne Coal, both near Steubenville, Ohio. According to fusulinid studies by Thompson (1936) and by Dunbar and Henbest (1942), the Ames Limestone could be placed as high as the Omega Limestone Member in the Illinois Basin.

Upper Pennsylvanian-Interval E

Relatively little detailed information is available about the swamp plant occurrences in Interval E. The coal balls from the coal in the Shumway Cyclothem of Illinois and those from the Redstone and/or Pittsburgh Coal are siliceous. The Ohio occurrence of coal balls reported by Good and Taylor (1974) indicates that they were not certain whether the coal is the Redstone or possibly the Pittsburgh Coal from which Cross (1952) had reported coal balls.

Newcastle Coal of Young County, Texas

Coal balls from Texas were briefly mentioned by Noe (1923), Feliciano (1924), and Reed (1941b, 1949). Schopf (1950) pointed out that the Newcastle Coal of northcentral Texas was from the Virgilian and the youngest source of coal balls in the United States. The Newcastle Coal was mined in Young County around Newcastle from 1908 into the early 1930s by the Belknap Coal Company (Evans,

1974). Cheney (1940, figs. 1, 2) placed the Newcastle in the Obregon Fm near the top of the Thrifty Group, Cisco Series; now it is placed in the Harperville Fm (Cisco Group) which extends across part of the late Virgilian and early Wolfcampian (Brown, Cleaves and Erxleben, 1973). The Newcastle is the major coal in the Harperville and occurs very near the undefined Pennsylvanian-Permian boundary within the Harperville. The coal balls are dolomitic, as are those from the lower Wolfcampian (Permian), which occur slightly higher (below the Saddle Creek Limestone) between Graham and Newcastle in Young County (Jizba, 1962). One species of *Medullosa* has also been described from the Moran Fm, Wichita Group, near the base of the Permian in Young County, 16 km W of Newcastle (Roberts and Barghoorn, 1952).

STRATIGRAPHIC OCCURRENCES AND RANGES OF PLANTS

It is fortunate in the development of systematic-evolutionary studies of coal-swamp plants from coal balls that the general historical trend of activities somewhat followed stratigraphic succession from Europe to North America; this has emphasized phylogenetic relationships and generated much interest in European coal-ball studies by the American workers. However, the taxonomy of American plants from coal balls, especially at the species level, developed under the constraints of comparisons largely with taxa from markedly older swamps (Westphalian A) of western Europe and from deposits from France (Stephanian and Autunian), which were not from coal balls. This combination of evolutionary and paleoecological differences between the American and the European sources, coupled with incomplete information and lack of specimens to directly compare from both, delayed the clear recognition that some species occurred on both continents. There are some duplications of specific epithets that only monographic studies can resolve; and there are some cases where different generic names have been used for the same kinds of plants, but these are exceptional. Such cases are mentioned, but only minor nomenclatural changes are within the scope of this study. Changes established in the literature have been adopted and references of "in press" articles are given for plant groups that are being taxonomically revised.

The principal sources of literature are quite scattered among journals and the bibliography given herein is limited to the most informative and more recent articles which, in turn, provide more comprehensive literature citations for occurrence data. Some of the European pioneers in coal-ball studies contributed so substantially to the field that it seemed unnecessary to cite all their pertinent papers; compiled bibliographies are already available: for W. C. Williamson's contributions on the English coal-ball plants see Williamson (1891, 1893); D. H. Scott's are listed by Oliver (1935); Bernard Renault (Renault, 1896) gives brief summaries of most of his own studies, including those of the Grand'Croix plants. Comprehensive reference lists of American literature are given by Andrews (1951); Baxter and Hornbaker (1965); and Phillips, Pfefferkorn, and Peppers (1973). All the pertinent Russian literature that was obtainable is cited. The available volumes of the ***Traité de Paléobotanique***, ***Fossilium Catalogus-Plantae***, and the ***Index of Generic Names of Fossil Plants, 1820-1965*** (Andrews, 1970) and its 1966-1973 ***Supplement***, (Blazer, 1975) were major supplemental guides to journal articles. However, taxa have no doubt been overlooked and it was not possible to check the compilation of plant species in the U. S. Geological Survey Index to Paleobotanical Species upon completion of the manuscript.

Slide and peel collections were examined from the following: Department of Paleontology, British Museum (Natural History), London; the University of London King's College and Imperial College; Cambridge University, Cambridge, England; Hunterian Museum, University of Glasgow, Scotland; University of Liége, Belgium; University of Montpellier, France; Museum National d'Histoire Naturelle, Paris, Müseum für Paläeontologie, Universität Humboldt zu Berlin, DDR; and the Komarov Botanical Institute, Leningrad, U.S.S.R. To the extent possible, published American coal-ball plant occurrences, largely from the Illinois Basin, have been supplemented from the Paleobotanical Herbarium at the University of Illinois. This constitutes the largest permanent coal-ball collection in Euramerica and the potential information available was not exhaustively extracted. The computerized holdings include more than 20,000 coal balls, which are systematically filed and cross-referenced according to taxa, locality and vertical profiles, and coal.

Taxonomic Groups

There are five major groups of vascular plants in coal-swamp peat floras: Lycophytina (lycopods), Sphenophytina (sphenopsids), Filicophytina (ferns), and the gymnosperm orders Cordaitales (cordaites) and Pteridospermales (seed ferns). More than 130 generic names are applied to these anatomically preserved Late Carboniferous plants, apart from the following: names such as *Megaloxylon, Kaloxylon, Dadoxylon, or Cordaioxylon,* form genera of fern and seed-fern foliage and those of *sporae dispersae* which are not included in the compilation. A relatively small number of genera described from coal-ball studies are not presently assignable to one of the five major taxa. These include plants with unknown affinities such as *Cyathotheca* (Taylor, 1972), *Mittagia* (Lignier, 1913b), *Sclerocelyphus* (Mamay, 1954b), and *Verticillaphyton* (Baxter, 1967) as well as seeds, cones, or axes of uncertain gymnospermous relationship.

No bryophytes are known from coal balls, but there is the report of one from the Grand'Croix flora, *Muscites bertrandii,* by Lignier (1914). The fungi and fungal-like structures are not included in the occurrence charts; Tiffany and Barghoorn (1974) provide references to the Pennsylvanian fungi, with more recent reports by Baxter (1975) and Stidd and Consentino (1975).

The taxa in each major vascular plant group are listed in Tables 2-1 through 2-21 according to their stratigraphic occurrences (indicated by X) in coal balls (or comparable anatomical preservation) by name of coal, equivalent coal, or associated limestone (Ames, Shumway); occurrences in roof nodules are indicated by an asterisk (★) and questionable occurrences (cf. plant identifications or uncertain stratigraphic provenance) are indicated by a question mark (?). Numerous genera and species have been subject to emendations by authors who are not indicated with the binomials on the charts except where the taxonomic concept has been expanded or modified in a very significant way by a particular author.

Lycophytina

Patterns of Occurrences (Tables 2.1-2.4)

The lycopod trees are the most important swamp plants of the Lower and Middle Pennsylvanian of Euramerica. The major taxa are long ranging (Westphalian); but, with the exception of *Selaginella* and the sigillarian trees, they do not extend into the coal-ball zones above the Westphalian-Stephanian transition. Coal palynology (Phillips, et al. 1974) indicates that *Polysporia,* as yet unknown anatomically from the coal-ball

TABLE 2.1 *Stratigraphic occurrences of Lycophytina—Selaginellales, Miadesmiales, and* Incertae sedis *and Lepidodendrales.*

	SELAGINELLA Beauv.	S. *fraipontii* (Leclercq) Schlanker & Leisman	*CARINOSTROBUS* Baxter	C. *foresmanii* Baxter	*MAIDESMIA* C. E. Bertrand	M. *membranacea* C. E. Bertrand	*MESOSTROBUS* Watson	M. *scottii* Watson	*SPENCERITES* Scott	S. *insignis* (Will.) Scott	S. *majusculus* Scott	S. *membranaceus* Kubart	S. *moorei* (Cridland) Leisman	*POLYSPORIA* Newberry	P. *sp.* ?	*SPORANGIOSTROBUS* Bode	S. *kansanensis* Leisman	**BOTHRODENDRACEAE**	*BOTHRODENDRON* Lindley & Hutton	B. *mundum* (Will.) Scott	*BOTHRODENDROSTROBUS* Hirmer	B. *watsonii* Chaloner
NEWCASTLE																						
SHUMWAY																						
GRAND'CROIX																						
CALHOUN		X																				
DUQUESNE		X																				
FRIENDSVILLE		X																				
AMES		X																				
UNNAMED																						
PARKER		X																				
DANVILLE																						
BAKER		X																				
HERRIN		X									X		X									
SPRINGFIELD		X																				
IRON POST		X																				
SUMMUM		X																				
BEVIER		X									X		X				X					
COLCHESTER																						
MINERAL-FLEMING		X		X							X		X				X					
TEBO																						
WEIR-PITTSBURG		X									X		X				X					
MURPHYSBORO		X																				
BUFFALOVILLE		X															X					
SECOR																						
ROCK ISLAND		X															X					
k8		X																				
AEGIR																						
COPLAND		X													X							
KATHARINA																						
h2/4																						
UPPER FOOT		X																		X		
UNION-FINEFRAU		X				X		X		X	X									X		X
"FIRST"		X				X				X										X		
HAUPTFLÖZ																						
KOKSFLÖZ												X										

TABLE 2.2 *Stratigraphic occurrences of Lycophytina — Lepidodendrales (cont.)*

	LEPIDODENDRON Sternberg	L. *dicentricum* C. Felix	L. *schizostelicum* Arnold	L. *scleroticum* Pannell	L. *serratum* C. Felix	L. *vasculare* Binney	*LEPIDODENDRONS ?*	L. *brevifolium* Williamson *sensu* Scott	L. *hallii* Evers	L. *hickii* Watson	L. *obovatum* Sternberg *sensu* Koopmans	L. *wilsonii* (Andrews & Baxter) Evers	*LEPIDOPHLOIOS* Sternberg	L. *harcourtii* (Witham) Seward & Hill,	L. *fuliginosus* (Will.) Seward	L. *kansanus* (C. Felix) Eggert	L. *pachydermatikos* Andrews & Murdy	*LEPIDOPHYLLOIDES* Snigirevskaya	L. *aciculum* (Reed) Snigirevskaya	L. *alatum* (Reed) Snigirevskaya	L. *angulatum* (Graham) Snigirevskaya	L. *equilaterale* (Graham) Chaloner	L. *latifolium* (Graham) Chaloner	L. *minor* (Graham) Chaloner	L. *papillonaceum* (Graham) Chaloner	L. *sewardii* (Graham) Chaloner	L. *thomasii* (Graham) Chaloner	L. *trichosulcata* (Reed) Chaloner
NEWCASTLE																												
SHUMWAY																												
GRAND'CROIX																												
CALHOUN																												
DUQUESNE																												
FRIENDSVILLE																												
AMES																												
UNNAMED																												
PARKER																												
DANVILLE																												
BAKER	X		X	X							X		X			X	X	X										
HERRIN	X		X	X	?			X	X	X			X			X	X	X										
SPRINGFIELD	X		X	X	X			X		X			X			X	X	X	X	X			X				X	X
IRON POST	X		X	X									X				X	X										
SUMMUM	X																	X										
BEVIER													X			X	X	X										
COLCHESTER	X																											
MINERAL-FLEMING	X	X	X	X	X			X			X		X			X	X	X										
TEBO													X			X	X											
WEIR-PITTSBURG																												
MURPHYSBORO	X					X							X				X	X										
BUFFALOVILLE	X		X			X							X			X		X										
SECOR	X				?	?												X										
ROCK ISLAND	X											X	X			X		X										
k8	X					X												X	X	X	X							
AEGIR	X					X				X	X							X										
COPLAND								X					X				X	X										
KATHARINA	X					X		X		X			X		X			X										
h2/4										X			X															
UPPER FOOT	X					X							X		X			X						X	X		X	
UNION-FINEFRAU	X				X	X		X		X	X		X		X			X			X	X	X			X	X	
"FIRST"	X					X				X			X		X													
HAUPTFLÖZ	X					X																						
KOKSFLÖZ																												

TABLE 2.3 *Stratigraphic occurrences of Lycophytina—Lepidodendrales (cont.)*

	ACHLAMYDOCARPON Schumacker-Lambry	A. *belgicum* Schumacker-Lambry	A. *takhtajanii* (Snigirevskaya) Schumacker-Lambry	A. *varius* (Baxter) Taylor & Brack-Hanes	*LEPIDOSTROBUS* Brongniart	L. *arberii* Jongmans	L. *bertrandii* Zalessky	L. *binneyanus* Arber	L. *diversus* C. Felix	L. *maslenii* Jongmans	L. *minor* Leisman & Rivers	L. *schopfii* Brack	L. *oldhamius* Williamson *sensu* Balbach	*LEPIDOCARPON* Scott	L. *lomaxii* Scott *sensu* Balbach
NEWCASTLE															
SHUMWAY															
GRAND'CROIX															
CALHOUN															
DUQUESNE															
FRIENDSVILLE															
AMES															
UNNAMED															
PARKER															
DANVILLE															
BAKER	X		X	X	X								X		X
HERRIN	X		X	X	X				X				X		X
SPRINGFIELD	X		X	X	X				X				X		X
IRON POST					X										X
SUMMUM	X			X											X
BEVIER					X								X		X
COLCHESTER															
MINERAL-FLEMING	X			X	X				X		?		X		X
TEBO															
WEIR-PITTSBURG															X
MURPHYBORO	X			X											
BUFFALOVILLE	X			X											
SECOR					X				X						X
ROCK ISLAND					X								X		X
k8	X		X	X											
Undesignated below K5					X		X								
AEGIR															
COPLAND					X							X	X		X
KATHARINA					X								X		X
h2/4															
UPPER FOOT	X		X	X	X					X			X		X
UNION-FINEFRAU	X	X	X	X	X			X		X			X		X
"FIRST"					X	X							X		X
HAUPTFLÖZ					X										
KOKSFLÖZ															

TABLE 2.4 *Stratigraphic occurrences of Lycophytina—Lepiodendrales (cont.)*

	MAZOCARPON (Scott) Benson	M. *cashii* Benson	M. *oedipternum* Schopf	M. *shorense* Benson	*SIGILLARIA* Brongniart	S. *brardii* Brongniart	S. *elegans* (Sternberg) Brongniart	S. *ichthyolepis* Presl in Sternberg sensu Weiss	S. *mamillaris* Brongniart	S. *rugosa* Brongniart	S. *tessellata* (Steinhauer) Brongniart	*SIGILLARIOPSIS* Scott non Renault	S. *cordata* Reed	S. *halifaxensis* Graham	S. *laevis* Koopmans	S. *sulcata* Scott	*STIGMARIA* Brongniart	S. *arachnoidea* Koopmans	S. *bacupensis* Lang	S. *ficoides* (Sternberg) Brongniart	S. *lohestii* Leclerq	S. *radiculosa* (Hick) Hirmer	S. *weissiana* Leclercq
NEWCASTLE																							
SHUMWAY																							
GRAND'CROIX					X	?																	
CALHOUN	X		X		X			X									X						
DUQUESNE	?				X							X					X						
FRIENDSVILLE	X		X		X			X									X						
AMES					X												X						
UNNAMED					X												X						
PARKER	X		X		X																		
DANVILLE					X												X			X			
BAKER	X				X												X			X			
HERRIN	X				X												X			X			
SPRINGFIELD	X				X							X	X				X			X			
IRON POST																	X			X			
SUMMUM																	X			X			
BEVIER																	X			X			
COLCHESTER																	X			X			
MINERAL-FLEMING					X												X			X			
TEBO																							
WEIR-PITTSBURG																							
MURPHYSBORO																	X			X			
BUFFALOVILLE																	X			X			
SECOR																	X			X			
ROCK ISLAND																	X			X			
k8																	X			X			
AEGIR																	X	X	X	X			
COPLAND	X																X			X			
KATHARINA	X																X			X			
h2/4																							
UPPER FOOT	X			X	X		X		X		X	X				X	X			X			
UNION-FINEFRAU	X	?		X	X		X		X	X		X		X	X	X	X	X	X	X	X	X	X
"FIRST"	X	X		X	X				X								X			X			
HAUPTFLÖZ																							
KOKSFLÖZ																							

peats, also extended into the Upper Pennsylvanian where it was dominant in some coal-swamp environments. *Endosporites*, the spores found in *Polysporia*, has been described from some American coal balls (Brack and Taylor, 1972; Good and Taylor, 1974). *Sigillaria* was more abundant in the lower Westphalian and in the Stephanian than in between, but it has not been established as the dominant tree type in any of the swamps known from coal balls. A parallel pattern of relative abundance of calamitalean trees also occurs.

The lycopod trees that were apparently the most highly adapted to aquatic environments in their vegetative growth and reproductive cycle with seed-like dispersable units are *Lepidophloios* (with *Lepidocarpon)* and several species of *Lepidodendron (L. serratum, L. schizostelicum,* and *L. scleroticum)* apparently producing *Achlamydocarpon.* All of the common species of these seed-like units are present in lower Westphalian A coal swamps and they disappeared from the swamps during the Westphalian-Stephanian transition, as well as most of those bisporangiate types of lycopods that were part of the same community structure. Heterosporous reproduction and its relationships to the changing patterns of arborescent lycopods in the Late Carboniferous are discussed by Phillips (1979), and a review of the Carboniferous Lepidodendraceae and Lepidocarpaceae is given by Thomas (1978).

Taxonomy

Numerous described species of *Lepidophloios, Lepidocarpon,* and extremely large cones of *Lepidostrobus* (microsporangiate) of complementary plant assemblages in the Westphalian apparently represent a fairly unchanging morphological type of lycopod tree which was important in many swamps. Balbach (1965, 1967) reduced the coal-ball lepidocarps to one species, *Lepidocarpon lomaxii*, and the giant lepidostrobi to one species, *Lepidostrobus oldhamius;* both species were originally based on lower Westphalian A specimens. *Lepidostrobus bertrandii* probably can be synonymized with *L. oldhamius* (Zalessky, 1908; Snigirevskaya, 1972). The taxonomic treatments by Balbach (1965, 1967) of described species scattered across some 20 million years may be viewed as rash, but she has documented the basis for them. We simply lack a means of distinguishing between Westphalian forms of *Lepidocarpon lomaxii sensu lato* and of *Lepidostrobus oldhamius sensu lato* because they are morphologically so similar. It should be noted that the same situation exists in the genus *Achlamydocarpon* across the Westphalian, with two very distinctive types of reproductive structures known as *A. varius* and *A. takhtajanii* (Snigirevskaya, 1964; Balbach, 1966; Leisman and Phillips, 1979). The *Lepidophloios* trees from American coal balls have recently been studied by DiMichele (1979) and one widespread species is recognized; the European forms have not yet been reexamined, but *L. harcourtii* and *L. fulginosus* seem indistinguishable.

In contrast to the above, the species of *Mazocarpon* in the lower Westphalian A (Benson, 1918) and in the Missourian of the Illinois Basin (Schopf, 1941a) are quite distinct and are indicative of subtle differences in reproductive morphology and strategy (Phillips, 1979) relevant to survival in dwindling swamp environments of the Stephanian. The occurrences of species of *Sigillaria* that are known anatomically (Lemoigne, 1960) are not well documented in coal balls in this recent monograph; consequently, the diversity of such trees in the lower Westphalian is uncertain.

Recent studies of *Stigmaria* (Frankenberg and Eggert, 1969; Eggert,

1972) provide résumés of the described species, detail occurrences of *S. ficoides*, and illustrate the anatomy of sigillarian stigmarias from the Upper Pennsylvanian of the Illinois Basin. Both of these studies point out that most of the specimens of *S. weissiana* and that of *S. lohestii* (Leclercq, 1925, 1928b) are probably that of a main aerial stem of *Sigillaria* from near the base.

Sphenophytina

Patterns of Occurrences (Tables 2.5-2.7)

Sphenophyllum parallels *Selaginella* as a small plant occurring through the Pennsylvanian in varied swamp environments. The greatest abundance of *Sphenophyllum*, encountered thus far, is in the Baker Coal near Providence, Kentucky. The calamitalean trees occur more abundantly in the lower and upper parts of the Upper Carboniferous than in the middle (Des Moinesian), paralleling the occurrences of sigillarian trees. One of the longest ranging species is the monosporic *Calamocarpon insignis*. At least locally in the Namurian C (Teichmuller and Schonefield, 1955) and in coal-ball layers of the Katharina Seam some peats were primarily calamitalean; in each case they represent only a peat layer or zone in the seam. The range of maximum calamites apparently extends into the Lower Block Coal of Indiana (Westphalian B). Calamitean trees are usually the third most important group in the Upper Pennsylvanian and some of their very large trunks (45 cm diameter) have been found in the Friendsville and Calhoun Coals (Andrews and Agashe, 1965). Calamites were apparently even more important elements in the Redstone Coal or Pittsburgh Coal (Good and Taylor, 1974) and, at least locally (one site), in the Newcastle Coal.

Taxonomy

The patterns of occurrences for the Sphenophytina are drawn largely from the literature and they reflect both relative abundance and extensive studies at certain stratigraphic intervals. Much variation has been encountered in the sporangiophores of cones and in calamitean wood anatomy.

Species concepts in *Sphenophyllum* are quite varied; *S. plurifoliatum* includes considerable variation in kinds of shoots as well as decorticated stems. Some sphenophylla were dimorphic with *S. constrictum* constituting the rarer type of shoot. The latter, along with *S. insigne* Williamson and Scott, occurs in the Lower Carboniferous as well as in the Middle Pennsylvanian. Revision of the Sphenophyllales has been undertaken by Good (1973, 1978), who has synonymized American species of *Lithostrobus* and *Sphenophyllum multirame* (cone), placing them in *Sphenostrobus*.

Filicophytina

Patterns of Occurrences (Tables 2.8-2.12).

The ferns of the Upper Carboniferous are divisible into two major groups in terms of stratigraphic patterns. The small true ferns, *Anachoropteris* (and in part *Tubicaulis)*, *Ankyropteris*, *Botryopteris* and *Zygopteris* extend across the Westphalian and Stephanian. Botryopterid and zygopterid ferns were well established in the Visean or earlier (Phillips, 1974). *Botryopteris* shows considerable evolutionary change from the Westphalian A to the Westphalian C; *B. forensis* is a cosmopolitan species, occurring in all the known swamp environments from the lower Des Moinesian upward. B. *tridentata* first occurs in the Katharina Seam and is subsequently abundant in the swamps rich in cordaites, up to but not including the Summum (No 4) Coal. The occurrences of both species are well documented in Europe and North America. *Anachoropteris* is often abundant where there is a paucity of other small ferns, even *Botryopteris;* this is par-

TABLE 2.5 *Stratigraphic occurrences of Sphenophytina—Sphenophyllales.*

	SPHENOPHYLLUM Brongniart (N.C.)	S. *constrictum* Phillips	S. *multirame* E. Darrah	S. *plurifoliatum* Williamson & Scott	S. *reedae* Good	S. *renaultii* Phillips	*PSEUDOSPHENOPHYTON* Baxter	P. *höegii* Baxter	*BOWMANITES* Binney	B. *bifurcatus* Andrews & Mamay	B. *dawsoni* (Williamson) Weiss	B. *fertilis* (Scott) Hoskins & Cross	B. *jablokovii* Snigirevskaya	B. *moorei* Mamay	B. *pterosporous* Snigirevskaya	B. *scottii* Hoskins & Cross	B. *simplex* Hoskins & Cross	B. *trisporangiatus* Hoskins & Cross	*LITOSTROBUS* Mamay	L. *iowensis* Mamay	L. *novikae* Snigirevskaya	L. *paulus* (Leisman) Baxter	*PELTASTROBUS* Baxter	P. *reedae* Baxter	*SPHENOSTROBUS*	S. *thompsonii* Levittan & Barghoorn
NEWCASTLE																										
SHUMWAY																										
GRAND'CROIX	X					X			X								X									
CALHOUN	X			X					X	X																
DUQUESNE	X								X																	
FRIENDSVILLE	X			X																						
AMES	X																									
UNNAMED	X			X																						
PARKER	X			X																						
DANVILLE																										
BAKER	X	X	X	X					X															X		
HERRIN	X	X	X	X	X			X																X		
SPRINGFIELD	X	X	X	X																				X		
IRON POST	X																									
SUMMUM	X	X		X																				X		
BEVIER	X			X																				X		
COLCHESTER																										
MINERAL-FLEMING	X	X	X	X					X					X										X		
l_6	X																		X		X					
TEBO																										
WEIR-PITTSBURG																			X			X				
MURPHYSBORO	X			X																				X		
BUFFALOVILLE	X	X		X																						
SECOR	X			X																						
ROCK ISLAND	X			X					X									X	X	X						X
k_8	X	X							X				X		X											
AEGIR	X			X					?																	
COPLAND	X			X					X		X															
KATHARINA	X			X																						
h2/4	X			X																						
UPPER FOOT	X			X					X		X	X														
UNION-FINEFRAU	X			X					X		X	X				X										
"FIRST"	X			X					X		X															
HAUPTFLÖZ	X																									
KOKSFLÖZ																										

TABLE 2.6 *Stratigraphic occurrences of Sphenophytina — Equisetales*

	CALAMOCARPON Baxter	C. *insignis* Baxter	*CALAMOSTACHYS* Schimper	C. *americana* Arnold	C. *binneyana* (Carr.) Schimper	C. *bosselensis* Leggewie & Schonefeld	C. *casheana* Williamson	C. *inversibractis* Good	C. *magnae-crucis* Browne	C. *zeilleri* Renault	*PALAEOSTACHYA* Weiss	P. *andrewsii* Baxter	P. *decacnema* Delevoryas	P. *vera* Seward	*PENDULOSTACHYS* Good	P. *cingulariformis* Good	*WEISSISTACHYS* Rothwell & Taylor	W. *kentuckiensis* Rothwell & Taylor	*ARTHROPITYS* Goeppert	A. *approximata* (Schlotheim) Renault	A. *bistriatoides* Hirmer & Knoell	A. *communis* (Binney) Renault	A. *communis* (Binney) var. *interlignea* Hirmer & Knoell	A. *communis* (Binney) var. *septata* Andrews	A. *felixi* Hirmer & Knoell	A. *grigorievii* Zalessky	A. *herbacea* Hirmer & Knoell	A. *hirmeri* Knoell	A. *hirmeri* var. *intermedia* Knoell	A. *illinoensis* Anderson
NEWCASTLE																			X											
SHUMWAY																			X											
GRAND'CROIX			X						X	X									X	X										
CALHOUN		X	X	X												X			X					X						X
DUQUESNE		X	X																X											
FRIENDSVILLE		X	X	X															X											
AMES																			X											
UNNAMED																														
PARKER		X	X																											
DANVILLE																														
BAKER		X	X								X		X																	
HERRIN		X	X					X			X	X																		
SPRINGFIELD		X																												
IRON POST		X																												
SUMMUM		X																												
BEVIER		X									X		X																	
COLCHESTER																														
MINERAL-FLEMING		X									X		X						X			X								
l_6																														
TEBO																														
WEIR-PITTSBURG		X									X	X																		
MURPHYSBORO		X	X																											
BUFFALOVILLE																														
SECOR		X																												
ROCK ISLAND		X									X	X							X			X								
k_8																			X									X		
AEGIR																														
COPLAND		X	X		X													X	X			X								
KATHARINA			X		X														X		X	X	X		X		X	X	X	
h2/4																			X							X				
UPPER FOOT			X		X		X				X			X					X			X								
UNION-FINEFRAU			X		X		?				X			X					X		X	X	X				X		X	
"FIRST"			X		X														X			X								
HAUPTFLÖZ			X		X	X																								
KOKSFLÖZ																														

TABLE 2.7 *Stratigraphic occurrences of Sphenophytina—Equisetales (cont.)*

	A. jongmansii Hirmer	*A. kansana* Andrews	*A. variabilis* Snigirevskaya	*A. versifoveata* Anderson	*ARTHROXYLON* Reed	*A. oldhamium* Reed	*A. werdensis* Haas	*A. williamsonii* Reed	*CALAMODENDRON* Brongniart	*C. americanum* Andrews	*C. congenium* Grand'Eury	*ANNULARIA* Sternberg	*A. hoskinsii* Good	*ASTEROPHYLLITES* Brongniart (N.C.)	*A. multifolia* Reed	*CALAMITES* Brongniart (N.C.)	*C. retangularis* Anderson	*ASTHENOMYELON* Leistikow	*A. adversale* Leistikow	*A. contractum* Leistikow	*A. intervallare* Leistikow	*A. tenuitrabeculatum* Leistikow	*ASTROMYELON* Williamson	*A. castrense* Leistikow	*A. cauloides* Anderson	*A. dadoxylinum* Renault	*A. pluriradiatum* Anderson	*A. williamsonii* (Hick & Cash) Williamson	*MYRIOPHYLLOIDES* Hick & Cash	*M. williamsonii* Hick & Cash	*ZIMMERMANNIOXYLON* Leistikow	*Z. multangulare* Leistikow
NEWCASTLE																																
SHUMWAY																																
GRAND'CROIX									X		X												X			X						
CALHOUN																	X						X		X		X					
DUQUESNE									X					X																		
FRIENDSVILLE																																
AMES																																
UNNAMED																																
PARKER																							X		X							
DANVILLE																																
BAKER																																
HERRIN									X	X			X	X	X																	
SPRINGFIELD					X			X						X	X								X		X							
IRON POST																							X									
SUMMUM																																
BEVIER																																
COLCHESTER																																
MINERAL-FLEMING		X		X	X			X	X	X																						
l_6																																
TEBO																																
WEIR-PITTSBURG																																
MURPHYSBORO																																
BUFFALOVILLE																																
SECOR																																
ROCK ISLAND					X			X						X	X																	
k_8			X																													
AEGIR																							X									
COPLAND					X			X																								
KATHARINA	X																						X					X				X
h2/4																																
UPPER FOOT					X	X												X	X				X	X				X	X	X		
UNION-FINEFRAU					X			X										X	X	X	X	X	X					X	X	X		
"FIRST"																																
HAUPTFLÖZ					X		X																X									
KOKSFLÖZ																																

TABLE 2.8 *Stratigraphic occurrences of Filicophytina—Coenopteridales or Filicales*

	ANACHOROPTERIDACEAE	*ANACHOROPTERIS* Corda	A. *clavata* Graham	A. *gillotii* Corsin	A. *involuta* Hoskins *sensu lato*	*w/lateral shoots*	*w/adaxial shoots*	*w/dichotomy*	A. *pautetii* Corsin	A. *pulchra* Corda	A. *robusta* Corsin	A. *williamsonii* Koopmans	*APOTROPTERIS*	*A. minuta* Morgan & Delevoryas	*PSALIXOCHLAENA*	P. *cylindrica* (Will.) Holden	*TUBICAULIS* Cotta	T. *multiscalariformis* Delevoryas & Morgan	T. *scandens* Mamay	T. *stewartii* Eggert	T. *sutcliffii* Stopes	SERMAYACEAE	*SERMAYA Eggert & Delevoryas*	S. *biseriata* Eggert & Delevoryas
NEWCASTLE																								
SHUMWAY																								
GRAND'CROIX		X		X	X				X		X													
CALHOUN		X	X		X	X			X					X			X		X	X			X	X
DUQUESNE		X			X																		X	
OPDYKE		X			X									X										
FRIENDSVILLE		X	X		X												X							
AMES					X																			
UNNAMED																								
PARKER		X	X		X				X	X							X							
DANVILLE																								
BAKER		X	X		X		X		X								X						X	X
HERRIN		X	X		X		X	X	X								X							
SPRINGFIELD		X	X																					
IRON POST		X																						
SUMMUM		X			X												X							
BEVIER																								
COLCHESTER																								
MINERAL-FLEMING		X															X	X						
RADNITZ?		X								X														
WEIR PITTSBURG																								
MURPHYSBORO		X			X				X								X							
BUFFALOVILLE		X			X												X							
SECOR																								
ROCK ISLAND		X			X																			
k8		X																						
AEGIR		X										?												
COPLAND		X																						
KATHARINA		X										X												
h2/4																								
UPPER FOOT		X										X				X	X				*			
UNION-FINEFRAU		X										X				X								
"FIRST"												X												
HAUPTFLÖZ																								
KOKSFLÖZ																								

TABLE 2.9 *Stratigraphic occurrences of Filicophytina—Coenopteridales or Filicales (cont.) and* Incertae sedis.

	BOTRYOPTERIDACEAE	*BOTRYOPTERIS* Renault	B. *sp.* "pseudoantiqua"	B. *ramosa* (Will.) Scott	B. *tridentata* (Felix) Scott	B. *hirsuta* (Will.) Scott	B. *sp.* "intermediate"	B. *forensis* Renault	B. *globosa* W. Darrah	B. *mucilaginosa* Kraentzel	B. *sp.* "intermediate"	B. *renaultii* Bertrand & Cornaille	B. *fecunda* Mamay	B. *illinoensis* Mamay	B. *spinosa* Mamay	INCERTAE SEDIS	*CATENOPTERIS* Phillips & Andrews	C. *simplex* Phillips & Andrews	*EOPTERIDANGIUM* Andrews & Agashe	E. *dictyosporum* Andrews & Agashe	*NORWOODIA* Rothwell	N. *angustum* Rothwell	*PTERIDOTHECA* Scott	P. *williamsonii* Scott	*RHABDOXYLON* Holden	R. *americanum* Dennis	R. *dichotomum* Holden	*STURIELLA* Weiss	S. *intermedia* (Renault) Weiss
NEWCASTLE																													
SHUMWAY																													
GRAND'CROIX		X	X					X				X																	X
CALHOUN		X	X					X						X				X		X						X			
DUQUESNE		X																											
OPDYKE																													
FRIENDSVILLE		X						X																					
AMES		X																											
UNNAMED		X						X																					
PARKER		X	X					X																					
DANVILLE																													
BAKER								X			X																		
HERRIN		X	X					X			X																		
SPRINGFIELD		X	X					X			X																		
IRON POST																													
SUMMUM		X	X					X			X																		
BEVIER		X	X		X			X			X																		
COLCHESTER		X	X					X																					
MINERAL-FLEMING		X	X		X			X			X											X							
TEBO		X						X																					
WEIR-PITTSBURG																													
MURPHYSBORO		X	X		X			X			X																		
BUFFALOVILLE		X			X			X			X																		
SECOR		X						X			X																		
ROCK ISLAND		X			X			X	X		X		X		X														
k8		X			X		?																						
AEGIR		X			?																								
COPLAND		X	X		X		X																						
KATHARINA		X			X		X																						
h2/4																													
UPPER FOOT		X		X		X																		X					
UNION-FINEFRAU		X		X		X				X														X			X		
"FIRST"		X		X		X																							
HAUPTFLÖZ		X		X																									
KOKSFLÖZ																													

TABLE 2.10 *Stratigraphic occurrences of Filicophytina —* Ankyropteris, *Zygopteridales, and* Stauropteris

	ANKYROPTERIS Stenzel	A. *corrugata* (Will.) P. Bertrand	A. *(Tedelea) glabra* Baxter	A. *grayi* (Will.) P. Bertrand	A. *westphaliensis* P. Bertrand	**ZYGOPTERIDACEAE**	*ZYGOPTERIS* Corda	Z. *berryvillensis* Dennis	Z. *illinoiensis* (Andrews) Baxter	Z. *lacattei* Renault	Z. *paradoxum* (Scott) Baxter	*ETAPTERIS* P. Bertrand	E. *bertrandii* Scott	E. *diupsilon* (Will) P. Bertrand	E. *leclerqiae* Smoot & Taylor	E. *renieri* Leclercq	E. *scottii* P. Bertrand	E. *shorensis* P. Bertrand	*BISCALITHECA* Mamay	B. *musata* Mamay	*CORYNEPTERIS* Bailey	C. *involucrata* Baxter & Baxendale	C. *scottii* Galtier & Holmes	*NOTOSCHIZAEA* Graham	N. *robusta* Graham	*Corynepteris/Notoschizaea*	**STAUROPTERIDACEAE**	*STAUROPTERIS* Binney	S. *oldhamia* Binney
NEWCASTLE																													
SHUMWAY																													
GRAND'CROIX										X																			
CALHOUN	X						X	X				X								X					X	X			
DUQUESNE	X							X				X								X									
OPDYKE																													
FRIENDSVILLE								X				X								X									
AMES																													
UNNAMED							X	X																					
PARKER							X	X				X								X						X			
DANVILLE																													
BAKER												X								X						X			
HERRIN	X		X				X		X			X														X			
SPRINGFIELD	X		X									X														X			
IRON POST												X																	
SUMMUM							X		X			X																	
BEVIER							X		X																				
COLCHESTER							X		X																				
MINERAL-FLEMING	X		X				X		X			X		X			X									X			
TEBO																													
WEIR PITTSBURGH																													
MURPHYSBORO	X		X									X														X			
BUFFALOVILLE												X																	
SECOR												X														X			
ROCK ISLAND	X		X									X																	
k8																													
AEGIR																	X												
COPLAND												X			X														
KATHARINA												X					X												X
h2/4																													
UPPER FOOT	X			X	X		X				X	X	X	X			X	*					X						X
UNION-FINEFRAU	X	X		X	X		X				X	X		X		X	X						X			X			X
"FIRST"					X												X												X
HAUPTFLÖZ																													
KOKSFLÖZ																													

TABLE 2.11 *Stratigraphic occurrences of Filicophytina — Marattiales*

	PSARONIUS Cotta	monocyclic *sp.*	*P. simplicicaulis* DiMichele & Phillips	P. *renaultii* Williamson	polycyclic *sp.*	P. *blicklei* Morgan	P. *chasei* Morgan	P. *giffordii* Gillette	P. *melanedrus* Morgan	P. *pertusus* Blickle	*STEWARTIOPTERIS* Morgan & Delevoryas	S. *singularis* Morgan & Delevoryas	*STIPITOPTERIS* Grand'Eury	S. *americana* Lenz	S. *gracilis* Morgan & Delevoryas	*ACAULANGIUM* Millay	A. *bulbaceus* (Graham) Millay	*ANDREWSIOPTERIS* Baxter	A. *revoluta* Baxter	*CYATHOTRACHUS* Watson	C. *altus* Watson	C. *altissimus* Mamay	*EOANGIOPTERIS* Mamay	E. *andrewsii* Mamay
NEWCASTLE					X																			
PITTSBURGH					X					X														
SHUMWAY, IL, KS					X			X																
GRAND'CROIX																								
CALHOUN					X	X	X			X			X	X			X					X		
DUQUESNE							X						X							X			?	
OPDYKE					X	X				X			X											
FRIENDSVILLE					X								X											
AMES							?						X							X				
UNNAMED													X											
PARKER					X		X					X	X		X									
DANVILLE																								
BAKER					X					X			X											
HERRIN					X				X				X							X		X		
SPRINGFIELD					X					X			X											
IRON POST													X											
SUMMUM					X								X											
BEVIER																								
COLCHESTER																								
MINERAL-FLEMING					X				X				X						X					
TEBO																								
WEIR-PITTSBURG										X														
MURPHYSBORO					X								X											
BUFFALOVILLE					X																			
SECOR																								
ROCK ISLAND					X															X		X	X	X
k8					X							X												
AEGIR																								
COPLAND																								
KATHARINA																								
h2/4																								
UPPER FOOT		X		X	*															X	X			
UNION-FINEFRAU													X											
"FIRST"																								
WESTERN ILLINOIS		X	X																					

TABLE 2.12 *Stratigraphic occurrences of Filicophytina—Marattiales (cont.) and* Incertae sedis.

	SCOLECOPTERIS Zenker	*S. illinoensis* Ewart	*S. incisifolia* Mamay	*S. iowensis* Mamay	*S. latifolia* Graham	*S. macrospora* Jennings & Millay	*S. major* Mamay	*S. minor* Hoskins	*S. minor* var. parvifolia Mamay	*S. monothrix* Ewart	*CYATHOTHECA* Taylor	*C. tectata* Taylor	*SCLEROCELYPHUS* Mamay	*S. oviformis* Mamay	*VERTICILLAPHYTON* Baxter	*V. paradoxum* Baxter
NEWCASTLE																
SHUMWAY																
GRAND'CROIX																
CALHOUN	X	X			X				X	X						
DUQUESNE	X															
OPDYKE																
FRIENDSVILLE	X															
AMES	X															
UNNAMED																
PARKER	X								X							
DANVILLE								X								
BAKER	X				X											
HERRIN	X				X	*		X	X			X				
SPINGFIELD	X				X				X							
IRON POST	X															
SUMMUM	X															
BEVIER																
COLCHESTER																
MINERAL-FLEMING	X				X			X	X					X		
TEBO	X															
WEIR-PITTSBURG																
MURPHYSBORO	X															
BUFFALOVILLE	X															
SECOR																
ROCK ISLAND	X		X	X	X		X									X
k_8																
AEGIR																
COPLAND																
KATHARINA																
h2/4																
UPPER FOOT	X								X							
UNION-FINEFRAU																
"FIRST"																
HAUPTFLÖZ																
KOKSFLÖZ																

ticularly the case in the Baker Coal, which contains an unusual amount of herbaceous peat compared to other Westphalian coal swamps. In the Missourian and Virgilian the true ferns are relatively abundant among the peats dominated by marattiaceous root systems.

The patterns of occurrence of *Psaronius* are of particular interest because this tree fern genus was the last major vegetational type to expand into coal swamps. The earliest known specimens of *Psaronius* from coal balls or a similar kind of preservation in the Upper Carboniferous are monocyclic and distichous (DiMichele and Phillips, 1977). They represent, probably in large part, the anatomically preserved form of *Megaphyton* (see Remy and Remy, 1977, p. 153). The small stems of these late Namurian-early Westphalian psaronii were quite rare. Evidence of larger polycyclic stems is even rarer; the oldest such specimen known with certainty is from a roof nodule above the Upper Foot from Dearnley near Littleborough in Lancashire (Scott Collection 3238-3240, British Museum). Robust polycyclic and typically polystichous *Psaronius* trees have not been anatomically discovered below the Westphalian B-C transition thus far, but coal palynology indicates that *Psaronius* trees became important in coal swamps in the Illinois Basin in about the middle of the Abbott Fm (Phillips et al., 1974). *Psaronius* was abundant in coal swamps thereafter and became the dominant tree form in many of the Upper Pennsylvanian coal swamps in the Illinois Basin.

One genus that has been included among the ferns for convenience is *Stauropteris*. While the affinities of *Stauropteris* are not known, the stratigraphic range of *S. oldhamia* is of particular interest. It coincides with that of the gymnosperm *Lyginopteris* in the Westphalian A coal balls, with both genera extending to the Katharina Seam.

Taxonomy

Many of fern groups are currently being studied intensively and some of the research has been completed but has not yet been published. A review of the vegetative evolution of coenopterid ferns was given by Phillips (1974), who illustrated many of them, including the "intermediate" forms indicated for *Botryopteris*. *Anachoropteris* still lacks taxonomic delineation because of the vegetative similarities of the "involuta types"--hence the use of morphological criteria on branching and positions of frond-borne shoots of the *Tubicaulis*-type. Zygopterid sporangia in the Upper Carboniferous listed as *Corynepteris*/*Notoschizaea* reflects uncertainty about the distinctions between the two genera. This in large part is because of the limitations of the type material of *Notoschizaea*. The very close similarities between *Biscalitheca musata* and *Zygopteris lacattei* have been noted by Galtier and Grambast (1972).

Most of the detailed studies of *Psaronius* trees from coal balls are based on specimens from the Upper Pennsylvanian of the Illinois Basin. (Morgan, 1959; Stidd, 1971) and similar studies are needed for some of those in the Middle Pennsylvanian to enable precise identification of species. The marattiaceous fructifications have recently been monographed by Millay (1976, 1977) and his studies of *Scolecopteris* will soon be published.

Cordaitales

Patterns of Occurrences (Tables 2.13-2.15)

The cordaites occur in coal balls through the entire Upper Carboniferous but they expanded (ca 10%) from the Katharina Seam up to and including the Summum (No. 4) Coal. The zenith of cordaites is in the coal swamps known from

TABLE 2.13 *Stratigraphic occurrences of gymnosperms—Cordaitales*

	AMYELON Williamson	A. *iowense* (Pierce & Hall) Cridland	A. *radicans* (Will.) Williamson	*RADICULITES* Lignier *non* Zalessky	R. *reticulatus* Lignier	*MESOXYLON* Scott & Maslen	M. *birame* Baxter	M. *demetrianum* Zalessky	M. *lomaxii* Scott & Maslen	M. *multirame* Scott & Maslen	M. *nauterianum* Andrews	M. *platypodium* Scott & Maslen	M. *poroxyloides* Scott & Maslen	M. *sutcliffii* (Scott) Scott & Maslen	M. *thompsonii* Traverse	*MESOXYLOPSIS* Scott	M. *arberae* Scott	*CORDAITES* Unger (stems)	C. *iowensis* Wilson & Johnston	C. *validus Cohen & Delevoryas*	*CORDAITES* Unger (leaves)	C. *angulosostriatus* Grand'Eury	C. *crassus* Renault	C. *duplicinervis* Grand'Eury	C. *filicis* Benson	C. *kansanus* Huertas	C. *lingulatus* Grand'Eury	C. *loculosus* Felix	C. *principalis* (Germar) Geinitz	C. *pyramidales* Huertas	C. *robustus* Felix	C. *rhombinervis* Grand'Eury	C. *rotundinervis* Grand'Eury
NEWCASTLE																																	
SHUMWAY																																	
GRAND'CROIX	X	X			X													X			X	?	X	?			?		?			?	?
CALHOUN	X																	X		X	X												
DUQUESNE	X					X												X			X												
FRIENDSVILLE	X																	X															
AMES	X					X												X			X												
UNNAMED	X																	X			X												
PARKER	X																	X			X												
DANVILLE	X																																
BAKER	X																	X			X												
HERRIN	X	X				X					X							X			X									X			
SPRINGFIELD	X																	X			X												
IRON POST	X																				X												
SUMMUM	X																	X			X												
BEVIER	X	X																X			X												
COLCHESTER	X																	X			X												
MINERAL-FLEMING	X	X				X	X											X			X					X			X				
TEBO																																	
WEIR-PITTSBURGH	X	X																															
MURPHYSBORO	X																	X			X												
BUFFALOVILLE	X	X																X			X												
SECOR	X																	X			X												
ROCK ISLAND	X	X	?			X	X								X			X	X		X		?		?				X				
k8																																	
AEGIR	X		X			X															X				X								
COPLAND	X		X															X			X				X								
KATHARINA	X		X															X			X				X			X			X		
h2/4						X		X																									
UPPER FOOT	X		X			X			*	X		*	X	*			*	X															
UNION-FINEFRAU	X		X			X				X								X			X				X								
"FIRST"	X		X															X			X				X								
HAUPTFLÖZ																		X			X				X								
KOKSFLÖZ																																	

TABLE 2.14 *Stratigraphic occurrences of gymnosperms—Cordaitales (cont.)*

	C. *tenuistriatus* Grand'Eury	C. *wedekindii* Felix	C. *weristeri* Leclercq	*GOTHANIA* Hirmer	G. *westfalica* Hirmer	*CARDIOCARPUS* Brongniart (N.C.)	C. *leclercqiae* Snigirevskaya	C. *magnicellularis* Baxter & Roth	C. *spinatus* Graham	C. *oviformis* Leisman	C. *bigibbosus* C. E. Bertrand	C. *drupaceus* Brongniart	C. *orbicularis* Brongniart	C. *sclerotesta* Brongniart	*MITROSPERMUM* A. Arber	M. *compressum* (Williamson) Arber	M. *leeanum* (Kern & Andrews) Baxter	M. *florinii* (Darrah) Baxter	*NUCELLANGIUM* Andrews	N. *glabra* (Darrah) Andrews	*CORDAIANTHUS* Grand'Eury (N.C.)	C. *compactus* Fry	C. *concinnus* Delevoryas	C. *shuleri* Darrah	C. *thompsonii* Fry	C. *williamsonii* Renault	C. *zeilleri* Renault
NEWCASTLE																											
SHUMWAY																											
GRAND'CROIX	?					X					X	X	X	X							X					X	X
CALHOUN						X			X	X											X						
DUQUESNE						X				?					?						X						
FRIENDSVILLE						X				X											X						
AMES																					X						
UNNAMED						X				X											X						
PARKER						X				?											X						
DANVILLE																											
BAKER						X				X																	
HERRIN						X				X											X						
SPRINGFIELD						X				X											X						
IRON POST																											
SUMMUM						X				X											X						
BEVIER						X			X	X											X						
COLCHESTER						X				X											X						
MINERAL-FLEMING						X			X	X											X		X				
TEBO																											
WEIR-PITTSBURG						X			X															X			
MURPHYSBORO						X			X	X											X						
BUFFALOVILLE						X		X	X	X					X		X			X	X			X	X		
SECOR						X			X												X						
ROCK ISLAND						X		X	X						X		X	X		X	X	X		X			
k8						X	X														X						
AEGIR						X	?	?																			
COPLAND															X	X					X						
KATHARINA		X			X										X	X											
h2/4																											
UPPER FOOT						*?									X	X											
UNION-FINEFRAU			X												X	X											
"FIRST"															X	X											
HAUPTFLÖZ																											
KOKSFLÖZ																											

TABLE 2.15 *Stratigraphic occurrences of gymnosperms—Cordaitales (cont.) and* Incertae sedis *gymnosperms*

	CYCLOSPERMUM Seward	C. *nummalaris* (Brongniart) Seward	C. *tenuis* (Brongniart) Seward	*DIPLOTESTA* Brongniart non Cookson & Eisenack	D. *avellana* (Brongniart) C. E. Bertrand	D. *grand'euryana* Brongniart	*LEPTOCARYUM* Brongniart	L. *avellanum* Brongniart	*LEPTOTESTA* Loubiere	L. *grandeuryii* Loubiere	*RHABDOSPERMUM* Seward	R. *conicus* (Grand'Eury) Seward	R. *cyclocaryon* (Brongniart) Seward	R. *spinatus* Andrews	R. *subtunicatus* (Grand-Eury) Seward	*SARCOTAXUS* Brongniart	S. *angulosus* Brongniart	*Stephanospermum caryoides* Oliver	*CLADITES* Scott	C. *bracteatus* Scott	*LASIOSTROBUS* Taylor	L. *polysacci* Taylor	*POROXYLON* Renault *non* Andrae	P. *stephanense* C. E. Bertrand & Renault	*STELASTELLARA* Baxter	S. *baxteri* DiMichele & Phillips	S. *parvula* Baxter
NEWCASTLE																											
SHUMWAY																											
GRAND'CROIX	X	X	X	X	X	X		X		X	X	X	X		X		X	X						X			
CALHOUN																						X					
DUQUESNE																											
FRIENDSVILLE																											
AMES																											
UNNAMED																											
PARKER																											
DANVILLE																											
BAKER																									X	X	
HERRIN																									X		X
SPRINGFIELD																											
IRON POST																											
SUMMUM																											
BEVIER																											
COLCHESTER																											
MINERAL-FLEMING																											
TEBO																									X		X
WEIR-PITTSBURG																											
MURPHYSBORO																									X		X
BUFFALOVILLE																									X		X
SECOR																											
ROCK ISLAND											X			X											X		X
k_8																											
AEGIR																											
COPLAND																											
KATHARINA																											
h2/4																											
UPPER FOOT																				*							
UNION—FINEFRAU																											
"FIRST"																											
HAUPTFLÖZ																											
KOKSFLÖZ																											

coal-ball peats is in the lower part of the Des Moinesian (Rock Island equivalent of Iowa to the Bevier Coal of Kansas). In coal swamps of the Illinois Basin during that interval cordaites were secondary to lycopod trees, but in the Western Interior coal region the swamps were generally dominated by cordaites with lycopods secondary and even sometimes rare in Iowa. Cordaites were the only seed plants to dominate the swamp vegetation of the Pennsylvanian, except for perhaps one site in the Newcastle Coal, which has largely yielded medullosans. It is thought that many of the cordaites occupied coastal swamps, which were subjected to greater saline influences and perhaps higher energy environments. If this were the case in the early Westphalian, relatively limited environments of that type existed (with coal-ball formation). Cordaites are apparently absent from the Union Seam of the Burnley coal basin of Lancashire and most of the specimens described from England occur in the Upper Foot Seam at Shore-Littleborough and Oldham as well as in roof nodules. They are also at Dulesgate and in the "First" coal. A few have also been reported from the Boux-harmont (Leclercq, 1928a); but in the lower Westphalian A, cordaites were on the whole rare. The large number of taxa reported in the "Great Coal-ball Horizon" is regarded as misleading as to diversity, numbers, and distribution.

Above the Bevier Coal in the Western Interior coal region and above the Summum (No. 4) Coal of the Eastern Interior coal region cordaites are also rare. In the Illinois Basin, coals immediately above the Summum were derived from the vast and long-lasting deltaic swamps of the Springfield and Herrin. Cordaites nevertheless occur in all the coals sampled extensively from coal balls higher in the section in the Illinois Basin but they constitute less than 1% of the peat. Cordaites are quite abundant in the Grand'Croix flora which has a great diversity of ovule forms although the cordaitalean affinities are uncertain for a number of them. Rothwell (1976b) also indicated the importance of cordaites in the flora of the Duquesne Coal of the Appalachian region. It should be noted that the Grand'Croix flora is regarded as limnic.

The stratigraphic sequences of *Mitrospermum*, *Cardiocarpus*, and *Nucellangium* in the Westphalian and equivalent coals seem to coincide with aspects of the pattern described for the "cordaitean zone." *Mitrospermum* is the only cordaitean seed found in the lower Westphalian; and the maximum diversity of cordaitean ovules occurs at the Westphalian B-C transition, which coincides with the maximum known development of such swamps. *Cardiocarpus magnicellularis*, the very similar *C. leclercqiae*, and a comparable form from the Aegir occur in this interval along with *Nucellangium*. Two additional species of *Cardiocarpus* appear in the coal-swamp peats during that peak development: *C. spinatus* extends upward to the Bevier Coal, the end of the cordaite zone in the Western Interior coal region (it does not extend as high in the Des Moinesian of the Illinois Basin); and *C. oviformis* is apparently the cordaitean ovule in the Colchester, Summum and higher coals, including those of the Upper Pennsylvanian. One anomaly in the range of *C. spinatus* is that of the rather poorly preserved type specimen from the Calhoun Coal.

Taxonomy

Variation in anatomy is obviously a major problem in dealing with cordaitalean stems and leaves. The taxonomy of stems has been commented on by Traverse (1950); a review of much of the foliage is included in the work of Harms and Leisman (1961); and Arnold (1967) has discussed the designations of foliage and stems of the Cordaitales. The affinities of many of the ovules tentatively attributed to cordaites need to be reexamined. This includes species of *Cardiocarpus*, *Mitrospermum*, and others, especially in the Stephanian, and seeds such as *Stephanospermum caryoides*, which have been misassigned to the pteridosperms (Florin, 1937, p. 310).

Pteridospermales

Patterns of Occurrences (Tables 2.16-2.21).

Medullosan trees occur through the Westphalian and Stephanian, somewhat paralleling the abundance pattern of tree ferns but with a steady and less dramatic rise. *Medullosa* is not very abundant in the early Westphalian coal balls but much more frequent in occurrence than *Psaronius*. It shows a steady rise in both cordaitalean and lycopod-dominated swamps of the Middle Pennsylvanian and is the second most important group in the Upper Pennsylvanian swamps of the Illinois Basin. Medullosan peats and ephemeral communities characterized by *Medullosa* are usually less persistent laterally than those of *Psaronius* except where the two occur together. In the Springfield and Herrin Coals, tree ferns and pteridosperms tend to occur as distinctive communities in certain zones, somewhat portending the *Psaronius-Medullosa* swamps of the Upper Pennsylvanian. At one site in the Newcastle Coal near the Virgilian-Wolfcampian boundary, medullosan trees were more abundant than *Psaronius*. *Sutcliffia* is fairly rare with scattered stratigraphic occurrences on each side of the DesMoinesian-Missourian transition; some have been reported from roof nodules above the Upper Foot of Shore-Littleborough in the lower Westphalian A.

The small pteridosperms known from shoots show a varied distribution pattern. *Lyginopteris* was very common in the Koksflöz and in coal balls of the Westphalian A, including the Katharina Seam; but there are no reports of the genus in coal balls from higher strata in the Westphalian B. None was found in the Hauptfloz. *Heterangium* is apparently quite abundant in the Koksfloz, as it is in petrifactions from Calciferous Sandstone Series (Visean). On the whole, *Heterangium* is usually a fairly rare genus in coal balls (often preserved in a fusinized state); but the genus occurs in coal balls through the entire Upper Carboniferous. *Microspermopteris* is also rare and occurs at sites and in coals with cordaitean vegetation from the Upper Foot Seam of Shore to the top of the cordaitean zone in the Summum (No. 4) Coal. A single specimen of *Microspermopteris* is known from the Upper Foot; it was originally described as *Syncrama liratum* (Holden, 1955). *Schopfiastrum* is also fairly rare and known only from the DesMoinesian. A questionable occurrence of the genus is recorded for the Aegir based on Koopmans' (1934) illustration of a rachis attributed to *Heterangium*.

Callistophyton is quite abundant above the cordaitean zone, both in the DesMoinesian and Missourian. The lowest stratigraphic, and a questionable occurrence, of fusinized *Callistophyton* is from the "First" coal at the Hough Hill Colliery, Stalybridge (Scott Collection 2050-2052, British Museum).

Taxonomy

Although the pteridosperms or seed ferns did not constitute dominant vegetational types in the extensive swamp lands of the Upper Carboniferous, more taxa have been described in this group than in any other. Extensive studies have been carried out on *Medullosa* (Delevoryas, 1955; Stewart and Delevoryas, 1956), *Lyginopteris* (Blanc-Louvel, 1966),*Heterangium* (Kubart, 1914; Scott, 1917; Hirmer 1933b; Shadle and Stidd, 1975), *Callistophyton* (Delevoryas and Morgan, 1954b; Rothwell, 1975), *Microspermopteris* (Baxter, 1949; Taylor

TABLE 2.16 *Stratigraphic occurrences of gymnosperms — Pteridospermales*

	CALLISTOPHYTON Delevoryas & Morgan	C. *boyssetii* (Renault) Rothwell	C. *poroxyloides* Delevoryas & Morgan	*HETERANGIUM* Corda	(Euheterangium Group)	H. *alatum* Kubart	H. *andrei* Kubart	H. *minimum Scott*	*H. paradoxum* Corda	H. *polystichum* Kubart	H. *schusteri* Kubart	H. *sturii* Kubart	(Polyangium Group)	H. *americanum* Andrews	H. *hoppstaedteri* Hirmer	H. *intermedium* Kubart	H. *kukukii* Hirmer	H. *lomaxii* Williamson *in litt.*	*H. shorense* Scott	H. *tiliaeoides* Williamson
NEWCASTLE																				
REDSTONE OR PITTSBURGH	X																			
GRAND'CROIX	X																			
CALHOUN	X		X	X										X						
DUQUESNE	X		X	X																
OPDYKE	X		X	X																
FRIENDSVILLE	X		X	X										X						
AMES	X		X																	
UNNAMED																				
PARKER	X		X	X										X						
DANVILLE																				
BAKER	X	X		X										X						
HERRIN	X	X		X										X						
SPRINGFIELD	X	X		X										X						
IRON POST				X																
SUMMUM	X	X		X																
BEVIER	X	X																		
COLCHESTER																				
MINERAL-FLEMING	X	X		X																
RADNITZ?				X					X											
WEIR-PITTSBURG																				
MURPHYSBORO				X																
BUFFALOVILLE																				
SECOR																				
ROCK ISLAND																				
k8																				
AEGIR				X																
COPLAND				X																
KATHARINA				X											X	X	X			
h2/4																				
UPPER FOOT				X													X		X	
UNION-FINEFRAU				X				X										X		X
"FIRST"																				
HAUPTFLÖZ																				
KOKSFLÖZ				X		X	X			X	X	X								

TABLE 2.17 *Stratigraphic occurrences of gymnosperms—Pteridospermales (cont.)*

	LYGINOPTERIS H. Potonié	L. *corsinii* Blanc-Louvel	L. *kubartii* Blanc-Louvel	L. *oldhamia* (Binney) H. Potonié	*LYGINORACHIS* Kidston	L. *taitiana* Kidston	*MICROSPERMOPTERIS* Baxter	M. *aphyllum* Baxter	M. *lyrata* (Holden) *comb. nov.*	*SCHOPFIASTRUM* Andrews	S. *decussatum* Andrews	*MEDULLOSA* Cotta	M. *anglica* Scott	M. *anglica* var. *thiessenii* Schopf	M. *centrofilis* de Fraine	M. *endocentrica* Baxter	M. *geriensis* (Scott) Boureau	M. *noei* Steidtmann	M. *primaeva* Baxter	M. *pusilla* Scott	M. *thompsonii* Andrews	*MYELOXYLON*	M. *americanum* (Hoskins) Seward	M. *bendixenii* Andrews	M. *landriotii* Renault	*SUTCLIFFIA* Scott	S. *insignis* Scott	S. *insignis* var. *tuberculata* Phillips & Andrews
NEWCASTLE																												
NEBRASKA												X		X														
GRAND'CROIX												X					X					X			X			
CALHOUN																X		X				X						
DUQUESNE												X										X						
OPDYKE												X						X				X						
FRIENDSVILLE												X				X		X				X						
AMES												X										X						
UNNAMED																						X						
PARKER												X										X				X		X
DANVILLE																												
BAKER										X	X	X										X				X		X
HERRIN										X	X	X				X		X				X				X		X
SPRINGFIELD										X	X	X				X						X	X			X		X
IRON POST												X										X						
S U M M U M							X	X				X										X						
BEVIER							X	X														X						
COLCHESTER																						X						
MINERAL-FLEMING							X	X				X							X			X						
TEBO																						X						
WEIR-PITTSBURGH																												
MURPHYSBORO							X	X		X	X	X										X						
BUFFALOVILLE							X	X				X							X			X						
SECOR												X										X						
ROCK ISLAND							X	X		X	X	X							X		X	X		X				
k8																						X						
AEGIR										?												X						
COPLAND							X	X				X	?															
KATHARINA	X	X		X								X	X									X						
h2/4	X			X																								
UPPER FOOT	X			X			X		X			X	X		X							X				X	*	
UNION-FINEFRAU	X	X	X	X		X						X	X							?		X						
"FIRST"	X			X								X	X									X						
HAUPTFLÖZ																												
KOKSFLÖZ	X	X	X	X																								

TABLE 2.18 *Stratigraphic occurrences of gymnosperms—Pteridospermales (cont.)*

	AETHEOTESTA Brongniart	A. *elliptica* Renault	A. *subglobosa* Brongniart	*ALBERTLONGIA* Taylor	A. *incostata* Taylor	*CALLOSPERMARION* Eggert & Delevoryas	C. *pusillum* Eggert & Delevoryas	*CODONOSPERMUM* Brongniart	C. *anomalum* Brongniart	*COMPSOTESTA* C. E. Bertrand	C. *brongniartii* C. E. Bertrand	*CONOSTOMA* Williamson	C. *anglo-germanicum* Oliver & Salisbury	C. *kestospermum* Taylor & Leisman	C. *leptospermum* Taylor	C. *oblongum* Williamson	C. *platyspermum* Graham	C. *quadratum* Graham	C. *villosum* Rothwell & Eggert	*CORONOSTOMA* Neely	C. *quadrivasatum* Neely	*GNETOPSIS* Renault & Zeiller	G. *elliptica* Renault & Zeiller	*HEXAPTEROSPERMUM* Brongniart	H. *delevoryii* Taylor	H. *pachypterum* Brongniart	H. *stenopterum* Brongniart	*LAGENOSTOMA* Williamson	L. *lomaxii* Oliver & Scott	L. *ovoides* Williamson
NEWCASTLE																														
SHUMWAY																														
GRAND'CROIX		X	X						X		X	X											X	X		X	X			
CALHOUN						X	X					X					X	X	X											
DUQUESNE						X											X	X	X		X			?						
OPDYKE																														
FRIENDSVILLE												X																		
AMES						X												X												
UNNAMED												X																		
PARKER												X																		
DANVILLE																														
BAKER																														
HERRIN					X							X		X	X		X							X	X					
SPRINGFIELD												X				X														
IRON POST												X																		
SUMMUM																														
BEVIER												X		X																
COLCHESTER																														
MINERAL-FLEMING																														
TEBO																														
WEIR-PITTSBURG																														
MURPHYSBORO																														
BUFFALOVILLE												X																		
SECOR																														
ROCK ISLAND												X				X														
k8																														
AEGIR																														
COPLAND												X	X			X														
KATHARINA												X	X			X														X
h2/4																														
UPPER FOOT												X	X			X														
UNION-FINEFRAU												X	X			X													X	X
"FIRST"																													X	X
HAUPTFLÖZ																													X	X
KOKSFLÖZ																														

TABLE 2.19 *Stratigraphic occurrences of gymnosperms—Pteridospermales (cont.)*

	PACHYTESTA Brongniart	P. *berryvillensis* Taylor & Eggert	P. *composita* Stewart	P. *gigantea* Brongniart *ex* Renault	P. *hexangulata* Stewart	P. *hoskinsii* Taylor	P. *illinoense* (Arnold & Steidtmann) Stewart	P. *incrassata* Brongniart	P. *noei* Hoskins & Cross	P. *olivaeformis* (Williamson) Hoskins & Cross	P. *pusilla* (Brongniart) Hoskins & Cross	P. *saharasperma* Taylor	P. *shorensis* (Salisbury) Hoskins & Cross	P. *stewartii* Taylor & Delevoryas	P. *vera* Hoskins & Cross	*PHYSOSTOMA* Williamson	P. *calcaratum* Leisman	P. *elegans* (Williamson) Oliver	P. *stellatum* Holden	*POLYLOPHOSPERMUM* Brongniart	P. *stephanense* Brongniart	*POLYPTEROSPERMUM* Brongniart	P. *renaultii* Brongniart	*PTYCHOTESTA* Brongniart	P. *tenuis* Brongniart	*SARCOSPERMUM* Deevers	S. *ovale* Deevers
NEWCASTLE																											
SHUMWAY																											
GRAND'CROIX	X			X				X			X										X		X		X		
CALHOUN	X	X			X		X																				
DUQUESNE	X	X			?									?													
OPDYKE																											
FRIENDSVILLE	X																										
AMES	X	X																									
UNNAMED																											
PARKER	X						X																				
DANVILLE																											
BAKER	X																										
HERRIN	X			X		X	X					X		X	X												
SPRINGFIELD	X						X		X			X															
IRON POST	X																										
SUMMUM	X																										
BEVIER			X			X	X									?	?										
COCHESTER																											
MINERAL-FLEMING	X		X												X	?	?										*
TEBO																											
WEIR PITTSBURG	X													X		?	?										
MURPHYSBORO	X																										
BUFFALOVILLE	X													X													
SECOR	X														X												
ROCK ISLAND																											
k8																											
AEGIR	X																										
COPLAND																											
KATHARINA																											
h2/4																											
UPPER FOOT	X									X			X			X		X									
UNION-FINEFRAU	X									X						X		X	X								
HAUPTFLÖZ										X																	
KOKSFLÖZ																											

TABLE 2.20 *Stratigraphic occurrences of gymnosperms—Pteridospermales (cont.)*

	STEPHANOSPERMUM Brongniart	S. *akenioides* Brongniart	S. *elongatum* Hall	S. *ovoides* Hall	*TAXOSPERMUM* Brongniart	T. *gruneri* Brongniart	T. *undulatum* Neely	*TRIPTEROSPERMUM* Brongniart	T. *rostratum* Brongniart	*TYLIOSPERMA* Mamay	T. *orbiculatum* Mamay	*CALLANDRIUM* Stidd & Hall	C. *callistophytoides* Stidd & Hall	*DOLEROTHECA* Halle (N.C.)	D. *fertilis* (Renault) Halle	D. *formosa* Schopf	D. *reedana* Schopf	D. *schopfii* Baxter	D. *sclerotica* Baxter	D. *villosa* Schopf	*FERAXOTHECA* Millay & Taylor	F. *culcitaus* Millay & Taylor	*HALLETHECA* Taylor	H. *reticulatus* Taylor	*IDANOTHEKION* Millay & Eggert	I. *gladulosum* Millay & Eggert	*POTONIEA* Zeiller	P. *sp.*
NEWCASTLE																												
SHUMWAY																												
GRAND'CROIX	X	X			X	X			X					X	X													
CALHOUN	X		X										X	X		X				X				X				
DUQUESNE	X															?										?		
OPDYKE	X		X																							X		
FRIENDSVILLE														X												X		
AMES																												
UNNAMED																												
PARKER	X													X														
DANVILLE																												
BAKER	X													X												X		
HERRIN	X		X	X	X		X						X	X												X		X
SPRINGFIELD	X				X		X							X			X									X		
IRON POST											X																	
SUMMUM	X													X														
BEVIER	X		X																									
COLCHESTER																												
MINERAL-FLEMING	X		X								X			X				X										
TEBO																												
WEIR—PITTSBURG																												
MURPHYSBORO														X														
BUFFALOVILLE	X																									X		
SECOR														X														
ROCK ISLAND	X		X											X				X	X									
k8																												
AEGIR																												
COPLAND																						X						
KATHARINA																												
h2/4																												
UPPER FOOT																												
UNION-FINEFRAU																												
"FIRST"																												
HAUPTFLÖZ																												
KOKSFLÖZ																												

TABLE 2.21 *Stratigraphic occurrences of gymnosperms—Pteridospermales (cont.)*

	RHETINOTHECA Leisman & Peters	R. *tetrasolenata* Leisman & Peters	*SULLITHECA* Stidd, Leisman & Phillips	S. *dactylifera* Stidd, Leisman & Phillips	*TELANGIUM* Benson	T. *pygmaeum* Graham	T. *scottii* Benson
NEWCASTLE							
SHUMWAY							
GRAND'CROIX							
CALHOUN						X	
DUQUESNE							
OPDYKE							
FRIENDSVILLE							
AMES							
UNNAMED							
PARKER							
DANVILLE							
BAKER							
HERRIN		X					
SPRINGFIELD							
IRON POST							
SUMMUM							
BEVIER							
COLCHESTER							
MINERAL-FLEMING							
TEBO							
WEIR-PITTSBURG							
MURPHYSBORO				X			
BUFFALOVILLE				X			
SECOR							
ROCK ISLAND				X			
k8							
AEGIR							
COPLAND							
KATHARINA							
h2/4							
UPPER FOOT							X
UNION-FINEFRAU							X
"FIRST"							
HAUPTFLÖZ							
KOKSFLÖZ							

and Stockey, 1976), and *Schopfiastrum (Rothwell and Taylor, 1972; Stidd and Phillips, 1973)*. Major monographs exist on *Pachytesta* (Hoskins and Cross, 1946b; Taylor, 1965), and *Dolerotheca* (Schopf, 1948; Baxter, 1949). References are given for most of the additional taxa based on fructifications.

Occurrences in Roof Nodules of the Upper Foot Seam

In the work of Stopes and Watson (1909) the then known permineralized plants from roof nodules were listed: *Sutcliffia insignis*, *Medullosa anglica*, *Myeloxylon*, *Pachytesta olivaeformis*, *Mesoxylon sutcliffii*, *Cordaites*, *Psaronius renaultii*, *Ankyropteris grayi*, *Etapteris diupsilon*, *E. shorensis*, *Tubicaulis sutcliffii*, *Sigillaria elegans*, *Sigillariopsis sulcata*, *Lepidostrobus oldhamius*, *Lepidophloios harcourtii*, *Lepidodendron hickii*, and calamite stems.

To these may be added the following: cf. *Cardiocarpus* (Scott Collection 2753, British Museum), *Mesoxylon lomaxii*, *M. platypodium*, *Mesoxylopsis arberae*, *Cladites bracteatus*, polycyclic *Psaronius*, *Botryopteris*, and *Lepidophylloides*. It was noted by Stopes and Watson (1909) that some of the common coal-ball plants such as *Lyginopteris* were not found in the roof nodules and that certain of the roof-nodule plants were not known from coal balls at all. In the latter category are *Sutcliffia*, several *Mesoxylon* species, *Mesoxylopsis*, *Cladites*, polycyclic *Psaronius*, *Etapteris shorensis*, and *Tubicaulis sutcliffii*. They concluded that the roof-nodule flora was different from the coal-ball flora and was formed from plants growing on some higher ground and that they had been transported by streams. It also seems significant that the occurrences of some of these plants, *Sutcliffia*, *Cordaites* or *Mesoxylon*, and polycyclic psaronii subsequently became more abundant in swamps higher in the Upper Carboniferous. At the time of the Stopes and Watson (1909) study, the idea had been expressed that the plants had the appearance of younger floras, which is consistent with the sequential introduction from nonswamp habitats and the expansion of cordaites and robust psaronii in later coal swamps. DiMichele and Phillips (1977, 2523) have pointed out that, "The increase in anatomical complexity concomitant with further increase in size in *Psaronius* may not have taken place initially in coal swamps."

BIOSTRATIGRAPHY

While coal balls are far more common than any of us suspected twenty years ago, data on the occurrences of the plants derived from them are still too meager from most coals and there are too few stratigraphic intervals well enough known to attempt correlations. From the data presently available one could attempt to bracket stratigraphically some new coal-ball occurrences; it does seem quite possible in many cases to determine the general part of the Pennsylvanian section of the seam source. This is only a very rough guide, and one ultimately has to turn to other sources for precise stratigraphic placement. The potential use of plants in coal balls for stratigraphic purposes obviously increases as more complete data become available from known sources and gaps are filled in the present stratigraphic distribution of coal balls. However, most of the well-known plant species that are common and widespread in many swamp habitats are long-ranging stratigraphically. Many of those that are potentially useful stratigraphically must be viewed with caution because of environmental factors.

The evolutionary history of plant introductions into the coal swamps is also of importance both in dealing with the biostratigraphy and in utilizing stratigraphic sequences as a guide for probable phylogenies. Plants could have been introduced into swamp environments anytime during the Upper Carboniferous, and some of the patterns mentioned later may well involve the introduction and expansion of new forms alongside swamp plants that have evolved considerably within swamps even from Early Carboniferous time. Our knowledge of the sequential entries of plants into coal swamps is derived from swamp assemblages in the Westphalian A and then from successively higher well-known swamp floras.

Biostratigraphic studies based on plants from coal balls differ significantly in their potential and emphasis from other paleobotanical data. The stratigraphic positions of coals bearing coal balls have been determined by means that do not rely on the botanical contents of the in situ coal balls. Consequently, the biostratigraphic use of coal-balls data is a separate but dependent means of establishing the stratigraphic and geographic patterns of in situ swamp-plant occurrences. To the extent that the precise stratigraphic position of a coal is unknown or that specific correlations between coal regions are uncertain, the biostratigraphic zones of swamp plants from coal balls are subject to change. The construction of a stratigraphic chart of coal-ball occurrences for Euramerica with the published data available, including the previous charts by Schopf (1941a), by Andrews (1951), and by Cross (1969), leaves some uncertaintly in the precise placement of certain coals of about the same age in different coal regions. This is particularly a problem in the upper part of the Pennsylvanian. The stratigraphic positions of some entries have required approximation because of the lack of precise stratigraphic placement within the region (particularly in Iowa) and in the relative positions of coals of Europe to those of North America.

As evolutionary and ecological indicators, swamp plants provide a means to establish the general patterns of swamp vegetation through portions of the Upper Carboniferous, where available data are stratigraphically close enough. It is desirable to document stratigraphic trends in prevalent swamp vegetation as well as to detect geographical differences. Major general patterns are already evident (Phillips et al., 1974), but how they differ from one region to another is known in only a few portions of the stratigraphic column.

Because differing plant communities of variable duration and frequency occur in many individual seam profiles, we need a measure of their relative importances, numerically and volumetrically; these determinations can be made from peat profiles of coal balls obtained in situ. The superposition of successive plant communities permits their study as with other peat beds, except the quality of anatomical preservation is, on the whole, better than that of most Holocene deposits. To a limited extent, the ephemeral communities of coal-ball deposits in a seam provide a preview of potential major vegetation types of later coals; where these are not so evident at one site, they may be at another site of the same coal. Plant occurrences, or lack thereof, readily call attention to changes or differences in swamps; but, as in coal palynology, they are most useful when accompanied by quantitative data. Because the analysis of plant occurrences and their inferred patterns are dependent on the extent of our knowledge of each swamp, the distribution of coal balls within coal seams is relevant.

Distribution Within Coal Seams

Coal-ball occurrences vary greatly in frequency and vertical distribution within and among coals. These factors have an important bearing on evaluations of reported occurrences of plant taxa.

There are marked differences in coal-ball occurrences among major commercial coals which are heavily mined. The Colchester Coal and equivalent coals and the Pittsburgh Coal have yielded relatively few coal balls thus far; these two coals constitute the most widespread and the most heavily mined coals, respectively, in the American Pennsylvanian. In the Illinois Basin the two most economically important coals (Springfield and Herrin Coal Members) have yielded by far the largest number of coal-ball occurrences.

The European coals from which coal balls have been obtained at numerous localities include the Union (Halifax Hard) and its Upper Foot Seam in England and equivalent coals across western Europe (Great Coal-ball Horizon), the Katharina Seam in the Aachen and especially the Ruhr, and Coals k_8, l_3, l_4, l_6 and m_3 in the Donets (Snigirevskaya, pers. comm., 1976). In the United States, coals in which there have been three or more coal-ball localities include the following: Copland (Taylor) coal bed of eastern Kentucky; an unnamed coal from Iowa (equivalent to the Rock Island of Illinois) along with the Secor and Blue Jacket of the Western Interior coal region; an unnamed Iowa coal approximately equivalent to the Buffaloville; the Mineral Coal of Kansas and its equivalent in western Kentucky, the DeKoven Coal; the Fleming Coal of southeastern Kansas; the Bevier Coal of Kansas; the Springfield and Herrin Coals of the Illinois Basin; and the Calhoun Coal of Illinois. These include about one-half of the coals from which numerous plant occurrences are known, although the available information from each of them varies a great deal.

It is a reasonable assumption that differing plant communities existed through the temporal span of a given coal swamp and the vertical distribution of coal balls within seams with a thickness of 0.5 m or more is pertinent, where known, to the relative completeness of plant occurrences. This has been established in the Herrin Coal (Phillips, Kunz, and Mickish, 1977). Some taxa occur frequently in numerous peat zones and others are much more restricted. Some seams have such different floras vertically that restricted coal-ball occurrences, often in the upper half, record only part of the swamp vegetation. Acknowledging that we have no published information on vertical coal-ball distribution for most of the coals, it is still desirable to mention those for which we do. At some sites in some coals, coal balls do occur throughout the entire vertical extent of the seam. In the "Great Coal-ball Horizon," Stopes and Watson (1909) established this in the Lower Mountain and Upper Foot components of the Union Seam which were separated by 20 cm at the Old Meadows Pit near Bacup in Lancashire. In the 30-cm thick Upper Foot Seam at the Shore Mine near Littleborough, massive coal balls extended up to a thickness of 2 m (Stopes and Watson, 1909; Lomax, 1915). Koopmans (1928) also observed coal balls occuping the entire seam at places in the Finefrau-Nebenbank of The Netherlands. In the first sampling of coal swamp plants by coal-ball zones, Leclercq (1930a) noted numerous layers in the 95-cm thick Bouxharmont seam but not vertically contiguous coal balls. In the Katharina Seam, Mentzel (1904) and Kukuk (1909) diagrammed coal-ball occurrences from the Ruhr indicating that the distribution was variable, occurring largely in the upper and rarely in the lower parts (separated by a clastic band) but not vertically throughout at any given exposure. According to Snigirevskaya (pers. comm., 1976), coal balls occur in three or more zones in the Coals k_8, l_3, l_6, and m_3 but there are no known solid coal-ball occurrences from floor to roof in the Donets shaft mines (Zaritsky and Snigirevskaya, pers. comm., 1976). American occurrences of almost complete peat profiles preserved in coal balls have been found in the Mineral, Fleming, and Bevier Coals of Kansas and the Calhoun Coal of Illinois by Mamay and Yochelson (1962); in an unnamed coal (equivalent to the Rock Island of Illinois) at the Angus Mine in Iowa (Stewart, Kosanke, and Delevoryas, mine notes dated 1951); in the unnamed coal approximately equivalent to the Buffaloville at the Star Mine in Iowa; in the Springfield Coal of Illinois and Indiana; in the Herrin Coal of Illinois and the equivalent coal in Kentucky (Kosanke, Simon and Smith, 1958; Evans and Amos, 1961; Phillips, Kunz and Mickish, 1977); and the Friendsville Coal of Illinois at the abandoned Allendale Coal Company Mine. There are other occurrences where coal balls occupy the entire thickness of coals, but the seam is relatively thin in that area except for the coal balls; these are the Copland coal bed of eastern Kentucky, the Parker Coal of Indiana, the Ames (coal below Ames Limestone) and Duquense Coals of Ohio (Rothwell, 1976b), and the coal below the Shumway Limestone in Illinois (Scheihing, 1978). There are a few coals in which coal balls were consistently encountered in the strip mines, but from frequent visits to in situ exposures it was determined that they were restricted to about the upper one-half of the seam. These were in the unnamed coal in Indiana, equivalent to the Murphysboro Coal of Illinois, at the Maple Grove Strip Mine near Cayuga, and in the Summum (No. 4) Coal, Pit 14 of the Peabody's Northern Illinois Mine near South Wilmington, Illinois. In the Briar Hill (No. 5) of southern Illinois, a core indicates coal-ball material occupying slightly more than the upper one-half. The thickest coal-ball deposits recorded are from different mines in the Herrin Coal of Illinois with coal-ball material constituting 2.7 to 2.9 m. of a 3-m seam thickness (Kosanke, Simon and Smith, 1958; DeMaris and Bauer, 1978). The seam in such areas is two to three times thicker that the adjacent coal where coal balls were absent.

Biostratigraphic Zones

While it is premature to draw biostratigraphic zones of coal-swamp vegetation, some further suggestions can be made on the patterns of plant occurrences from coal balls to augment those given by Phillips et al. (1974) for the Illinois Basin. These are given with the realization that a given swamp changes spatially and temporally and that one pattern of vegetation overlaps another through time. However, there are patterns based on quantitative occurrences of plants that can be further analyzed and such research is continuing.

The two oldest sources of coal balls included in this compilation are not adequately known but it is noted that the Koksflöz contains *Lyginopteris* and the Hauptflöz contains *Calamostachys binneyana* and *Lepiododendron vasculare*. All three taxa occur in the Westphalian A interval which includes the "First," Union-Bouxharmont-Finefrau, h2/4, and Katharina Seam. This interval is also marked by the uppermost occurrences of *Lyginopteris* and *Stauropteris oldhamia* and the lowest occurrence of *Botryopteris tridentata* and abundant cordaites.

Numerous coal-ball taxa extend through or part way through the Westphalian from the Katharina Seam; and there are subpatterns that need to be delineated in what is termed the "cordaitean zone," which extends from the Katharina Seam to the Bevier Coal in Kansas and the Summum (No. 4) Coal in Illinois. This spans about one-half of the Westphalian and includes major geographic changes in the relationships of the dominant forms — cordiates and lycopods. This has been noted also in the Donets basin in Coal l_3 (Phillips and Snigirevskaya, unpublished data). The cordaitean zone includes many different kinds of cordaites along the Westphalian B-C transition, which apparently represents the zenith of such swamp development. The differences in occurrences of some cordaitean taxa have been observed from site to site in the same coal (Buffaloville equivalent) as well as between the Western Interior coal region and the Illinois Basin. It was during this interval that the polycyclic psaronii became abundant in swamps and some fern genera, such as *Botryopteris*, evolved markedly. Some of the smaller plants that seem to be consistently associated with cordaitean rich (dominant or subdominant) communities include *Botryopteris tridentata*, *Microspermopteris*, and *Stelastellara parvula*, which was probably a gymnosperm (DiMichele and Phillips, 1979). *Botryopteris tridentata* is not known to occur above the Bevier Coal; *Microspermopteris* has not been reported from above the Summum Coal.

The generalization can be made that lycopod trees formed the basic swamp vegetation of the Westphalian coal swamps; thus, we are seeking the changes in vertical distribution of the other major groups. Above the cordaitean zone the importance of *Psaronius* and *Medullosa* is most evident as the second and third major groups. This pattern apparently extends upward from the Iron Post Coal of Oklahoma and the Springfield Coal of the Illinois Basin to the hiatus in coal-ball occurrences. During the DesMoinesian-Missourian (Westphalian-Stephanian) transition, the major lycopods disappear form the coal swamps of the Illinois Basin. While this is the greatest change in the Upper Carboniferous coal-swamp vegetation, there were apparently transitional aspects in the occurrences of the taxa that may have been regarded as strictly Upper Pennsylvanian. Some of the ferns found more commonly in the Parker, Friendsville, and Calhoun Coals do appear in small numbers in the Baker Coal or below. These include *Anachoropteris clavata*, *Sermaya*, and *Biscalitheca musata*. In the few coal balls available from the Danville (No. 7) sigillarian periderm is apparently more abundant than in coals immediatley below. *Sigillaria* was one of the few lycopods to survive the transition into the Upper Pennsylvanian.

While detailed interregional comparisons of coal-swamp vegetation in the Upper Pennsylvanian or equilvalent strata can not be made at this time, there are known differences among the peat floras from the Illinois Basin, the Ames and Duquesne Coals of Ohio (Rothwell, 1976b), the Redstone or Pittsburgh Coal of Ohio (Good and Taylor, 1974), and the Grand'Croix flora of France. Some to many of the same taxa occur in all of these but the dominant tree types differ among them. The limnic Grand'Croix flora differs the most because cordaites are dominant and exceedingly diverse; pteridosperms and ferns, however, are subdominant. In the coal balls from the coal below the Ames Limestone and from the Duquense Coal, psaroniaceous tree ferns and medullosan seed ferns are the most abundant, but there is a significant amount of cordaitean material. Rothwell (1976b) stated that the majority of taxa in the Duquense also occur in the Calhoun Coal of Illinois and that most of the plants reported from the Redstone or Pittsburgh Coal by Good and Taylor (1974) also occur in the Duquesne. If the Ames Limestone Member of Ohio were considered equivalent to the Omega Limestone Member of Illinois, then the coal-ball floras in the Duquesne Coal would be slightly younger than the Calhoun.

In the Redstone or Pittsburgh Coal, Sphenophytina is the most important group, with ferns second (Good and Taylor, 1974). Pteridosperms were relatively low in abundance, with *Callistophyton* being the most common pteridosperm. Coal balls from the Newcastle Coal of Texas contain more abundant medullosans or calamites, depending on the site, than *Psaronius*. These dominant vegetational types reported from the Virgilian of Texas and the Monongahelan of Ohio indicate stratigraphically higher patterns of distribution than those observed in the Illinois Basin. Thus, in the Virgilian or equivalent strata the dominant tree forms may be *Psaronius* (Shumway), medullosan (Newcastle), or calamites (Redstone or Pittsburgh and Newcastle).

ACKNOWLEDGMENTS

It is a pleasure to thank colleagues for their generous help and advice in the preparation of the manuscript. The stratigraphic chart was an expansion on charts prepared by James M. Schopf. U.S.G.S., Columbus, Ohio, and by Aureal T. Cross, Michigan State University, East Lansing, who kindly helped with references and advice, as did Robert M. Kosanke, U.S.G.S., Denver, Colorado, and Russel A. Peppers, Illinois State Geological Survey, Urbana, on palynological correlations. The maps were prepared by Philip J. DeMaris, Illinois State Geological Survey. Plant occurrences, peels-slides, and other regional information were provided by: Matthew J. Avcin, Iowa Geological Survey, Iowa City; Manfred Barthel, Museum fur Palaontologie, Humboldt-Universität zu Berlin, DDR; Robert W. Baxter, University of Kansas, Lawrence; Muriel Farion-Demoret and Maurice Streel, University of Liege, Belgium; Jean Galtier and John Holmes, University of Montpellier, France; Donald L. Eggert and Harold C. Hutchinson, Indiana Geological Survey, Bloomington; Robert E. McLaughlin, University of Tennessee, Knoxville; Hermann W. Pfefferkorn, University of Pennsylvanian, Philadelphia; Jeffry Schabilion and Nancy Brotzman, Univeristy of Iowa, Iowa City; Natasha S. Snigirevskaya, Komarov Botanical Institute, Leningrad, U.S.S.R.; and Benton M. Stidd, Western Illinois University, Macomb. For access to slide collections I thank Madame Blanc, Muséum National d'Histoire Naturelle, Paris; J.D.R. Fryer, University of London King's College; Majorie D. Muir and K. L. Alvin, University of London Imperial College; Cedric Shute, British Museum (N.H.), London; W. D. Ian Rolfe, Hunterian Museum, Glasgow, Scotland; and Kenneth R. Sporne, Cambridge University, England.

Christine Y. Bethin, Department of Slavic Languages, translated the Russian articles and William A. DiMichele, Botany Department, helped in the compilation of lycopod occurrences. The chart was made by Alice Prickett, School of Life Sciences, University of Illinois.

This study was supported in part by NSF Grant DEB 75/13695 for the Quantitative Analysis of Pennsylvanian Coal-Swamp Vegetation in Relation to Coal and by a 1975-1976 Fellowship Grant from the John Simon Guggenheim Memorial Foundation.

REFERENCES

Abernathy, G. E. 1946. Strip-mined areas in the southeastern Kansas coal field. *Kansas Geo. Surv. Bull.* 64:125-144.

Absolom, R. G. 1929. Lower Carboniferous coal-ball flora of Haltwhistle, Northumberland. *Proc. Univ. Durham Phil. Soc.* 8:73-87.

Aisenverg, D. E.; N. E. Brazhnikova; K. O. Novik; A. P. Rotay (ed.); and P.L. Shulga. 1960. *4th Cong. Intern. Stratigr. Géol. Carbonifère, Compte Rendu* (Heerlen 1958) 1:1-12.

Aisenverg, D. E.; V. V. Lagutina; M. L. Levenshtein; and V. S. Popov (eds).

1975. *Field excursion guidebook for the Donets Basin, 8th Intern. Cong. Carboniferous Stratigr. Geol.* (Moscow 1975), 360 p.

Anderson, B. R. 1954. A study of American petrified calamites. *Missouri Bot. Garden Annals* 41:395-418.

Andrews, H. N., Jr. 1942. Contributions to our knowledge of American Carboniferous floras. I. *Scleropteris*, gen. nov., *Mesoxylon* and *Amyelon*. *Missouri Bot. Garden Annals* 29:1-19.

Andrews, H. N., Jr. 1945. Contributions to our knowledge of American Carboniferous floras. VII. Some pteridosperm stems from Iowa. *Missouri Bot. Garden Annals* 32:323-360.

Andrews, H. N., Jr. 1949. *Nucellangium*, a new genus of fossil seeds previously assigned to *Lepidocarpon*. *Missouri Bot. Garden Annals* 36:479-504.

Andrews, H. N., Jr. 1951. American coal-ball floras. *Bot. Rev.* 17:431-469.

Andrews, H. N., Jr. 1952. Some American petrified calamitean stems. *Missouri Bot. Garden Annals* 39:189-218.

Andrews, H. N., Jr. 1970. Index of generic names of fossil plants, 1820-1965. *U.S. Geol. Surv. Bull. 1300*, 354 p.

Andrews, H. N., and S. N. Agashe. 1965. Some exceptionally large calamite stems. *Phytomorphlogy* 15:103-108.

Andrews, H. N.; C. A. Arnold; E. Boureau; J. Doublinger; and S. Leclercq. 1970. *Filicophyta IV. Fasc. I. Traite de Paleobotanique*. Paris: Masson et Cie, 519 p.

Andrews, H. N., and C. J. Felix. 1952. The gametophyte of *Cardiocarpus spinatus* Graham. *Missouri Bot. Garden Annals* 39:127-135.

Andrews, H. N., and J. A. Kernen. 1946. Contributions to our knowledge of American Carboniferous floras. VIII. Another *Medullosa* from Iowa. *Missouri Bot. Garden Annals* 33:141-146.

Andrews, H. N., and W. H. Murdy. 1958. *Lepidophloios*-and ontogeny in arborescent lycopods. *Am. Jour. Bot.* 45:552-560.

Andrews, H. N., and E. Pannell. 1942. Contributions to our knowledge of American Carboniferous floras. II. *Lepidocarpon*. *Missouri Bot. Garden Annals* 29:19-28.

Arber, A. 1910. On the structure of the Palaeozoic seed *Mitrospermum compressum* (Will.). *Ann. Bot.* 24:491-509.

Arber, A. 1914. An anatomical study of the Palaeozoic cone-genus *Lepidostrobus*. Trans. Linn. Soc. London (Botany ser. 2) 7:205-238.

Arber, E.A.N., and H. H. Thomas. 1909. A note on the structure of the cortex of *Sigillaria mamillaris*, Brongn. *Ann. Bot.* 23:513-514.

Arnold C. A. 1958. Petrified cones of the genus *Calamostachys* from the Carboniferous of Illinois. *Univ. Michigan Contrib. Mus. Paleont.* 14:149-165.

Arnold, C. A. 1960. A lepidodendrid stem from Kansas and its bearing on the problem of cambium and phloem in Paleozoic lycopods. *Univ. Michigan Contrib. Mus. Paleont.* 15:249-267.

Arnold, C. A. 1967. The proper designations of the foliage and stems of the Cordaitales. *Phytomorphology* 17:346-350.

Balbach, M. K. 1965. Paleozoic lycopsid fructifications. I. *Lepidocarpon* petrifactions. *Am. Jour. Bot.* 52:317-330.

Balbach, M.K. 1966. Paleozoic lycopsid fructifications. II. *Lepidostrobus takhtajanii* in North America and Great Britain. *Am. Jour. Bot.* 53:275-283.

Balbach, M. K. 1967. Paleozoic lycopsid fructifications. III. Conspecificity of British and North American *Lepidostrobus* petrifactions. *Am.*

Jour. Bot. 54:867-875.

Barnard, P.D.W. 1962. Revision of the genus *Amyelon* Williamson. *Palaeontology* 5:213-224.

Barss, M. S., and P.A. Hacquebard. 1967. Age and stratigraphy of the Pictou Group in Maritime Privinces as revealed by fossil spores. *Geol. Assoc. Canada, Spec. Vol. 4,* 267-282.

Baxendale, R. W. 1977. A study of the morphology and anatomy of cordaitean organ genera from Middle Pennsylvanian Kansas and Iowa coal balls. Ph.D. thesis, Lawrence: University of Kansas, 259 p.

Baxter, R.W. 1949. Some pteridosperm stems and fructifications with particular reference to the Medullosae. *Missouri Bot. Garden Annals* 36:287-352.

Baxter R. W. 1951. Coal balls — new discoveries in plant petrifactions from Kansas coal beds. *Trans. Kansas Acad. Sci.* 54:526-535.

Baxter, R. W. 1952. The coal-age flora of Kansas. II. On the relationships among the genera *Etapteris, Scleropteris* and *Botrychioxylon. Am. Jour. Bot.* 39:263-274.

Baxter, R. W. 1955. *Palaeostachya andrewsii,* a new species of calamitean cone from the American Carboniferous. *Am. Jour. Bot.* 42:342-351.

Baxter, R. W. 1959. A new cordaitean stem with paired axillary branches. *Am. Jour. Bot.* 46:163-169.

Baxter, R. W. 1960. A first report of coal balls from the Pennsylvanian of New Brunswick, Canada. *Canadian Jour. Bot.* 38:697-699.

Baxter, R. W. 1963. *Calamocarpon insignis,* a new genus of heterosporous petrified calamitean cones from the American Carboniferous. *Am. Jour. Bot.* 50:469-476.

Baxter, R. W. 1967. A revision of the sphenopsid organ genus, *Litostrobus. Univ. Kansas Sci. Bull.* 47:1-23.

Baxter, R. W. 1971. *Carinostrobus foresmani,* a new lycopod cone genus from the Middle Pennsylvanian of Kansas. *Palaeontographica* 134B: 124-130.

Baxter, R.W. 1972. A comparative study of nodal anatomy in *Peltastrobus reedae* and *Sphenophyllum plurifoliatum. Rev. Palaeobot. Palynol.* 14:41-47.

Baxter, R. W. 1975. Fossil fungi from American Pennsylvanian coal balls. *Univ. Kansas Paleont. Contrib. Paper* 77, 6 p.

Baxter, R. W., and R. W. Baxendale. 1976. *Corynepteris involucrata,* sp. nov., a new fertile fern of possible zygopterid affinities from the Pennsylvanian of Kansas. *Univ. Kansas Paleont. Contrib. Paper 85,* 8 p.

Baxter, R. W., and A. L. Hornbaker. 1965. Pennsylvanian fossil plants from Kansas coal balls. *Field Conference Guidebook for Annual Meetings.* Geol. Soc. Am. and associated societies, Kanasa City: 34 p.

Baxter, R. W., and G. A. Leisman. 1967. A Pennsylvanian calamitean cone with *Elaterites triferens* spores. *Am. Jour. Bot.* 54:748-754.

Baxter, R. W., and E. A. Roth. 1954. *Cardiocarpus magnicellularis* sp. nov., a prelminiary report. *Trans. Kansas Acad. Sci.* 57:458-460.

Benninghoff, W. S. 1942. Preliminary report on a coal ball flora from Indiana. *Proc. Indiana Acad. Sci.* 52:62-68.

Benson, M. 1904. *Telangium Scotti,* a new species of *Telangium (Calymmatotheca)* showing structure. *Ann. Bot.* 18:161-177.

Benson, M. 1907. *Miadesmia membranacea,* Bertrand: a new Palaeozoic lycopod with a seed-like structure. *Phil. Trans. Roy. Soc. London* 199B: 409-425.

Benson, M. 1912. *Cordaites Felices,* sp.

nov., a cordaitean leaf from the Lower Coal Measures. *Ann. Bot.* 26:201-207.

Benson, M. 1918. *Mazocarpon* or the structural *Sigillariostrobus*. *Ann. Bot.* 32:569-589.

Bertrand, C. E., and B. Renault. 1886. Recherches sur les Poroxylons. Gymnospermes fossiles des Terrains Houillers Superieurs. *Archives Bot. Nord France* 11:243-337.

Blanc-Louvel, C. 1966. Etude anatomique comparée des tiges et des pétioles d'une Pteridospermée du Carbonifère du genre *Lyginopteris* Potonié. *Mém Mus. Nat. Historie Nat. Sér. C. Sci. Terra,* Paris 18, fasc. 1, 103p.

Blazer, A.M. 1975. Index of generic names of fossil plants, 1966-1973. *U.S. Geol. Surv. Bull. 1396.* Washington: U.S. Gov. Print. office, 54 p.

Boureau, E. 1952. Sur un nouveau *Medullosa* du Stephanien de Rive-de-Gier. *Bull. Soc. Geol. France* (1951) ser. 6, 1:419-423.

Boureau, E. 1964. Sphenophyta-Noeggerathiophyta, III. *Traite de Paleobotanique*. Paris: Masson et Cie, 544 p.

Brack, S. D. 1970. On a new structurally preserved arborescent lycopsid fructification from the Lower Pennsylvanian of North America. *Am. Jour. Bot.* 57:317-330.

Brack, S. D., and T. N. Taylor. 1972. The ultrastructure and organization of *Endosporites*. *Micropaleontology* 18:101-109.

Bradley, W. H. 1956. Use of series subdivisions of the Mississippian and Pennsylvanian Systems in reports by members of the U.S. Geological Survey. *Bull. Am. Assoc. Petrol. Geologists* 40, (pt. II): 2284-2285.

Branson, C. C., and G. C. Huffman. 1965. Geology and oil and gas reserves of Craig County, Oklahoma. Part 1. Geology of Craig County. *Oklahoma Geol. Surv. Bull. 99,* 58p.

Brant, R. A., and R. M. DeLong. 1960. Coal resoures of Ohio. *Ohio Geol. Surv. Bull. 58,* 245p.

Brongniart, A. 1874. Les graines fossiles trouvees a l'etat silicifie dans le terrain houiller de Saint-Etienne. *Ann. Sci. Nat., Bot.* 20:234-265. (1881, Recherches sur les Graines fossiles silicifees, Imprimerie Nationale, Paris.)

Brown, L. F., Jr.; A. W. Cleaves,II; and A. W. Erxleben. 1973. Pennsylvanian depositional systems in north-central Texas. *Guidebook No. 14.* Austin: Bur. Econ. Geol., Univ. Texas, 117p.

Brush, G. S., and E. S. Barghoorn. 1955. *Kallostachys scottii:* A new genus of sphenopsid cones from the Carboniferous. *Phytomorphology* 5:346-356.

Brzyski, B. 1965. A petrified Carboniferous lepidodendrid-*Lepidophloios fulginosus* from the vicinity of Rybnik (Upper Silesian Coal Basin). *Acta Palaeobot.* Kraków 6:3-13.

Cady, G. H. 1915. Coal resources of District I (Longwall). *Illinois State Geol. Surv. Coop. Coal Mining Invest. Bull. 10,* 149p.

Calver, M.A. 1969. Westphalian of Britian *6th Congr. Intern. Stratigr. Géol. Carbonifère Compte Rendu* (Sheffield, 1967) 1:233-254.

Calver, M.A. 1969. Westphalian of Britian. *6th Congr. Intern. Stratigr. Geol. Carbonifere Compte Rendu* (Sheffield, 1967) 1:233-254.

Canright, J. E. 1959. Fossil plants of Indiana. *Indiana Geol. Surv. Rept. Prog. 14,* 45p.

Chaloner, W. G., and E. Boureau. 1967. Lycophyta. In *Traite de Paleobotanique,* E. Boureau, ed. Paris: Masson et Cie, 845 p.

Cheney, M.G. 1940. Geology of north-central Texas. *Bull. Am. Assoc. Petrol. Geologists* 24:65-118.

Clayton, G.; R. Coquel; J. Doubinger; K.

J. Gueinn; S. Loboziak; B. Owens; and M. Streel. 1977. Carboniferous miospores of Western Europe: illustration and zonation. *Meded. Rijks Geol. Dienst* 29:1-71.

Cohen, L.M., and Theodore Delevoryas. 1959. An occurrence of *Cordaites* in the Upper Pennsylvanian of Illinois. *Am. Jour. Bot.* 46:545-549.

Corda, A.J. 1845. *Beitrage zur Flora der Vorwelt.* Prague: J. G. Calve'scheg Buchhandlung.

Corsin, Paul. 1937. Contribution a l'étude des fougères anciennes du groupe des Inversicaténales. Thesis, Univ. Lille, Lille. 247p.

Cridland, A. A. 1964. *Amyelon* in American coal-balls. *Palaeontology* 7:186-209.

Cross, A. T. 1952. The geology of the Pittsburgh Coal: Stratigraphy, petrology, origin and composition, and geologic interpretation of mining problems. In: Origin and Constitution of Coal. Nova Scotia Dept. Mines, June 1952. pp. 32-99.

Cross, A. T., 1967. A coal-ball flora from Pennsylvania (Abs.). *Am. Jour. Bot.* 54:652.

Cross, A. T. 1969. Coal balls: Their origin, nature, geologic significance and distribution (Abs.). *Geol. Soc. America Program,* Atlantic City, N.J. p. 42.

Damberger, H.H. 1970. Petrographic character of the Colchester (No. 2) Coal Member at the Banner Mine, Peoria and Fulton Counties, Illinois. In: *Depositional Environments in Parts of the Carbondale Formation—Western and Northern Illinois, Francis Creek Shale and Associated Strata and Mazon Creek Biota,* Smith, W. H.; R. B. Nance; M. E. Hopkins; R. G. Johnson; and C. W. Shabica. Urbana; Illinois Geol. Surv., Guidebook Ser. 8, pp. 99-105.

Darrah, E. L. 1968. A remarkable branching *Sphenophyllum* from the Carboniferous of Illinois. *Palaeontographica* 121B: 87-101.

Darrah, W. C. 1939. The fossil flora of Iowa coal balls. I. Discovery and occurrence. *Bot. Leaflets Harvard Univ.* 7:125-136.

Darrah, W. C. 1940. The fossil flora of Iowa coal balls. III. *Cordaianthus. Bot. Mus. Leaflets Harvard Univ.* 8:1-20.

Darrah, W. C. 1941a. The fossil flora of Iowa coal balls. IV. *Lepidocarpon. Bot. Mus. Leaflets Harvard Univ.* 9:85-100.

Darrah, W. C. 1941b. Studies of American coal balls. *Am. Jour. Sci.* 239:33-53.

Darrah, W. C. 1967. The structure of *Cardiocarpus Florini* (Darrah), A Pennsylvanian cordaite seed from Iowa. *Proc. Pennsylvania Acad. Sci.* 40:80-86.

Deevers, C. L. 1937. Structure of Paleozoic seeds of the Trigonocarpales. Contributions from the Hull Botanical Laboratory 476. *Bot. Gaz.* 98:572-585.

Delevoryas, T. 1953. A new male cordaitean fructification from the Kansas Carboniferous. *Am. Jour. Bot.* 40:144-150.

Delevoryas, T. 1955. The Medullosae-structure and relationships. *Palaeontographica* 97B:114-167.

Delevoryas, T. 1957. The anatomy of *Sigillaria approximata. Am. Jour. Bot.* 44:654-660.

Delevoryas, T., and J. Morgan. 1952. *Tubicaulis multiscalariformis:* A new American coenopterid. *Am. Jour. Bot.* 39:160-166.

Delevoryas, T., and J. Morgan. 1954a. A further investigation of the morphology of *Anachoropteris clavata. Am. Jour. Bot.* 41:192-198.

Delevoryas, T., and J. Morgan. 1954b. A new pteridosperm from Upper Pennsylvanian deposits of North America.

Palaeontographica 96B:12-23.
DeMaris, P.J., and R.A. Bauer. 1978. Geology of a longwall mining demonstration at Old Ben No. 24: Roof lithologies and coal balls. *Proc. Illinois Min. Inst.* (1977), pp. 80-91.
Dennis, R. L. 1974. Studies of Paleozoic ferns: *Zygopteris* from the Middle and Late Pennsylvanian of the United States. *Palaeontographica*148B:95-136.
DiMichele, W. A. 1979. Arborescent lycopods of Pennsylvanian age coals: *Lepidophloios. Palaeontographica* 171B: 57-77.
DiMichele, W. A., and T. L. Phillips. 1977. Monocyclic *Psaronius* from the lower Pennsylvanian of the Illinois Basin. *Canadian Jour. Bot.* 55:2514-2524.
DiMichele W. A., and T. L. Phillips. 1979. *Stelastellara* Baxter, axes of questionable gymnosperm affinity with unusual habit — Middle Pennsylvanian. *Rev. Palaeobot. Palynol.* 27:103-117.
Dopita, M., and J. Králik. 1971. Mineralized cordaitic tissues from the seams of the Jaklovec series (Namur A) of the Ostrava-Karviná basin. *Collected Scientific Works of Mining Institute in Ostrava, Mining* — Geology series (Article 266) 17:75-91. (Czech).
Dunbar, C. O., and L. G. Henbest. 1942. Pennsylvanian Fusulinidae of Illinois. *Illinois Geol. Surv. Bull. 67*, 218p.
Dutcher, R.R.; J. C. Ferm; N.K. Flint; and E. G. Williams. 1959. The Pennsylvanian of western Pennsylvania. In *Guidebook for Field Trips*, T. V. Buckwalter; S. C. Addison; C. Gray, and R. B. Carter. Pittsburgh Meeting (1959), Geol. Soc. Am., pp. 61-76.
Eggert, D. A. 1959. Studies of Paleozoic ferns: *Tubicaulis stewartii* sp. nov. and evolutionary trends in the genus. *Am. Jour. Bot.* 46:594-602.
Eggert, D. A. 1972. Petrified *Stigmaria* of sigillarian origin from North America. *Rev. Palaeobot. Palynol.* 14:85-99.
Eggert, D. A., and T. Delevoryas. 1960. *Callospermarion*— a new seed genus from the upper Pennsylvanian of Illinois. *Phytomorphology* 10: 131-138.
Eggert, D. A., and T. Delevoryas. 1967. Studies of Paleozoic ferns: *Sermaya*, gen. nov. and its bearing on filicalean evolution in the Paleozoic *Palaeontographica* 120B:169-180.
Evans, T. J. 1974. *Bituminous coal in Texas.* Austin: Bur. Econ. Geol., Univ. Texas, Handbook 4, 65 p.
Evans, W. D., and D. H. Amos. 1961. An example of the origin of coal balls. Proc. Geologists Assoc. 72:445-454.
Ewart, R. B. 1961. Two new members of the genus *Scolecopteris. Missouri Bot. Garden Annals* 48:275-289.
Feliciano, J. M. 1924. The relation of concretions to coal seams. *Jour. Geology* 32: 230-239.
Felix, C. J. 1952. A study of the arborescent lycopods of southeastern Kansas. *Missouri Bot. Garden Annals* 39: 263-288.
Felix, C. J. 1954. Some American arborescent lycopod fructifications. *Missouri Bot. Garden Annals* 41: 351-394.
Felix, J. 1886. Untersuchungen uber den inneren Bau westfalischer Carbon-Pflanzen. *Abh. Geol. Specialkarte Preuss. Thuring. Staaten* 7: 153-225.
Ferm, J. C., and V. V. Cavaroc, Jr. 1969. A field guide to Allegheny deltaic deposits in the upper Ohio Valley with a commentary on deltaic aspects of Carboniferous rocks in the northern Appalachian plateau. Ohio Geol. Soc. and Pittsburgh Geol. Soc., 21 p.
Fiebig, H. E. R., and W. Leggewie. 1974. Die Namurflora des Ruhrgebietes und ihre stratigraphische Bedeutung. *7th Cong. Intern. Stratigr. Géol. Carbonifère,* Compte Rendu (Krefeld 1971) 3: 45-53.
Florin, R. 1937. On the morphology of the pollen-grains in some Palaeozoic

pteridosperms. *Svensk Bot. Tidskrift* 31: 305-338.

Foster, W. D. and F. L. Feicht. 1946. Mineralogy of concretions from Pittsburgh coal seam, with special reference to analcite. *Am. Mineralogist* 31: 357-364.

deFraine, E. 1914. On *Medullosa centrofilis*, a new species of *Medullosa* from the Lower Coal Measures. *Ann. Bot.* 28: 251-264.

Frankenberg, J. M., and D. A. Eggert. 1969. Petrified *Stigmaria* from North America, I. *Stigmaria ficoides*, the underground portions of Lepidodendraceae. *Palaeontographica* 128B: 1-47.

Frezon, S. E., and G. H. Dixon. 1975. Texas Panhandle and Oklahoma. *U.S. Geol. Surv. Prof. Pap. 853-J*, pp. 177-195.

Fry, W. L. 1956. New cordaitean conco from the Pennsylvanian of Iowa. *Jour. Paleontology* 30: 35-45.

Galtier, J. and L. Grambast. 1972. Observations nouvelles sur les structures reproductrices attribuées à *Zygopteris lacattei* (Coenopteridales de l'Autuno-Stephanien Francais). *Rev. Palaeobot. Palynol.* 14: 101-111.

Galtier, J., and John Holmes. 1976. Un *Corynepteris* à structure conservée du Westphalien d'Angleterre. *C. R. Acad. Sci.*, Paris, Ser. D. 282: 1265-1268.

Galtier, J., and T. L. Phillips. 1977. Morphology and evolution of *Botryopteris*, a Carboniferous age fern. Part 2. Observations on Stephanian species from Grand'Croix, France. *Palaeontographica* 164B: 1-32.

Good, C. W. 1973. Studies of *Sphenophyllum* shoots: Species delimitation within the taxon *Sphenophyllum*. *Am. Jour. Bot.* 60: 929-939.

Good, C. W. 1978. Taxonomic characteristics of sphenophyllalean cones. *Am. Jour. Bot.* 65: 86-97.

Good, C W., and T. N. Taylor. 1970. On the structure of *Cordaites felicis* Benson from the Lower Pennsylvanian of North America. 1970. *Palaeontology* 13: 29-39.

Good, C. W., and T. N. Taylor. 1974. Structurally preserved plants from the Pennsylvanian (Monongahela Series) of southeastern Ohio. *Ohio Jour. Sci.* 74: 287-290.

Graham, R. 1934. Pennsylvanian flora of Illinois as revealed in coal balls. *Bot. Gaz.* 95: 453-476.

Graham. R. 1935a. Pennsylvanian flora of Illinois as revealed in coal balls. Part 2. *Bot. Gaz.* 97:156-168.

Graham, R., 1935b. An anatomical study of the leaves of the Carboniferous lycopods. *Ann. Bot.* 49: 587-608.

Grand'Eury, F.C. 1877. Mémoire sur la flore Carbonifère du Départment de la Loire et du centre de la France. *Mém. Acad. Sci. Inst. France* 24: 1-624.

Haas, H. 1975. *Arthroxylon werdensis* n. sp. — ein Calamit aus dem Namur C. des Ruhrkarbons mit vollstandig erhaltenen Geweben. *Argumenta Palaeobotanica* 4: 139-154.

Hacquebard, P. A. 1972. The Carboniferous of Eastern Canada. *7th. Congr. Intern. Stratigr. Géol. Carbonifère, Compte Rendu* (Krefeld 1971) 1: 69-90.

Halket, A. C. 1930. The rootlets of *Amyelon radicans* Will.; their anatomy, their apices and the endophytic fungus. *Ann. Bot.* 44: 865-905.

Hall, J. W. 1954. The genus *Stephanospermum* in American coal balls. *Bot. Gaz.* 115: 347-360.

Hall, J. W. 1961. *Anachoropteris involuta* and its attachment to a *Tubicaulis* type of stem from the Pennsylvanian of Iowa. *Am. Jour. Bot.* 48: 731-737.

Harms, V. L., and G. A. Leisman. 1961. The anatomy and morphology of certain *Cordaites* leaves. *Jour. Paleon-*

tology 35: 1041-1064.

Havlena, V. 1967. Stratigraphische tabelle des Karbon und Perm. Munster: Fachschaft Geologie, 1 p.

Hendricks, T. A. 1937. The McAlester District, Pittsburg, Atoka and Latimer counties. *U.S. Geol. Surv. Bull. 874-A*, 90 p.

Hirmer, M. 1928. Über Vorkommen und Verbreitung der Dolomitknollen und deren Flora. 1st *Congr. Stratigr. Carbonifère, Compte Rendu* (Heerlen 1927) pp. 289-312.

Hirmer, M. 1933a. Zur Kenntnis der strukturbietenden Pflanzenreste des jungeren Palaeozoikums. *Palaeontographica* 77B: 121-140.

Hirmer, M. 1933b. Zur Kenntnis der strukturbietenden Pflanzenreste des jüngeren Paläeozoikums. 2: Über zwei neue, im mittleren Oberkarbon Westdeutschlands gefundene Arten von *Heterangium* Corda nebst Bemerkungen über *Heterangium shorense* Scott. *Palaeontographica* 78B: 57-113.

Holden, H. S. 1930. On the structure and affinities of *Ankyropteris corrugata*. *Phil. Trans. Roy. Soc. London* 218B:78-114.

Holden, H.S. 1955. Some features in the morphology of a hitherto undescribed stem from the Lancashire Coal-Measures. *Jour. Linn. Soc. Botany* 55: 313-317.

Holmes, J. C. 1977. The Carboniferous fern *Psalixochlaena cylindrica* as found in Westphalian A coal balls from England. Part I. Structures and development of the cauline system. *Palaeontographica* 164B:33-75.

Hooker, J. D., and E. W. Binney. 1855. On the structure of certain limestone nodules enclosed in seams of bituminous coal, with a description of some Trigonocarpons contained in them. *Phil. Trans. Roy. Soc. London* 145: 149-156.

Hopkins, M. E., and J. A. Simon. 1975. Pennsylvanian System.In Handbook of Illinois Stratigraphy Illinois Surv. Bull. 95 H. B. Willman et al., 1975, pp. 163-201.

Hoskins, J. H. 1926. Structure of Pennsylvanian plants from Illinois. I. *Bot. Gaz.* 82: 427-437.

Hoskins, J. H. 1934. *Psaronius illinoensis*. *Am. Midland Naturalist* 15: 358-362.

Hoskins, J. H., and A. T. Cross. 1941. A consideration of the structure of *Lepidocarpon* Scott based on a new strobilus from Iowa. *Am. Midland Naturalist* 25: 523-547.

Hoskins, J. H., and A. T. Cross. 1943. Monograph of the Paleozoic cone genus *Bowmanites* (Sphenophyllales). *Am. Midland Naturalist* 30: 113-163.

Hoskins, J. H., and A. T. Cross. 1946a. Studies in the Trigonocarpales Part II. Taxonomic problems and a revision of the genus *Pachytesta*. *Am. Midland Naturalist* 36:331-361.

Hoskins, J. H., and A. T. Cross, 1946b. Studies in the Trigonocarpales Part Ii. Taxonomic problems and a revision of the genus *Pachytesta*. *Am. midland Naturalist* 36: 331-361.

Huertas, G. 1960. A study of the leaf cuticles from Carboniferous deposits of North America. *Univ. Ind. Santander, Bol. Geol.* (Colombia, S. A.) no. 4, 5-18.

Humblet, E. 1941. Le bassin houiller de Liege. *Rev. Univers. Mines*, ser. 8, 17: 1-21.

Jennings, J. R, and M. A. Millay. 1978. A new permineralized marattialean fern from the Pennsylvanian of Illinois. *Palaeontology* 21: 709-716.

Jewett, J. M.; C. K. Bayne; E. D. Goebel; H. G. O'Connor; A. Swineford; and D. E. Zeller. 1968. The stratigraphic succession in Kansas. *Kansas Geol. Surv. Bull. 189*, 81p.

Jizba, K. M. M. 1962. Late Paleozoic

bisaccate pollen from the United States midcontinent area. *Jour. Paleontology* 36: 871-887.

Judd, R. W., and J. J. Nisbet. 1969. Pennsylvanian coal-ball flora of Indiana. *Proc. Indiana Acad. Sci.* (1968) 78:120-138.

Kern, E. M., and H. N. Andrews. 1946. Contributions to our knowledge of American Carboniferous floras. IX. Some petrified seeds from Iowa. *Missouri Bot. Garden Annals* 33: 291-307.

Knoell, H. 1935. Zur Kenntnis der strukturbietenden Pflanzenreste des jungeren Palaeozoikums. (Herausgegeben von Max Hirmer), Teil 4: Zur Systematik der strukturbietenden Calamiten der Gattung *Arthropitys* Goeppert aus dem mittleren Oberkarbon Westdeutschlands und Englands. *Palaeontographica* 80B: 1-51.

Koopmans, R. G. 1928. *Researches on the flora of the coal-balls from the "Finefrau-Nebenbank" horizon in the province of Limburg (The Netherlands).* Heerlen: Geologisch Bureau voor het Nederlandsche Mijngebied, 53p.

Koopmans, R. G. 1934. Researches on the flora of coal balls from the "Aegir" horizon in the province of Limburg (The Netherlands) *Jaarverslag over 1933.* Heerlen: Geologisch Bureau voor het Nederlandsche Mijngebied, pp. 45-46.

Kosanke, R. M. 1973. Palynological studies of the coals of the Princess Reserve District in northeastern Kentucky. *U.S. Geol. Surv. Prof. Pap. 839,* 22 p.

Kosanke, R. M.; J. A. Simon; and W. H. Smith. 1958. Compaction of plant debris—forming coal beds. *Geol. Soc. Am. Bull.* 69: 1599-1600.

Kosanke, R. M.; J. A. Simon; H. R. Wanless; and H. B. Willman. 1960. Classification of the Pennsylvanian strata of Illinois. *Illinois Geol. Surv. Rept. Inv. 214,* 84p.

Kubart, Bruno. 1908. Pflanzenversteinerungen enthaltende Knollen aus dem Ostrau-Karwiner Kohlenbecken. *Sitzungsber.* Akad. *Wiss. Wien, Math. — Naturwiss. Kl.* 117: 573-578.

Kubart, B. 1910. Untersuchungen uber die Flora des Ostrau-Karwiner Kohlenbeckens. I. Die spore von *Spencerites membranaceus* nov. spec. *Denkschr. Akad. Wiss. in Wien, Matth. — Naturwiss. Kl.* 85:83-89.

Kubart, B. 1911. Corda's Sphaerosiderite aus dem Steinkohlenbecken Radnitz — Břaz in Böhmen nebst Bemerkungen über *Chorinopteris gleichenioides* Corda. *Sitzungsber. Akad. Wiss. in Wien. Math — Naturwiss. KL.*120: 1035-1048.

Kubart, B. 1914. Über die Cycadofilicineen *Heterangium* und *Lyginodendron* aus dem Ostrauer Kohlenbecken. *Österreichische Bot. Zeitschr., Wien,* 64: 8-19, (continued as *Plant Systematics and Evolution, Wien).*

*Kubart, B. 1931. Untersuchungen über die Flora des Ostrau-Karwiner Kohlenbekens. II. Ein Lyginodendron-*Stämmchen mit zwei Zuwachszonen. *Denkschr. Akad. Wiss. Wein. Math — Naturwiss Kl.* 102: 369-372.

Kukuk, P. 1909. Über Torfdolomite in den Flözen der niederrheinisch—westfälischen Steinkohlenablagerung. *Gluckauf* 45 (32):1137-1150.

Kukuk, P. 1938. *Geologie des Niederrheinisch-Westfalischen Steinkohlengebietes* Berlin: Julius Springer, 706p.

Landis, E. R., and O. J. Van Eck. 1965. Coal Resources of Iowa. *Iowa Geol. Surv. Tech. Pap. No. 4, 141 p.*

Leclercq, S. 1925. Introduction à l'Etude anatomique des Végétaux houillers de Belgique: Les coal balls de la couche

Bouxharmont des Charbonnages de Wérister. *Mém. Soc. Geol. Belgique.* 79p.

Leclercq, S. 1928a. Les végétaux à structure conservée du Houiller Belge. Note I: Feuilles et racines de Cordaitales des coal-balls de la couche Bouxharmont. *Ann. Soc. Geol. Belgique Bull.* 51: 1-14.

Leclercq, S. 1928b. Les vegetaux a structure conservee du Houiller Belge. Note II. Sur un *Stigmaria* a bois primary centripete des coal-balls de la couche Bouxharmont. *Ann. Soc. Geol. Belgique 51:* B89-B93.

Leclercq, S. 1930a. Etude d'une coupe verticale dans une couche a coal-balls du Houiller de Liege. *Ann Soc. Geol. Belgique Bull.* 54: B63-B67.

Leclercq, S. 1930b. A monograph of *Stigmaria bacupensis* Scott et Lang. *Ann. Bot.* 44: 31-54.

Leclercq, S. 1934. Les coal-balls et la formation des couches de houille qui les renferment. *Ann. Soc. Geol. Belgique Bull.* 58: B214-B220.

Leclercq, S. 1952. Sur la presence de coal-balls dans la couche Petit Buisson (Assise du Felnu) du bassin houiller de la Campine. *3rd Congr. Intern. Stratigr. Carbonifere Compte Rendu* (Heerlen 1951)2: 397-400.

Ledran, C. 1962. Sur la structure anatomique de quelques feuilles de Cordaites. *Bull. Soc. Bot. France* 109: 63-75.

Leggewie, W. and W. Schonefeld. 1957. Pteridophyten und Pteridospermen der Sprockhöveler (—Magerkohlen—) Schichten (Namur C) *Palaeontographica* 101B-1-29.

Leggewie, W., and W. Schonefeld. 1959. Die Steinkohlenflora der westlichen paralischen Steinkohlenreviere Deutschlands. II. Die Mesocalamiten der Sprockhöveler Schichten. *Geol. Jahrb., Beih.* 36:60-90.

Leggewie, W., and W. Schonefeld. 1961. Die Calamariaceen der Westfal-Schichten im Ruhrkarbon. *Palaeontographica* 109B: 1-44.

Leisman, G. A. 1961. A new species of *Cardiocarpus* in Kansas coal balls. *Trans. Kansas Acad. Sci.* 64:117-122.

Leisman, G. A. 1964. *Physostoma calcaratum* sp. nov., a tentacled seed from the middle Pennsylvanian of Kansas. *Am. Jour. Bot.* 51: 1069-1075.

Lesiman, G. A., and J. L. Bucher. 1971. Variability in *Calamocarpon insignis* from the American Carboniferous. *Jour. Paleontology* 45:494-501.

Leisman, G. A., and Charles Graves. 1964. The structure of the fossil Sphenopsid cone, *Peltastrobus reedae. Am. Midland Naturalist* 72:426-437.

Lesiman, G. A., and J. S. Peters. 1970. A new pteridosperm male fructification from the middle Pennsylvanian of Illinois. *Am. Jour. Bot.* 57: 867-873.

Leisman, G. A., and T. L. Phillips. 1979. Megasporangiate and microsporangiate cones of *Achlamydocarpon varius* from the Middle Pennsylvanian. *Paleontographica* 168B:100-128.

Leisman, G. A., and J. Roth. 1963. A reconsideration of *Stephanospermum. Bot. Gaz.* 124: 231-240.

Leisman, G. A., and B. M. Stidd. 1967. Further occurrences of *Spencerites* from the middle Pennsylvanian of Kansas and Illinois. *Am. Jour. Bot.* 54: 316-323.

Leistikow, K. U. 1962. Die Wurzeln der Calamitaceae. Inaug. dissertation, Univ. Tubingen. 67p.

Lemoigne, Y. 1960. Etudes analytiques et comparées des structures internes des sigillaries. *Ann. Sci. Nat. Bot. Biol. Veg.* ser. 12, 1: 469-639.

Lenz, L. W. 1942. Contributions to our knowledge of American Carboniferous floras. III. *Stipitopteris. Missouri Bot. Garden Annals* 29: 59-68.

Lignier, O. 1911. Les *"Radiculites*

reticulatus Lignier" sont probablement des radicelles de Cordaitales. *Assoc. Francaise Avanc. Sci. Compte Rendu* (40th session) *Notes et Memoires, Dijon.* 1: 509-513.

Lignier, O. 1913a. Differenciation des tissus dans le bourgeon vegetatif du *Cordaites lingulatus* B. Ren. *Ann. Sci. Nat. Bot.* ser. 9, 187 (17): 233-254.

Lignier, O. 1913b. Un nouveau sporange seminiforme, *Mittagia seminiformis,* gen. et sp. nov. Soc. *Linnéenne Normandie Mem.* 24:49-67.

Lignier, O. 1914. Sur une Mousse houillere a structure conservee. *Bull. Soc. Linnéenne Normandie* 7: 128-131.

Lomax, James. 1915. The formation of coal-seams in the light of recent microscopic investigations. Part I. *Trans. Inst. Min. Engineers* 50:1-32.

Long, A. G. 1977. Observations on Carboniferous seeds of *Mitrospermum, Conostoma* and *Lagenostoma. Trans. Royal Soc. Edinburgh* 70:37-61.

Loubiere, A. 1929. Etude anatomique et comparee du *Leptotesta grand 'euryi* n. gen., n. sp. graine silicifee du *Pecopteris pluckeneti* Schlot. *Rev. Gén. Bot.* 41: 593-605.

Maistre, J. de. 1963a. Description géologique du bassin houiller de la Loire. *Rev. Ind. Miner.* 45:541-600.

Maistre, J. de. 1963b. Caractères paléontologiques des subdivisions du Stéphanien dans le gisement type de Saint-Etienne 5th *Congr. Intern. Stratigr. Géol. Carbonifère Compte Rendu* (Paris) 2:569-580.

Mamay, S. H. 1950. Some American Carboniferous fern fructifications. *Missouri Bot. Garden Annals* 37:409-476.

Mamay, S. H. 1952. An epiphytic American species of *Tubicaulis* Cotta. *Ann. Bot.,* n.s. 16:145-163.

Mamay, S. H. 1954a. A new sphenopsid cone from Iowa. *Ann. Bot.,* n.s. 18:229-239.

Mamay, S. H. 1954b. Two new plant genera of Pennsylvanian age from Kansas coal balls. *U.S. Geol. Surv. Prof. Pap. 254-D,* pp. 81-95.

Mamay, S. H., and E. L. Yochelson. 1962. Occurrences and significance of marine animal remains in American coal balls. *U.S. Geol. Surv. Prof. Pap. 354-I,* pp. 193-224.

McCullough, L. A. 1977. Early diagenetic calcareous coal balls and roof shale concretions from the Pennsylvanian (Allegheny Series). *Ohio Jour. Sci.* 77:125-134.

McKee, E. D., and E. J. Crosby (coord.). 1975. Paleotectonic investigations of the Pennsylvanian System in the United States. Part I. Introduction and regional analyses of the Pennsylvanian System. *U.S. Geol. Surv. Prof. Pap. 853,* 349p.

McLaughlin, R. E., and A. B. Reaugh. 1976. Pennsylvanian coal-ball petrifactions in the Cumberland Plateau section, East Tennessee (Abs). *Geol. Soc. America Program* 2:228-229.

Mentzel, H. 1904. Beitrage zur Kenntnis der Dolomitvorkommen in Kohlenflozen. *Glückauf* 40 (36, 37): 1164-1171.

Millay, M. A. 1976. Synangia of North American Pennsylvanian petrified marattialeans. Ph.D. thesis, Univ. Illinois, Urbana, 483p.

Millay, M. A. 1977. *Acaulangium* gen. n., a fertile marattialean from the upper Pennsylvanian of Illinois. *Am. Jour. Bot.* 64: 223-229.

Millay, M. A., and D. A. Eggert. 1970. *Idanothekion* gen. n., a synangiate pollen organ with saccate pollen from the middle Pennsylvanian of Illinois. *Am. Jour. Bot.* 57: 50-61.

Millay, M. A., and T. N. Taylor. 1977. *Feraxotheca* gen. n., a lyginopterid pollen organ from the Pennsylvanian of North America. *Am. Jour. Bot.* 64: 177-185.

Moore, R. C. (chm). 1944. Correlation of Pennsylvanian formations of North America. *Geol. Soc. Am. Bull.* 55: 657-706.

Morgan. J. 1959. The morphology and anatomy of American species of the genus *Psaronius. Illinois Biol. Monograph 27,* Urbana: Univ. Ill. Press, 108p.

Morgan, J., and T. Delevoryas. 1952. *Stewartiopteris singularis:* a new psaroniaceous fern rachis. *Am. Jour. Bot.* 39: 479-484.

Morgan, J., and T. Delevoryas 1954. An anatomical study of a new coenopterid and its bearing on the morphology of certain coenopterid petioles. *Am. Jour. Bot.* 41: 198-203.

Morris, J. E. 1958. Studies of specimens of the genus *Dolerotheca* from Kansas. Masters thesis, Univ. Kansas, Lawrence, 48p.

Mullins, A. T.; R. E. Lounsbury; and D. L. Hodgson. 1965. *Coal reserves of northwestern Kentucky.* Washington: USGPO, Tennessee Valley Authority, 28p.

Neely, F. E. 1951. Small petrified seeds from the Pennsylvanian of Illinois. *Bot. Gaz.* 113: 165-179.

Noé, A. C. 1923. Coal balls. *Science* 57:385.

Noé, A. C. 1925. Coal balls here and abroad. *Trans. Illinois State Acad. Sci.* (1924) 17:179-180.

Oliver, F. W. 1904. On the structure and affinities of *Stephanospermum,* Brongniart, a genus of fossil gymnosperm seeds. *Trans. Linn. Soc. London, Ser. 2, Botany,* 6:361-400.

Oliver, F. W. 1909. On *Physostoma elegans,* Williamson, an archaic type of seed from the Palaeozoic rocks *Ann. Bot.* 23:73-116.

Oliver, F. W. 1935. Dukinfield Henry Scott, 1854-1934. *Ann. Bot.* 49:823-840.

Oliver, F. W., and E. J. Salisbury. 1911. On the structure and affinities of the Palaeozoic seeds of the *Conostoma* group. *Ann. Bot.* 25:1-50.

Osborn, T. G. B. 1909. The lateral roots of *Amyelon radicans,* Will., and their mycorhiza. *Ann. Bot.* 23: 603-611.

Outerbridge, W. F. 1976. The Magoffin Member of the Breathitt Formation, In "Changes in stratigraphic nomenclature by the U.S. Geological Survey, 1975," Cohee, G. V., and W. B. Wright. *U.S. Geol. Survey Bull. 1422-A,* pp. A64-A65.

Pannell, E. 1942. Contributions to our knowledge of American Carboniferous floras. IV. A new species of *Lepidodendron. Missouri Bot. Garden Annals* 29: 245-274.

Pant, D. D., and B. K. Verma. 1964. The cuticular structure of *Noeggerathiopsis* Feistmantel and *Cordaites* Unger. *Palaeontographica* 115B:21-44.

Peppers, R. A. 1970. Correlation and palynology of coals in the Carbondale and Spoon Formations (Pennsylvanian) of the northeastern part of the Illinois Basin. *Illinois Geol. Surv. Bull. 93,* 173p.

Perkins, T. W. 1976. Textures and conditions of formation of Middle Pennsylvanian coal balls, central United States. *Univ. Kansas, Kansas Paleontol. Contrib., Paper No. 82,* 13p.

Phillips, T. L. 1959. A new sphenophyllalean shoot system from the Pennsylvanian. *Missouri Bot. Garden Annals* 46:1-17.

Phillips, T. L. 1970. Morphology and evolution of *Botryopteris,* a Carboniferous age fern. Part 1. Observations on some European species. *Palaeontographica* 120B: 137-172.

Phillips, T. L. 1974. Evolution of vegetative morphology in coenopterid ferns. *Missouri Bot. Garden Annals* 61:427-461.

Phillips, T. L. 1979. Reproduction of heterosporous arborescent lycopods in the Mississippian-Pennsylvanian of Euramerica. *Rev. Palaeobot. Palynol.* 27:239-289.

Phillips, T. L., and H. N. Andrews. 1963. An occurrence of the medullosan seed-fern *Sutcliffia* in the American Carboniferous. *Missouri Bot. Garden Annals* 50:29-51.

Phillips, T. L.; M. J. Avcin; and Dwaine Berggren. 1976. Fossil peat of the Illinois Basin—a guide to the study of coal balls of Pennsylvanian age. *Illinois Geol. Surv. Ed. Series 11.* 39p.

Phillips, T. L., A. B. Kunz; and D. J. Mickish. 1977. Paleobotany of permineralized peat (coal balls) from the Herrin (No. 6) Coal Member of the Illinois Basin. In *Interdisciplinary Studies of Peat and Coal Origins,* P.N. Given, and A. D. Cohen (eds). Geol. Soc. America Microform Pub. 7, pp. 18 49.

Phillips, T. L.; H. W. Pfefferkorn; and R. A. Peppers. 1973. Development of paleobotany in the Illinois Basin. *Illinois Geol. Surv. Circ. 480,* 86p.

Phillips, T. L.; R. A. Peppers, M. J. Avcin; and P. F. Laughnan. 1974. Fossil plants and coal: Patterns of change in Pennsylvanian coal swamps of the Illinois Basin. *Science* 184:1367-1369.

Posthumus, O. 1931. *Catalogue of fossil remains described as fern stems and petioles* Malang, Java: N. V. Jahn, 234p.

Ramanujam, C. G. K.; G. W. Rothwell; and W. N. Stewart. 1974. Probable attachment of the *Dolerotheca* campanulum to a *Myeloxylon-Alethopteris* type frond. *Am. Jour. Bot.* 61:1057-1066.

Read, C. B., and S. H. Mamay. 1964. Upper Paleozoic floral zones and floral provinces of the United States. *U.S. Geol. Surv. Prof. Pap. 454-K,* pp. K1-K19.

Reed, F. D. 1941a. Coal flora studies: Lepidodendrales. *Bot. Gaz.* 102:663-683.

Reed, F. D. 1941b. Some fossil plants found in coal balls from Texas (Abs.). *Am. Jour. Bot.* 28:95.

Reed, F. D. 1946. On *Cardiocarpon* and some associated plant fragments from Iowa coal fields. *Bot. Gaz.* 108:51-64.

Reed, F. D. 1949. Notes on the anatomy of two Carboniferous plants, *Sphenophyllum* and *Psaronius. Bot. Gas.* 110:501-510.

Reed, F. D. 1952. *Arthroxylon,* a redefined genus of calamite. *Missouri Bot. Garden Annals* 39:173-187.

Reed, F. D., and M. T. Sandoe. 1951. *Cordaites affinis:* A new species of cordaitean leaf from American coal fields. *Bull. Torrey Bot. Club* 78:449-457.

Remy, W., and R. Remy. 1977. *Die Floren des Erdaltertums.* Essen: Verlag Gluckauf GmbH, 468p.

Renault, B. 1879. Structure comparée de quelques tiges de la flore Carbonifère. *Nouv. Archiv. Mus. Nation. Hist. Nat.* (Paris) sér. 2, 2:213-348.

Renault, B. 1896. *Notice sur les travaux scientifiques.* Autun: Imprimere Dejussieu Pere et Fils, 162p.

Renier, A. 1926a. Etude stratigraphique du Westphalien de la Belgique. *13th Congr. Geol. Intern. Compte Rendu* (Liege, Belgium, 1922) 3:1796-1841.

Renier, A. 1926b. Sur l'existence de coal balls dans le bassin houiller des Asturies. *Comptes Rendus Acad. Sci.* (Paris) 182.

Roberts, D. C., and E. S. Barghoorn. 1952. *Medullosa olseniae:* A Permian *Medullosa* from north central Texas. *Bot. Mus. Leaflets Harvard Univ.* 15: 191-200.

Robertson, C. E. 1971. Evaluation of Missouri's coal resources. *Missouri Geol. Surv. Water Resources Rept. Invest. No. 48,* 92p.

Roth, E. A. 1955. The anatomy and modes of preservation of the genus *Cardiocarpus spinatus* Graham. *Univ.*

Kansas Sci. Bull. 37 (pt. I):151-174.

Rothwell, G. W. 1971. Additional observations on *Conostoma anglo-germanicum* and *C. oblongum* from the lower Pennsylvanian of North America. *Palaeontographica* 131B:167-178.

Rothwell, G. W. 1972. Pollen organs of the Pennsylvanian Callistophytaceae (Pteridospermopsida). *Am. Jour. Bot.* 59:993-999.

Rothwell, G. W. 1975. The Callistophytaceae (Pteridospermopsida): I. Vegetative structures. *Palaeontographica* 151B:171-196.

Rothwell, G. W. 1976a. A new pteropsid fructification from the middle Pennsylvanian of Kansas. *Palaeontology* 19:307-315.

Rothwell, G. W. 1976b. Petrified Pennsylvanian age plants of eastern Ohio. *Ohio Jour. Sci.* 76:128-132.

Rothwell, G. W., and D. A. Eggert. 1970. A *Conostoma* with tentacular sarcotesta from the upper Pennsylvanian of Illinois. *Bot. Gaz.* 131:359-366.

Rothwell, G. W., and T. N. Taylor. 1971. Studies of Paleozoic calamitean cones: *Weissia kentuckiense* gen. et sp. nov. *Bot. Gaz.* 132:215-224.

Rothwell, G. W., and T. N. Taylor. 1972. Carboniferous pteridosperm studies: Morphology and anatomy of *Schopfiastrum decussatum. Canadian Jour. Bot.* 50:2649-2658.

Salisbury, E. J. 1914. On the structure and relationship fo *Trigonocarpus Shorensis*, sp. nov.: A new seed from the Palaeozoic rocks. *Ann. Bot.* 28:39-80.

Schabilion, J.; N. Brotzman; and T. L. Phillips. 1974. Two coal-ball floras from Iowa (Abs.). *Am. Jour. Bot.* 61:19.

Scheihing, M. H. 1978. A paleoenvironmental analysis of the Shumway cylclothem (Virgilian), Effingham County, Illinois. Masters thesis, Univ. Illinois, Urbana, 164p.

Schlanker, C. M., and G. A. Leisman. 1969. The herbaceous Carboniferous lycopod *Selaginella fraipontii* comb. nov. *Bot. Gaz.* 130:35-41.

Schopf, J. M. 1939. *Medullosa distelica*, a new species of the anglica group of *Medullosa. Am. Jour Bot.* 26:196-207.

Schopf, J. M. 1941a. Contribution to Pennsylvanian paleobotany: *Mazocarpon oedipternum* sp. nov., and sigillarian relationships. *Illinois Geol. Surv. Rept. Inv.* 75, 53p.

Schopf, J. M. 1941b. Contribution to Pennsylvanian paleobotany—notes on the Lepidocarpaceae. *Am. Midland Naturalist* 25:548-463 *(Illinois Geol. Surv. Cir. 73,* 17p.*)*.

Schopf, J. M. 1948. Pteridosperm male fructifications: American species of *Dolerotheca*, with notes regarding certain allied forms. *Jour. Paleontology* 22:681-725 (Illinois Geol. Surv. Rept. Inv. No. 142, 1949).

Schopf, J. M. 1950. Age of American coal-ball plants. *Bot. Gaz.* 111:356-357.

Schopf, J. M. 1961. Coal-ball occurrences in eastern Kentucky. *U.S. Geol. Surv. Prof. Pap. 424B*, pp. 228-230.

Schopf, J. M. 1975. Pennsylvanian climate in the United States. In "Paleotectonic investigations of the Pennsylvanian System in the United States," Part II. *U.S. Geol. Surv. Prof. Pap. 853*, pp. 23-31.

Scott, D. H. 1899. On the structure and affinities of fossil plants from the Palaeozoic rocks. III. On *Medullosa anglica*, a new representative of the Cycadofilices. *Phil. Trans. Roy. Soc. London* 191B:81-126.

Scott, D. H. 1914. On *Medullosa pusilla*. *Proc. Roy. Soc. (London)* 87B:221-228.

Scott, D. H. 1917. The Heterangiums of the British Coal Measures. *Jour. Linn.*

Soc. Botany. 44:59-105.

Scott, D. H. 1919. On the fertile shoots of *Mesoxylon* and an allied genus. *Ann. Bot.* 33:1-21.

Scott, D. H. 1930. *Cladites bracteatus*, a petrified shoot from the Lower Coal Measures. *Ann. Bot.* 44:333-348.

Scott, D. H., and A. J. Maslen. 1910. On *Mesoxylon*, a new genus of Cordaitales—Preliminary Note. *Ann. Bot.* 24:236-239.

Searight, W. V. and W. B. Howe, 1961. Pennsylvanian System. In The Stratigraphic Succession in Missouri. J. W. Koenig (Ed.). *Missouri Geol. Surv. Water Resources. 40* (2d ser.), pp.78-122.

Seward, A. C. 1893. On the genus *Myeloxylon* (Brong.) *Ann. Bot.* 7:1-20.

Seward, A. C. 1899. Notes on the Binney Collection of Coal Measure plants: Part II. *Megaloxylon*, gen. nov. *Proc. Cambridge Phil. Soc.* 10:173-174.

Seward, A. C. 1923. A supposed Paleozoic angiosperm. *Bot. Gaz.* 76:215.

Shadle, G. L., and B. M. Stidd. 1975. The frond of *Heterangium*. *Am. Jour. Bot.* 62:67-75

Shaver, R. H.; A. M. Burger; G. R. Gates; H. H. Gray; H. C. Hutchinson; S. S. Keller; J. B. Patton; C. B. Rexroad; N.N. Smith; W. J. Wayne; and C. E. Wier. 1970. Compendium of rock-unit stratigraphy in Indiana. *Indiana Geol. Surv. Bull.* 43, 229p.

Smith, W. H., and G. E. Smith. 1967. Description of late Pennsylvanian strata from deep diamond drill cores in the southern part of the Illinois Basin. *Illinois Geol. Sruv. Circ. 411*, 27p.

Snigirevskaya, N. S. 1958. Anatomical study of the remains of the leaves of some lycopods in coal balls of the Donbass basin. *Bot. Zhur. (Leningrad)* 43: 106-112. (in Russian).

Snigirevskaya, N. S. 1961. The genus *Botryopteris* in the coal balls from the Donets coal basin. *Bot. Zhur. (Leningrad)* 46: 1329-1335. (In Russian).

Snigirevskaya, N. S. 1962a. The morphology and systematics of the genus *Botryopteris*. *Paleontol. Zhur.* 2: 122-132. (in Russian).

Snigirevskaya, N. S., 1962b. Remains of the sporangiophores of the Sphenophyllaceae in the coal balls of the Donbass. *Bot. Zhur. (Leningrad)*47:546-552. (in Russian).

Snigirevskaya, N. S. 1964. An anatomical study of the plant remains from the Donets coal balls. 1. Lepidodendraceae. *Tr. Bot. Inst. Akad. Nauk. SSSR, ser.8, Paleobotanica* 5:5-37. (in Russian).

Snigirevskaya, N. S. 1967. The remains of calamites and psaroniaceous ferns in coal balls of the Donets Basin. *Tr. Bot Inst. Akad. Nauk. SSSR, ser. 8, Paleobotanica* 6:7-25. (in Russian).

Snigirevskaya, N. S. 1972. Studies of coal balls of the Donets Basin. *Rev. Palaeobot. Palynol.* 14:197-204.

Snigirevskaya, N. S.; O.P. Fisunenko; I. K. Ikonnikova; S. N. Kotlik; and A. O. Murashova. 1977. On the first discovery of *Litostrobus* in coal balls of the Carboniferous in the Donets Basin. *Geol. Zhur.* (Kiev) 37:140-147. (In Russian).

Steidtman, W. E. 1937. A preliminary report on the anatomy and affinities of *Medullosa noei* sp. nov. from the Pennsylvanian of Illinois. *Am. Jour. Bot.* 24:124-125.

Stewart, G. F. 1975. Kansas. *U.S. Geol. Surv. Prof. Pap. 853-H*, pp. 127-156.

Stewart, W. N., and T. Delevoryas. 1956. The medullosan pteridosperms. *Bot. Rev.* 22:45-80.

Stidd, B. M. 1971 Morphology and anatomy of the frond of *Psaronius*. *Palaeontographica* 134B: 87-125.

Stidd, B. M., and Karen Cosentino. 1975. *Albugo*-like oogonia from the American Carboniferous. *Science* 190:

1092-1093.

Stidd, B. M., and K. Cosentino. 1976. *Nucellangium:* Gametophytic structure and relationships to cordaites. *Bot. Gaz.* 137:242-249.

Stidd, B. M., and J. W. Hall. 1970a. *Callandrium callistophytoides,* gen. et sp. nov., the probable pollen-bearing organ of the seed fern, *Callisophyton. Am. Jour. Bot.* 57:394-403.

Stidd, B. M., and J. W. Hall. 1970b. The natural affinity of the Carboniferous seed, *Callospermarion. Am. Jour. Bot.* 57:827-836.

Stidd, B. M.; G. A. Leisman; and T. L. Phillips. 1977. *Sullitheca dactylifera* gen. et. sp. n.: A new medullosan pollen organ and its evoutionary significance. *Am. Jour. Bot.* 64:994-1002.

Stidd, B. M.; L. L. Oestry; and T. L. Phillips. 1975. On the frond of *Sutcliffia insignis* var. *tuberculata. Rev. Palaeobot. Palynol.* 20:55-66.

Stidd, B. M. and T. L. Phillips. 1973. The vegetative anatomy of *Schopfiastrum decussatum* from the middle Pennsylvanian of the Illinois Basin. *Am. Jour. Bot.* 60:463-474.

Stopes, M. C. 1903. On the leaf-structure of *Cordaites. New Phytologist* 2:91-98.

Stopes, M. C., and D. M. S. Watson. 1909. On the present distribution and origin of the calcareous concretions in coal seams, known as coal balls. *Phil. Trans. Royal Soc. London* 200B:167-218.

Stur, D. 1885. Ueber die in Flötzen reiner Steinkohle enthaltenen Stein-Rundmassen und Torf-Sphärosiderite. *Jahr. K. K. Geol. Reichsanstalt, Wien* 35:613-648.

Taylor, T. N. 1962. Additional observations on *Stephanospermum ovoides,* a middle Pennsylvanian seed. *Am. Jour. Bot.* 49:794-800.

Taylor, T. N. 1965. Paleozoic seed studies: a monograph of the American species of *Pachytesta. Palaeontographica* 117B:1-46.

Taylor, T. N. 1966. Paleozoic seed studies: on the genus *Hexapterospermum. Am. Jour. Bot.* 53:185-192.

Taylor, T. N. 1967. Paleozoic seed studies: on the structure of *Conostoma leptospermum* n. sp., and *Albertlongia incostata* n. gen. and sp. *Palaeontographica* 121B:23-29.

Taylor, T. N. 1970. *Lasiostrobus* gen. n., a staminate strobilus of gymnospermous affinity from the Pennsylvanian of North America. *Am. Jour. Bot.* 57:670-690.

Taylor, T. N. 1971. *Halletheca reticulatus* gen. et sp. n.: a synangiate Pennsylvanian pteridosperm pollen organ. *Am. Jour. Bot.* 58:300-308.

Taylor, T. N. 1972. A new Carboniferous sporangial aggregation. *Rev. Palaeobot. Palynol.* 14:309-318.

Taylor, T. N., and S. D. Brack-Hanes. 1976. *Achlamydocarpon varius* comb. nov.: Morphology and reproductive biology. *Am. Jour. Bot.* 63:1257-1265.

Taylor, T. N., and T. Delevoryas. 1964. Paleozoic seed studies: a new Pennsylvanian *Pachytesta* from southern Illinois. *Am. Jour. Bot.* 51:189-195.

Taylor, T. N., and D. A. Eggert. 1969. On the structures and relationships of a new Pennsylvanian species of the seed *Pachytesta. Palaeontology* 12:382-387.

Taylor, T. N., and G. A. Leisman. 1963. *Conostoma kestospermum,* a new species of Paleozoic seeds from the middle Pennysylvanian. *Am. Jour. Bot.* 50:475-580.

Taylor, T. N., and W. N. Stewart. 1964. The Paleozoic seed *Mitrospermum* in American coal balls. *Palaeontogrpahica* 115B: 51-58.

Taylor, T. N., and R. A. Stockey. 1976. Studies of Paleozoic seed ferns: Anatomy and morphology of *Microspermopteris aphyllum. Am. Jour. Bot.* 63:1302-1310.

Teichmüller, M. 1952. Vergleichende mikroskopische Untersuchungen versteinerter Torfe des Ruhrkarbons und der daraus entstandenen Steinkohlen. *3rd Congr. Internl. Stratigr. Géol. Carbonifère* Compte Rendu (Heerlen 1951) 2:607-613.

Teichmüller, M. and W. Schonefeld. 1955. Ein verkieselter Karbontorf im Naumr C von Kupferdreh. *Geol. Jahb.* 71:91-111.

Thomas, B. A., 1978. Carboniferous Lepidodendraceae and Lepidocarpaceae. *Bot. Rev.* 44:321-364.

Thomas, H. H. 1912. On the leaves of Calamites *(Calamocladus* section). Phil. Trans. Roy. Soc. London 202B: 51-92.

Thompson, M. L. 1936. Pennsylvanian fusulinids from Ohio. *Jour. Paleontology* 10:673-683.

Tiffney, B. H., and E. S. Barghoorn. 1974. The fossil record of fungi. *Occasional Papers Farlow Herbarium, Cryptogamic Botany, Harvard Univ. No.* 7, 42p.

Traverse, A. 1950. The primary vascular body of *Mesoxylon thompsonii,* a new American cordaitalean. *Am. Jour. Bot.* 37:318-325.

Vetter, P. 1971. Le Carbonifère supérieur et la Permien du Massif Central. *Symposium J. Jung, Clermont-Ferrand, pp. 169-213.*

Wanless, H. R. 1939. Pennsylvanian correlations in the Eastern Interior and Appalachian Coal Fields. *Geol. Soc. Am. Spec. Pap. No. 17,* 130p.

Wanless, H. R. 1957. Geology and mineral resources of the Beardstown, Glasford, Havana and Vermont *Quadrangles. Illinois Geol. Surv. Bull 82,* 233p.

Wanless, H. R. 1975a. Appalachian Region. *U.S. Geol. Surv. Prof. Pap, 853-C,* pp. 17-62.

Wanless, H.R. 1975b. Illinois Basin Region. *U.S. Geol. Surv. Prof. Pap. 853-E,* pp. 71-95.

Wanless, H. R. 1975c. Missouri and Iowa. *U.S. Geol. Surv. Prof. Pap. 853-F,* pp. 97-114.

Watson, D.M.S. 1909. On *Mesostrobus,* a new genus of lycopodiaceous cones form the Lower coal Measure, with a note on the systematic position of *Spencerites. Ann. Bot.* 23:379-397.

Williamson, W. C. 1891. General, morphological, and historical index to the author's collective memoris on the fossil plants of the Coal Measures. *Mem. Proc.* Manchester *Lit. Phil Soc. Ser.* IV 4:53-68.

Williamson, W. C. 1893. General, morphological, and histological index to the author's collective memoirs on the fossil plants of the Coal Measures. Part II. *Mem. Proc.* Manchester *Lit. Phil. Soc.* series IV 7:91-127.'

Willman, H. B.; E. Athorton; T. C. Buschbach; C. Collinson; J. C. Frye; M. E. Hopkins; J. A. Lineback; and J. A. Simon. 1975. Handbook of Illinois stratigraphy. *Illinois Geol. Survey Bull.* 95, 261p.

Wilson, L. R., and A. W. Johnston. 1940. A new species of *Cordiates* from the Pennsylvanian strata of Iowa. *Bull. Torrey Bot. Club* 67:117-120.

Wright, W. B.; R. L. Sherlock; D. A. Wray; W. Lloyd; and L. H. Tonks. 1927. The Geology of the Rossendale Anticline *Mem., Geol. Surv. Great Britain, England, Wales, Explan. Sheet 76* (Rochdale), 182p.

Zalessky, M. D. 1908. Végétaux fossiles du terrain carbonifère du bassin du Donetz, 2, Étude sur la structure anatomique d'un *Lepidostrobus. Mém. Com. Geol.* n.s. 46:1-33.

Zalessky, M. D. 1910a. On the discovery of well preserved plant remains in one type of rock under Limestone S(I3) of a general section of the Donets Carboniferous deposits. *Bull. Acad. Imp. Sci. St. Petersbourg* 4: 447-449. (in

Russian).

Zalessky, M. D. 1910b. On the discovery of the calcareous concretions known as coal balls in one of the coal seams of the Carboniferous strata of the Donetz basin. *Bull. Acad. Imp. Sci. St. Petersbourg* 4:477-480.

Zalessky, M. D. 1911. Structure du rameau du *Lepidodendron obovatum* Sternb. *Etud. Paleobot.* St. Petersbourg 1:1-12.

Zalessky, M. D. 1912a. Anatomy of *Lepidodendron dichotomum* Sternberg. *Etud. Paleobot. pt. II, St. Petersburg, Lett. Scientifique,* 1: 1-13. (in Russian).

Zalessky, M. D. 1912b. The anatomy of *Lepidophloios laricinus* Sternberg. *Etud. Paleobot. St. Pétersbourg, Lett. Scientifique* 1: 2-3. (in Russian).

Zalessky, M. D. 1913. The anatomy of *Lepidodendron dichotomum* Sternb. *Izv. Obscht. Issl. Prir. Orlovsk. Gub.* 23:97-99 (in Russian).

Zalessky, M. D. 1915. Stem structure of *Lepidodendron dichotomum* Sternb. and *Lepidophloios laricinus* Sternb. *Geol. Vestn.* 1:16-20 (in Russian).

Zalessky, M. D. 1923. On new representatives of the genera *Arthropitys* and *Mesoxylon* from the Donets coal basin. *Vesta. Mosk. Gorn. Akad* 2:73-77. (in Russian).

Zalessky, M. D. 1928. Essai d'une division du terrain houiller du bassin du Donetz d'apres sa flore fossile. *Congr. Stratigr. Carbonifere Compte Rendu* (Heerlen, 1927), pp. 805-820.

Zaritsky, P.V. 1959. *Concretions of Coal-bearing Deposits of the Donets Basin.* Kharkov: Kharkov State Univ., 237p. (in Russian).

Zaritsky, P. V. 1971. Concretions in the coals of the Donets Basin, their genesis and importance for regional stratigraphy, In *Sedimentation and Genesis of Carboniferous Coals in the USSR,* I.I. Gorsky; P.P. Timofeev; and L. I. Bogolyuboa. (eds.) Moscow: Nauka, pp. 173-812. (in Russian).

Zaritsky, P.V.; A. V. Makedonov; and L. L. Salnikova. 1971. Concretions in Carboniferous and Permian Coal Measures of the USSR, In *Sedimentation and Genesis of Carboniferous Coal in the USSR,* I.I. Gorsky; P.P. Timofeev; and L. I. Bogolyuboa (eds). pp. 163-173, (in Russian).

Zhemchuzhnikov, Y. A.; V. S. Yablokov; L. I. Bogolyuboa; L. N. Botvinkina; A. P. Feofilova; M. I. Ritenberg; P.P. Timofeev; and Z. V. Timofeeva. 1960. The structure and conditions of sedimentation of the basic coal-bearing formations and coal layers of the middle Carboniferous in the Donets Basin. *Trudy Geol. Inst.* Issue 15. Moscow: ANSSSR, (in Russian).

ADDENDUM

The Westphalian C-D boundary equivalent in the United States is now suggested in the Illinois Basin just above the Rock Island (No. 1) Coal Member (Peppers in press) and in the proposed Pennsylvanian stratotype section in the Appalachians just above the Stockton coal bed (Gillespie and Pfefferkorn 1979). Considerable doubt has been expressed about the traditional Westphalian B-C correlation from the Soviet Union by Wagner and Higgins (1979) who suggested that the Bashkirian-Moscovian boundary should be stratigraphically "lowered" into the Westphalian A. A revised stratigraphic correlation chart of coal-ball occurrences is given in Phillips (in press).

Gillespie, W.H., and H.W. Pfefferkorn, 1979. Distribution of commonly occurring plant megafossils in the proposed Pennsylvania System stratotype, In *Proposed Pennsylvanian System Stratotype-Virginia and West Virginia,* K.J. Englund, H.H. Arndt and T.W. Henry (eds.). Falls Church, Virginia. American Geological Institute pp. 87-96.

Peppers, R.A., in press. Comparison of miospore assemblages in the Pennsylvania System of the Illinois Basin with those in the Upper Carboniferous of western Europe. *9th Congr. Intern. Stratigr. Geol. Carbonifere, Compte Rendu* (Urbana, 1979).

Phillips, T.L. in press. Stratigraphic occurrences and vegetational patterns of Pennsylvanian pteridosperms in Euramerican coal swamps. *Rev. Palaeobot, Palynol.*

Wagner, R.H., and A.C. Higgins. 1979. The Carboniferous of the U.S.S.R.; Its stratigraphic significance and outstanding problems of world-wide correlation. In *The Carboniferous of the U.S.S.R.,* R.H. Wagner, A.C. Higgins, and S.V. Meyen (eds.). Leeds, England. Occasional Pub. 4, Yorkshire Geol. Soc. pp. 5-22.

3

BIOSTRATIGRAPHY AND BIOGEOGRAPHY OF PLANT COMPRESSION FOSSILS IN THE PENNSYLVANIAN OF NORTH AMERICA

Hermann W. Pfefferkorn
William H. Gillespie

SUMMARY

Compression-impression plant fossils of Pennsylvanian age occur in at least twenty-seven states of the United States, three Canadian provinces, and one Mexican state. More than 500 reports have been published on these plant fossils, which are found in the coal basins of the eastern United States and in terrestrial beds intercalated with marine sediments in the western states. The sequence of first occurrences of new species is largely identical throughout North America and is similar to the sequence in the European Upper Carboniferous. The differences are smallest in the Lower and Middle Pennsylvanian and the variations, although more pronounced in the Upper Pennsylvanian, do not mask the general similarity. Four paleobiographic provinces apparently existed in North America during the Pennsylvanian. These were separated by mountains or latitude and were differentiated by their topography and geotectonic position. The Acadian Province extended through eastern Canada and New England, the Interior-Appalachian Province occupied the larger part of the North American continent, the Cordilleran Province existed in the area of the present middle and southern Rocky Mountains, and the Oregonian Province included several microcontinents and/or island arcs, which attached to the American continent after the Pennsylvanian. The difference between the provinces is indicated by the presence of a small number of endemic genera and species, which are not present in large enough numbers to hinder biostratigraphic work. Pennsylvanian floras in northern South America (Venezuela and Columbia) belonged to the Amerosinian floral realm and are comparable with the North American floras.

INTRODUCTION

Biostratigraphy is the major tool for time stratigraphic correlations in interbasinal and other long-distance comparisons, and plant megafossils are one of the most useful groups in the Pennsylvanian for this purpose.

Plant compression-impression fossils have been used extensively for biostratigraphic work since the 1890s and to a certain extent since the 1850s. Their value for the solution of stratigraphic problems in areas with a complex tectonic situation has always been well accepted: although, quite often, only physical stratigraphy was used in areas with more or less flat-lying beds. Nevertheless, even today's correlation from basin to basin still depends to a degree on the plant megafossil comparisons made by David White in the early years of this century.

Plant megafossils are found much more frequently in Pennsylvanian rocks than is generally realized. They are known to occur in at least twenty-seven of the fifty states of the United States *(Figure 3.1)* in three Canadian provinces and in at least one Mexican state. They are not restricted to coal-bearing sequences in the major coal basins; they also occur in thin terrestrial sequences intercalated in the marine sequences found over large areas of the United States. Consequently, they are common enough to be stratigraphically useful over large areas. This fact has been recognized; and, in the last few years, there has been an increase in research activity in this field in North America. A survey of existing publications has shown that there is a wealth of information in the more than 500 papers published on the subject in this country.

This survey of the published data, in combination with recent first-hand observations, indicates that there were geographic differences in the floras growing on the North American continent in Pennsylvanian time. Therefore, the scope of the study had to be enlarged to include biogeographic considerations. It is an opportune time to do this because the knowledge of continental positions made possible by the theory of plate tectonics allows us to interpret the biogeographic results in a more meaningful context than has been possible before.

BIOSTRATIGRAPHY AND CHRONOSTRATIGRAPHY

Lesquereux was the first to use plant megafossils extensively for stratigraphic purposes in this country. His summary, published in volume three (1884) of his "Description of the Coal Flora of the Carboniferous Formation in Pennsylvania and Throughout the United States," established tentative correlations: but his "zonation" was not well enough defined to be used by geologists. Ward presented the principles for using fossil plants for geologic correlation in 1892.

Several other paleobotanists used plant megafossils for chronostratigraphic correlations, but they worked in restricted areas and often only for short periods of time. The first to use plant megafossils in the modern sense for biostratigraphy was David White. His scientific work spanned the time from 1890 to 1934. He developed a zonation of Pennsylvanian beds based on plant megafossils, and his conclusions were aimed directly at the solution of geologic problems. His often-cited conclusions form the basis for several correlations still used today.

David White's work was continued by C.B. Read who published a formal summary of his zonation in several papers (Read, 1944, 1947). The most complete model appeared in 1964 (Read and Mamay), with zones 4 through 12 being those in the Pennsylvanian *(Table 3.1)*. These zones were characterized by the dominance of one to a few species or genera and by the overall assemblages, which indicate the general character of the zones. They are a combination of acme and assemblage zones.

A local biostratigraphy for the maritime provinces of Canada was published by Bell in 1944 *(Table 3.2)*. He emphasized the similarity to the European scheme, but he named the zones independently after the more common plants in the units.

In 1949, Arnold used plant megafossils to correlate the strata of the Michigan Basin *(Table 3.3)* and to relate them to other areas of North America and Europe.

Cridland, Morris, and Baxter (1963) discussed the biostratigraphy of the Pennsylvanian of Kansas and drew comparisons directly with the published European chronostratigraphy *(Figure 3.2)*. The

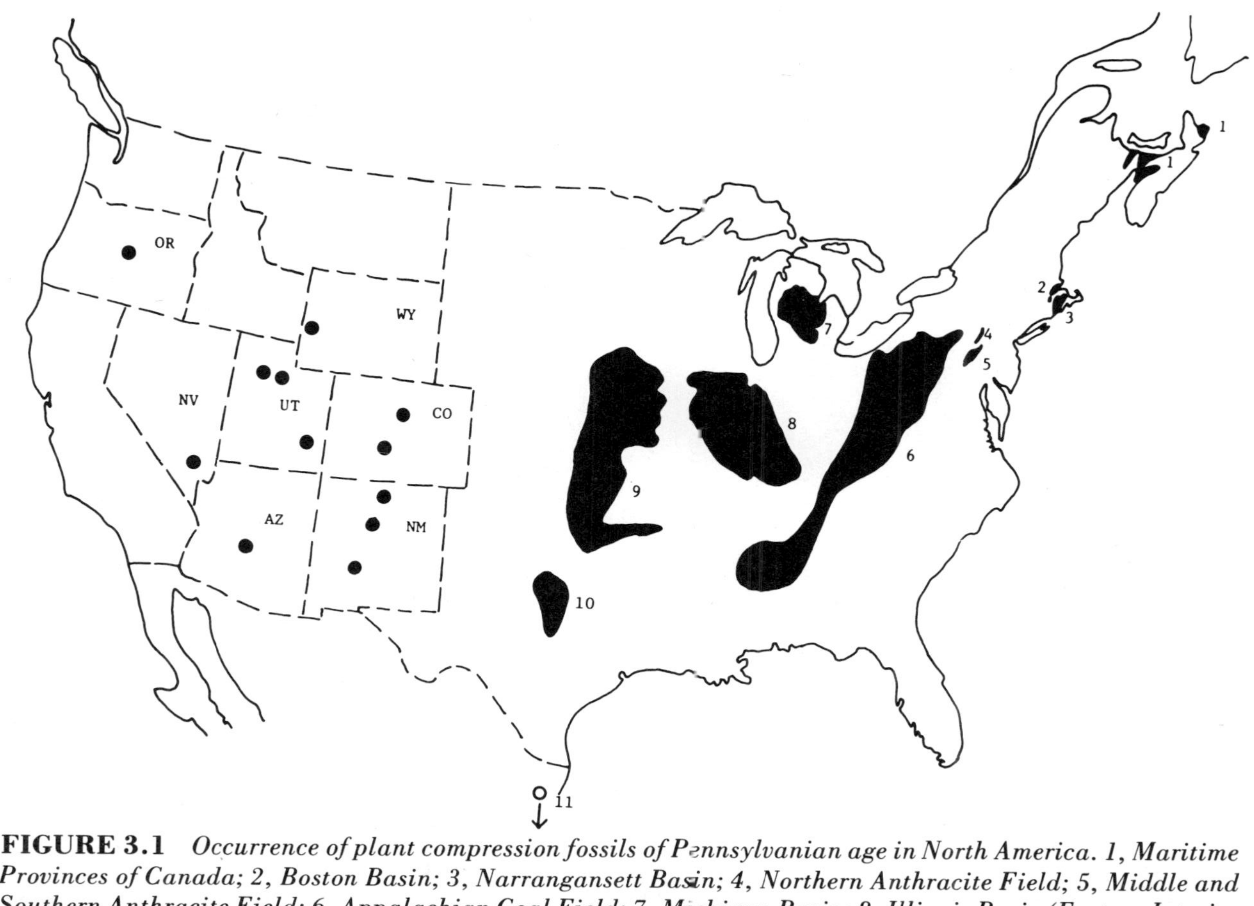

FIGURE 3.1 *Occurrence of plant compression fossils of Pennsylvanian age in North America. 1, Maritime Provinces of Canada; 2, Boston Basin; 3, Narragansett Basin; 4, Northern Anthracite Field; 5, Middle and Southern Anthracite Field; 6, Appalachian Coal Field; 7, Michigan Basin; 8, Illinois Basin (Eastern Interior Coal Field); 9, Western Interior Coal Field; 10, Southwestern Coal Field; 11, Puebla, Mexico; black dots denote areas of single or multiple localities. OR, Oregon; WY, Wyoming; NV, Nevada; UT, Utah; CO, Colorado; NM, New Mexico; AZ, Arizona.*

TABLE 3.1 *Floral zones of Read and Mamay (1964) for the Pennsylvanian of North America (reproduced with permission).*

Floral zone	Name	Appalachian region except for Southern anthracite field	Southern anthracite field	Midcontinent region
12	*Danaeites* spp.	Upper part Monongahela Formation.	Not known.	Missouri and Virgil Series.
11	*Lescuropteris* spp.	Lower part Monongahela Formation and upper part Conemaugh Formation.	Not known.	In midcontinent region, zones 11 and 12 are not separable and are together designated the zone of *Odontopteris* spp.
10	*Neuropteris flexuosa* and *Pecopteris* spp.	Lower part Conemaugh Formation and upper part Allegheny Formation.	Post-Pottsville rocks undifferentiated.	Upper part of Des Moines Series.
9	*Neuropteris rarinervis*	Lower part Allegheny Formation.	Upper part Sharp Mountain Member, Pottsville Formation.	Lower part of Des Moines Series.
8	*Neuropteris tenuifolia*	Major part Kanawha Formation.	Not known.	Major part of Atoka Series.
7	*Megalopteris spp.*	Base of Kanawha Formation.	Not known.	Base of Atoka Series.
6	*Neuropteris Tennesseeana* and *Mariopteris pygmaea.*	Upper part New River Formation and upper part Lee Formation.	Schuylkill Member, Pottsville Formation.	Bloyd Shale, Morrow Series.
5	*Mariopteris pottsvillea* and *Aneimites* spp.	Lower part New River Formation.	Lykens Valley No. 4 coal bed and adjacent strata of Tumbling Run Member, Pottsville Formation.	Locally, basal strata of Pennsylvanian System in midcontinent region.
4	*Neuropteris pocahontas* and *Mariopteris eremopteroides.*	Pocahontas Formation.	Lykens Valley No. 5 and No. 6 coal beds and ajdacent strata of Tumbling Run Member, Pottsville Formation.	No floras known.

TABLE 3.2 *Biostratigraphy of the Upper Carboniferous in northern Nova Scotia, Canada (from Bell, 1944, p. 30; reproduced with permission from the Geological Survey of Canada).*

Epoch	Age	Stages	Characteristic plants and animals
	D		*Annularia sphenophylloides, Asterophyllites equisetiformis, Asterotheca miltoni* forma *abbreviata, Asterotheca robbi, Mariopteris nervosa, Neuropteris rarinervis, Neuropteris scheuchzeri, Sphenopteris whitii, Annularia stellata, Sphenophyllum emarginatum* Zone of *Ptychocarpus unitus.* *Acitheca polymorpha, Dicksonites pluckeneti, Ptychocarpus unitus, Alethopteris friedeli, Sphenophyllum majus, Sphenophyllum oblongifolium, Neuropteris ovata, Mariopteris ? ribeyroni*
	C	Pictou and equivalent Morien and Stellarton groups	Zone of *Linopteris obliqua.* *Pecopteris plumosa* forma *dentata, Linopteris obliqua, Mariopteris latifolia, Mariopteris sphenopteroides, Sphenopteris striata, Telangium ? potieri, Zeilleria frenzli.* Acme of *Alethopteris serli, Neuropteris tenuifolia, Linopteris muensteri.* Disappearance of *Sphenophyllum cuneifolium*
West-phalian	B (late) or C (early)	(transgressive) Unconformity and	*Lonchopteris* zone. *Lonchopteris eschweileriana* Entrance of *Alethopteris serli, Linopteris muensteri, Neuropteris scheuchzeri, Sphenopteris striata* Entrance of *Anthracomya (Anthraconauta)* of *phillipsi* group Hemimylacrid blattoid insects
	B (early)	Disconformity Cumberland group (trangressive) Mainly	Disappearance of *Naiadites* Disappearance of *Neuropteris schlehani* Zone of *Andiantites adiantoides, Sphenopteris valida, Neuropteris obliqua, Neuropteris gigantea* or *pseudogigantea, Megalopteris,* abundant *Lepidodendra* and *Sigillaria, Pecopteris plumosa* forma *crenata, Pecopteris pilosa, Samaropsis* of *baileyi* group, *Dicranophyllum glabrum*. Entrance of *Boweria schatzlarensis, Obligocarpia brongniarti, Zeilleria frenzli, Mariopteris nervosa, Neuropteris tenuifolia.* Acme of *Naiadites* of *modiolaris* group
	A	Disconformity Riversdale group Unconformity and	*Sphenopteris obtusiloba, Sphenopteris polyphylla, Rhodea* cf. *sparsa, Sphenopteris rhomboidea, Sphenopteris schatzlarensis, Sphenopteris pseudo-furcata, Mariopteris acuta, Neurocardiopteris barlowi, Neuropteris smithsii, Whittleseya desiderata.* Small naiaditiform *Anthracomya.* Entrance of *Naiadites* of *modiolaris* group
Nam-urian	A	Disconformity Canso group	*Sphenopteridium* spp., *Telangium* of *affine* group, *Asterocalamites* of *scrobiculatus* group, *Mesocalamites, Lepidodendron praelanceolatum, Neuropteris* aff.? *N. smithsii*

TABLE 3.3 *Chronologic relation of the floras of the Saginaw Group, Michigan, to those of the eastern United States, the Maritime Provinces, and western Europe (from Arnold, 1949, p. 159; Reproduced with permission from the Museum of Paleontology, University of Michigan).*

Michigan	Eastern United States	Maritime Provinces	Western Europe
"Upper Flora" Bay City area and higher cyclical formations at Grand Ledge	Latest Kanawha, Upper Tradewater	Early Morien and late Cumberland	Late Yorkian, late Westphalian B, Assise d'Anzin, Mauritz Group
"Intermediate Flora" Cycle "A" and below at Grand Ledge and Corunna	Late Kanawha, Upper Tradewater	Cumberland	Late Yorkian, etc.
"Lower Flora" Saginaw Coal horizon	Upper Lee	Late Riversdale	Early Yorkian, Upper Westphalian A, Assise de Vicoigne, Baarlo - Wilhelmina Group.

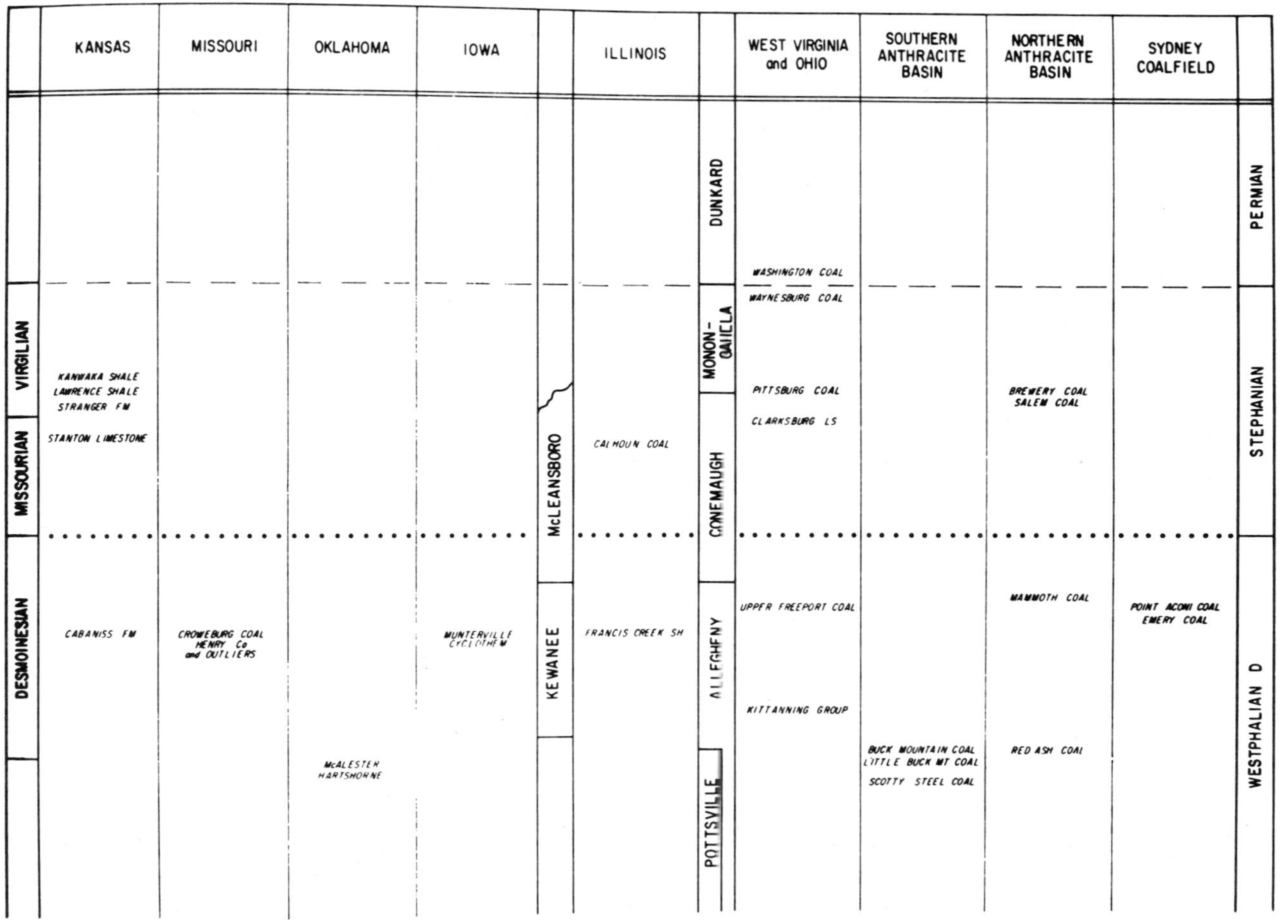

	KANSAS	MISSOURI	OKLAHOMA	IOWA		ILLINOIS		WEST VIRGINIA and OHIO	SOUTHERN ANTHRACITE BASIN	NORTHERN ANTHRACITE BASIN	SYDNEY COALFIELD	
							DUNKARD	WASHINGTON COAL				PERMIAN
VIRGILIAN	KANWAKA SHALE LAWRENCE SHALE STRANGER FM				McLEANSBORO		MONON-GAHELA	WAYNESBURG COAL PITTSBURG COAL		BREWERY COAL SALEM COAL		STEPHANIAN
MISSOURIAN	STANTON LIMESTONE				McLEANSBORO	CALHOUN COAL	CONEMAUGH	CLARKSBURG LS				STEPHANIAN
DESMOINESIAN	CABANISS FM	CROWEBURG COAL HENRY Co and OUTLIERS		MUNTERVILLE CYCLOTHEM	KEWANEE	FRANCIS CREEK SH	ALLEGHENY	UPPER FREEPORT COAL KITTANNING GROUP		MAMMOTH COAL	POINT ACONI COAL EMERY COAL	WESTPHALIAN D
			McALESTER HARTSHORNE				POTTSVILLE		BUCK MOUNTAIN COAL LITTLE BUCK MT COAL SCOTTY STEEL COAL	RED ASH COAL		WESTPHALIAN D

FIGURE 3.2 *Correlation of the Pennsylvanian of Kansas (from Cridland, Morris, and Baxter, 1963, Paleontographica B, v. 112, with permission of the publisher).*

biostratigraphy of the Pennsylvanian in Illinois has been studied by Pfefferkorn and preliminary results have been published (1970, 1975).

The publications of Lesquereux established that plant megafossils in the Pennsylvanian of North America were very similar to those in the Upper Carboniferous(=Silesian) of Europe and the British Isles, an observation used by Wegener (1915, p.64) as an argument for his hypothesis of continental drift.

In several instances, apparently identical fossils, already named on one continent, were independently described under different names on the other continent. This resulted from the slow communication between the areas, from language barriers, and from a lack of first-hand experience with the material from the other continent. Another major reason is that continental drift was not generally accepted. Many apparently believed that plants comprising the flora of such a vast area would show minor differences, even if contemporaneously developed. Still another reason was scientific isolation as evidenced by the publication of a number of scientific reports on both sides of the Atlantic that do not cite a single publication from the other continent.

Several attempts have been made to resolve the matter. Bertrand, Jongmans, and Bode, eminent European paleobotanists, visited and collected in the United States. They wrote several papers on the similarities of the compression flora on both continents (e.g.) Bertrand, 1934, 1935; Bode 1958; Jongmans 1937; Jongmans and Gothan 1934). Remy (1975) has recently published a literature-based correlation. Darrah visited several famous European localities and pointed out the floral similarities in several papers (1933, 1937a,b) and in a table (1970) (Table 3.4). It is obvious from a reading of these publications and from recent field work that similarities do indeed exist in the compression floras and that biostratigraphic experience gained on one continent may also apply to the other. There are several differences, but they are mostly minor biogeographic variations that do not restrict biostratigraphic comparison.

A tentative correlation between the American stratigraphic units and the European units is presented in *Table 3.5.* It is intended for general orientation only and no attempt has been made to develop a degree of precision that can be used with confidence. Several similar generalizations are in the literature, but research under way at the present time will result in a significant refinement of the table in the not too distant future.

There are three distinct stratigraphies *(Table 3.5)* in use at the present time in North America. The stratigraphy for the Appalachian region is a lithostratigraphy that is often used as a chronostratigraphy; that for the midcontinent is a chronostratigraphy based on marine invertebrates; and the letter designation in the *U. S. Geological Survey Professional Paper 853*(McKee et al., 1975) was adopted because the other two had not been correlated sufficiently well with each other and neither had been established as the standard chronostratigraphy.

The U. S.Geological Survey is presently establishing a stratotype for the Pennsylvanian System in West Virginia (Englund et al., 1974). The system and its series, as defined in this stratotype, will become the chronostratigraphic standard in the United States. The Pennsylvanian System stratotype will be a composite, with the individual sections being tied together by mapping and correlation through physical stratigraphy. Plant compression fossils have been collected in over 250 localities. They are presently used to establish local biostratigraphic zones. It is already apparent that many plant fossils appear and disappear in the approximate same sequence that they do in other coal

TABLE 3.4 *Stratigraphic correlation and stratigraphic distribution of major species (according to Darrah, 1970, p. 65; reproduced with permission).*

Appalachian Region	European Equivalent	Neuropteris	Alethopteris	Pecopteris	Sphenopteris	Mariopteris	Linopteris	Sphenophyllum	Other genera and species	Mid-Continent Region
Washington Monongahela Upper Conemaugh	Stephan.	*rogersi* *desorii*	*grandini*	*daubreei* *feminaeformis* *polymorpha* *schimperiana*	*minutisecta* *acrocarpa*	*pachyderma* *ribeyroni*		*filiculme* *longifolium* *oblongifolium*	*Callipteridium* spp. *Odontopteris* spp. *Lescuropteris moorii* *Daneites emersoni*	Virgil Missouri
Lower Conemaugh Allegheny	Westph. D.	*scheuchzeri*• *ovata*•	*serlii*•	*lamuriana* *arborescens*• *unita*	*chaerophylloides* *mixta* *obtusiloba*	*cordato-ovata*	*neuropteroides*	*emarginatum* *majus*	*Annularia stellata*• *Annularia sphenophylloides*•	Des Moines
Lowest Allegheny Upper Pottsville (Lykens I)	Westph. C	*vermicularis* *rarinervis* *tenuifolia*	*sullivantii* *lonchitica*	*dentata*• *miltoni* *abbreviata*	*furcata* *spinosa*	*nervosa* *neuropteroides* *muricata*	*munsteri*	*cuneifolium*		Des Moines
Kanawha Pottsville (Lykens) Kanawha	Westph. B	*tenuifolia* *heterophylla* *gigantea*	*lonchitica* *davreuxi*		*valida* *schazlarensis*	*latifolia*		*myriophyllum*		
Lower Pottsville (Lower Lykens)	Westph. A	*pocahontas* (=*schlehani*)		cf. *aspera*	*hoeninghausi*	*acuta*			"*Megalopteris*" spp.	Morrow

Certain species, especially those marked•, range beyond the limits suggested by this tabulation. The combined evidence or association of a number of forms should give good indication of chronologic correlation.

Note: The subdivisions do not show either relative thicknesses or relative durations of these stages or zones.

TABLE 3.5 *Tentative correlation of major chrono- and lithostratigraphies, and the floral zones of Read and Mamay (1964), with the European chronostratigraphies.*

U.S. G.S.	Midcontinent	Appalachians		Read & Mamay zones	Europe		
E	Virgil	Monongahela		11 + 12	Stephan	U.Carb.=Silesian	CARBONIFEROUS
D	Missouri	Conemaugh			Stephan		
				10	Westphal		
C	DesMoines	Allegheny		9	Westphal		
B	Atoka	Pottsville	Kanawha	8	Westphal		
A	Morrow		New River	6	Westphal		
				5	Namur		
			Pocahontas	4	Namur		
-	Chester	4 formations		X	Namur		
				3	Vise	LC	

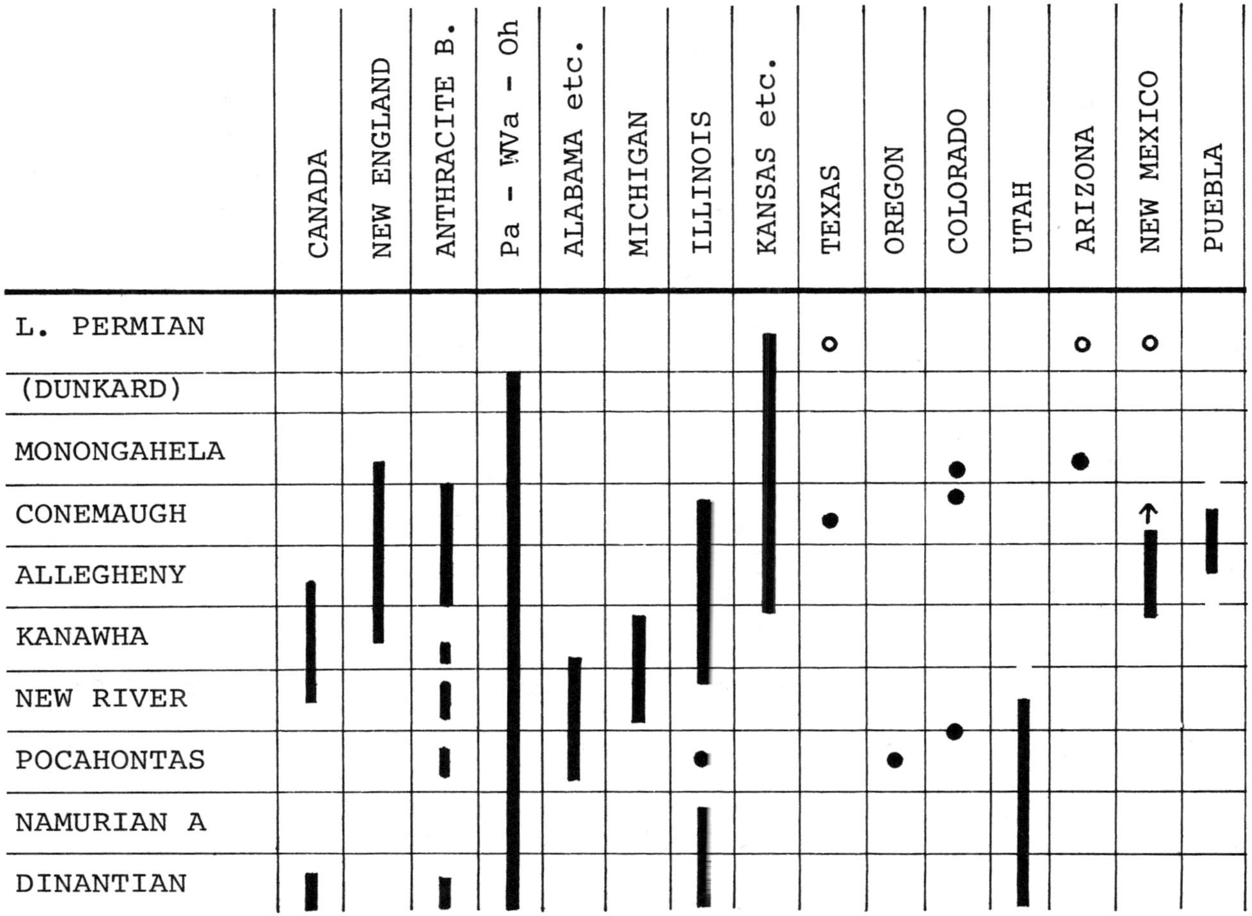

CANADA
NEW ENGLAND
ANTHRACITE B.
Pa - WVa - Oh
ALABAMA etc.
MICHIGAN
ILLINOIS
KANSAS etc.
TEXAS
OREGON
COLORADO
UTAH
ARIZONA
NEW MEXICO
PUEBLA

L. PERMIAN
(DUNKARD)
MONONGAHELA
CONEMAUGH
ALLEGHENY
KANAWHA
NEW RIVER
POCAHONTAS
NAMURIAN A
DINANTIAN

FIGURE 3.3 *Distribution in time of beds containing plant fossils in different parts of North America plotted against the Appalachian stratigraphic scale as a standard. (European terms Dinantian and Namurian A are used to denote possibilities of subdivisions best recorded with these stratigraphic terms.)*

districts of the United States and in Europe.

Using the section of the Appalachian area as exposed and studied in the stratotype as a standard, the distribution of plant-fossil-bearing beds in different parts of North America has been plotted in Figure 3.3. The figure shows quite clearly that the sequence in the stratotype area is the most complete sequence present. The European terms Dinantian and Namurian A have been used to show that a subdivision is possible between the two based on plant fossils and that no suitable terminology exists within the Chesterian to express this finding. The Dunkard is listed separately from the Lower Permian to show the uncertainty that still exists. The Permian (in its terrestrial facies) is accepted internationally as beginning with the *first* occurrence of *Callipteris*. In recent years, however, a controversy has arisen over this definition. This has happened because *C. conferta* has been found in so-called Stephanian rocks (Bouroz and Doubinger, 1977) and because at least one author (Remy, 1975; Remy and Remy, 1977) has synonymized *Dichophyllum* with *Callipteris*. Permian plant-fossil occurrences are plotted in Figure 3.3 only in those areas where unquestionable Permian material has been reported in print.

BIOGEOGRAPHY

North America was part of the Amerosinian floral realm (Havlena, 1970; Chaloner and Meyen, 1973; Chaloner and Lacey, 1973) during the Pennsylvanian and there is now enough evidence to subdivide this area into smaller provinces. Some of the differences and similarities have been recognized earlier. Bell (1944) emphasized the similarity of the Upper Carboniferous flora of the Maritime Provinces with Europe rather than with the flora of the United States. Read (in Read and Mamay, 1964) named the Cordilleran as a separate floral province.

Four floral provinces are now distinguishable in North America *(Figure 3.4)*. They are, from east to west; (1) the Acadian Province, which includes the maritime provinces of Canada, the Narragansett Basin, and other smaller occurrences in New England; (2) the Interior-Appalachian Province, including all of the large coal basins of the United States and part of New Mexico; (3) the Cordilleran Province, with the main occurrence of flora in Colorado and Utah; and (4) the Oregonian Province based on one occurrence in Oregon.

The endemic forms that allow us to separate the provinces are listed in *Table 3.6.* The reason for the differentiation of the four floral provinces can be found in their geotectonic positions, the topography of the area, and the latitude. Both the Acadian Province and the Interior-Appalachian Province were tropical and near the equator. However, they were separated by the Acadian Mountains and at least some coal basins of the Acadian Province were intermontaneous basins that had no connection to the sea and might have been situated at some altitude above sea level. The Interior-Appalachian Province, on the contrary, consisted of areas on the craton and in the fore-deep of the mountains and both were essentially at sea level and invaded at times by the sea. The two provinces represented different environments and were separated by mountains. The genus *Lonchopteris* in the Acadian Province was apparently not able to cross the barrier to the Appalachian Province; *Danaeites* apparently faced the same barrier, but from the opposite direction.

The Cordilleran Province was at least a number of degrees from the equator and was at least partly in a semi-arid climate as shown, for instance, by the Bear Gulch Limestone (lowermost Pennsylvanian) of central Montana (Melton, 1971) and other aridity indicators summarized in *Professional Paper 853* (McKee et al., 1975). In addition, a varied topography was produc-

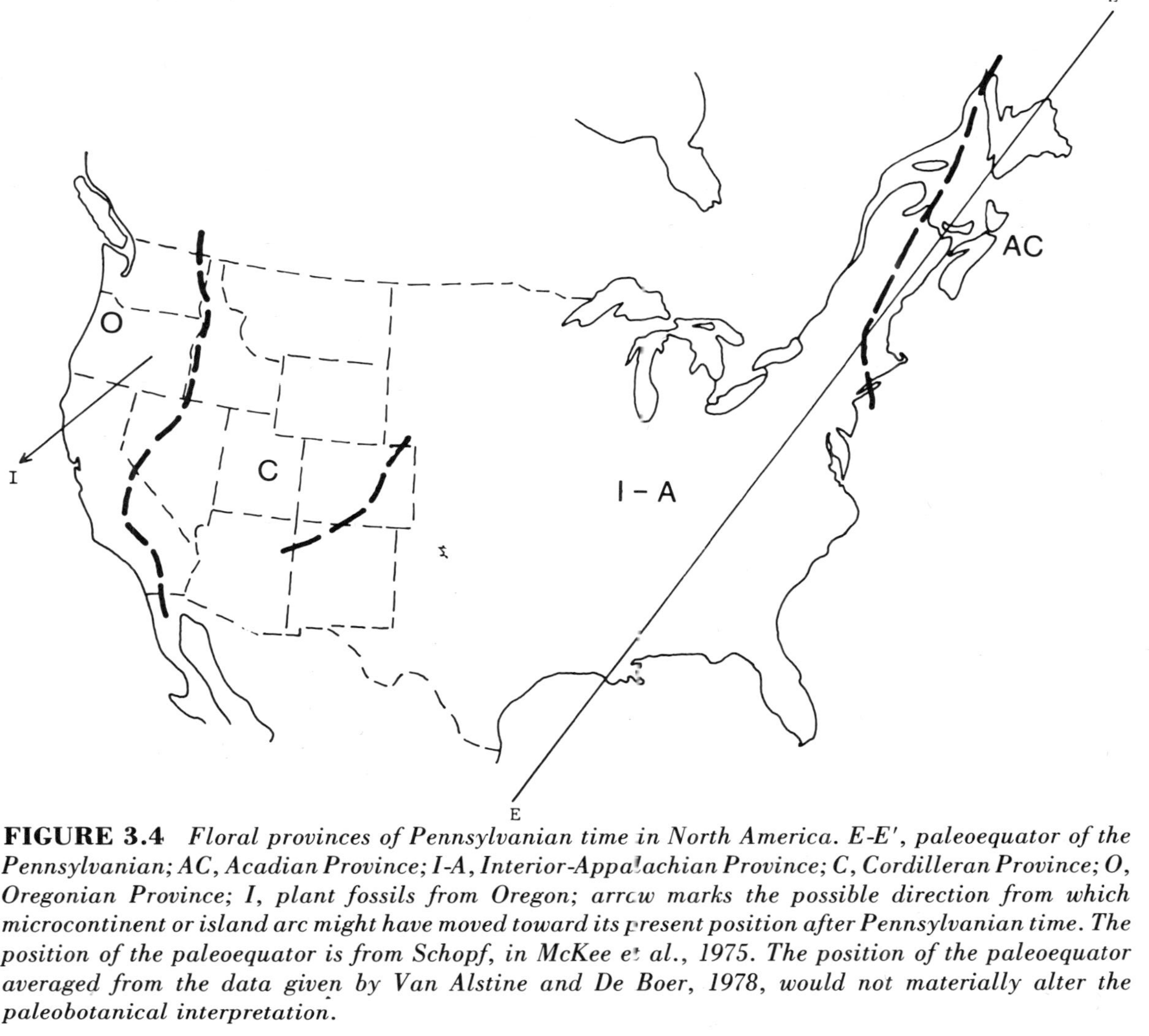

FIGURE 3.4 *Floral provinces of Pennsylvanian time in North America. E-E', paleoequator of the Pennsylvanian; AC, Acadian Province; I-A, Interior-Appalachian Province; C, Cordilleran Province; O, Oregonian Province; I, plant fossils from Oregon; arrow marks the possible direction from which microcontinent or island arc might have moved toward its present position after Pennsylvanian time. The position of the paleoequator is from Schopf, in McKee et al., 1975. The position of the paleoequator averaged from the data given by Van Alstine and De Boer, 1978, would not materially alter the paleobotanical interpretation.*

TABLE 3.6 *List of endemic species characterizing the North American floral provinces during the Pennsylvanian.*

ACADIAN FLORAL PROVINCE

genus *Lonchopteris*
(European form, absent in the other North American provinces.)

genus *Megalopteris*
(known only in the Acadian and Interior-Appalachian Provinces.)

Neuropteris agassizii

INTERIOR-APPALACHIAN PROVINCE

Danaeites emersoni

Lescuropteris moorii

genus *Megalopteris*

CORDILLERAN PROVINCE

genus *Crossopteris*

OREGONIAN PROVINCE

Phyllotheca paulensis

Dicranophyllum rigidum

ed by intensive block faulting in the region as shown in the maps of *Professional Paper 853*. Both factors would produce conditions suitable for intensive evolutionary pressure and the development of new forms. The floras preserved, however, were those growing in or near the area of deposition and contained many forms also found in more humid environments closer to the equator. Nevertheless, there is a clearly divergent evolutionary line in the group of *Neuralethopteris* (represented here by *Crossopteris)*. The presence of *Noeggerathia* (reported as *Tingia*) shows the presence of upland floras.

The Oregonian Province consists of areas that were not yet part of the American continent in Pennsylvania time. The border between the Cordilleran Province and the Oregonian Province as shown in *Figure 3.4* is a line derived from tectonic data. The areas west of that line contain pieces that were microcontinents and island arcs in Pennsylvanian time, and other areas that did not even exist then. Those pieces having Pennsylvanian beds were somewhere to the west (in our present orientation) and their precise position in the Pennsylvanian is not yet known. These old borderline terranes (Churkin and Eberlein, 1977) have to be lumped for our present purposes. It is possible that more floras will be found and that they might belong to several different provinces depending on the original position of their specific islands.

REVIEW OF LITERATURE ON COMPRESSION FLORAS

Research and publications dealing with compression-impression floras are presented below in a geographic arrangement. The discussion begins in eastern Canada and moves south through New England and the Appalachians to Alabama, it continues with Michigan, other Midwestern states to Texas, it moves to Oregon, and other western states, and ends with a discussion of Pennsylvanian paleobotany in Mexico and northern South America.

Canada

The Pennsylvanian flora of the maritime provinces of Canada has been studied and discussed by Bell (1938 to 1966), Bunbury (1846 a,b, 1847, 1852), Dawson (1860, 1891), Stopes (1914), and White (1901a,b, 1902). A number of their reports are monographs with numerous plates that allow a good evaluation of the plant fossils. Bell (1944) summarized the stratigraphic results for the area and showed that Namurian A and most of the Westphalian is present. A most remarkable occurrence is that of the genus *Lonchopteris*, which was common enough to be selected as a plant after which to name a zone. Its presence, although reported, has not been authenticated in other areas of North America.

New England

In New England, plant fossils are most common in the Narragansett Basin. Hitchcock (1841) was the first to describe the plant fossils and Lesquereux included them in his large monograph (1879-1884). Round (1921, 1927) revived the studies in this century and Lyons (1971 to 1978) and Oleksyshyn (1976) have done the most recent work. The plants occur in beds that were severely deformed and heated. The rank of the coals is that of a meta-anthracite and many of the plant fossils have been extensively deformed by tectonic forces. There are a few isolated occurrences in the area (Grew et al., 1970;Zen and Mamay, 1968) which are of great geologic importance. Most of the floras described by earlier workers are Stephanian in age, but Lyons, et al (1976), Oleksyshyn (1976), and Lyons and Darrah (1978) have recently shown the presence of different stages of the Westphalian.

Appalachia

Pennsylvania

Pennsylvania was an early center for the study of North American compression floras (Cist, 1822; Harlan, 1835; Goeppert, 1836; Lesquereux, 1854, 1858a,b). Due to the large number of active coal mines in Pennsylvania, plant fossils were more or less continuously available. Another factor was that a paleobotanical patron, Mr. R. D. Lacoe of Pittston, Pennsylvania, purchased thousands of specimens from collectors and made them available to Lesquereux and other scientists for study. The only other states with a similar intensity of research in the early years were Illinois, Ohio, and West Virginia.

The foremost workers who studied Pennsylvanian floras from Pennsylvania intensively were Lesquereux, David White, and Darrah. As Lesquereux was working for Henry Rogers and the Pennsylvania Survey, he centered his "Description of the Coal Floras of the Carboniferous Formation of Pennsylvania and throughout the United States" (1879-1884) around material from Pennsylvania, although he also studied specimens from other areas. This work is often cited simply as Lesquereux's "Coal Flora." David White's publications on the paleobotany of the State appeared between 1900 and 1943. Darrah's publications span the time from 1932 to 1970.

Ohio

The Pennsylvanian beds in Ohio are a continuation of the flat-lying beds of bituminous coal in western Pennsylvania and West Virginia. Among the scholars who did significant work were Andrews (1875a,b), Newberry (publications between 1853 and 1891), and Abbott (publications between 1950 and 1968). The stratigraphic problems are very similar to those of the two neighboring states and problems like the Dunkard Group have been studied in Ohio (Cross, 1958). A flora with very rare forms and upland elements was described by Andrews (1875a) from Rushville, Ohio.

West Virginia

Intensive paleobotanical research in West Virginia paralleled that of Pennsylvania. The first paper of any consequence was nearly two dozen plates of well-executed wood cuts of plant fossils appended to a comprehensive article (Hildreth, 1836) on the natural history of West Virginia and an adjacent strip of Ohio. Lesquereux made several collecting forays into West Virginia; he also studied the specimens collected there by Hildreth and by Lacoe. They are described in his early papers referred to above. Fontaine and I. C. White (1880), published their report on Upper Pennsylvania or Permian Flora in the 2nd Pennsylvanian Geological Survey, but the work was centered in West Virginia and many of the localities are undoubtedly in Pennsylvanian rocks. In 1892, Lacoe published the first catalog of fossil plants from West Virginia. David White worked extensively in West Virginia from 1884-1905 and proved the usefullness of plant megafossils as biostratigraphic guides. He revised Lacoe's Catalog in 1913. Cross started a program of paleobotanical research at West Virginia University in 1950 and Gillespie has continued the research.

Other States

The remaining states belonging to the Appalachian area are Maryland, Virginia, eastern Kentucky, Tennessee, Georgia, and Alabama. Bunbury (1846a) and Bassler (1916) published about plant fossils from Maryland, David White (1904), 1905) from Virginia, Lyell (1846a,b), Lesquereux (1876), and Mamay (1955) from Alabama. Little has been published on the paleobotany of the other states.

Midwest

Michigan

The Pennsylvanian flora of the Michigan Basin has been studied by Arnold who published on this area between 1930 and 1966. He treated the flora in two monographs (Arnold, 1934, 1949). In these works, he correlated the Pennsylvanian beds of Michigan, on the basis of their contained plant fossils, with the Appalachian area and with Europe and could show that beds of late Westphalian A to late Westphalian B were present.

Illinois

Many well-known workers in paleobotany have published on the compression-impression floras of the Pennsylvanian in the Illinois Basin (Illinois, Indiana, western Kentucky). The historical development has been described by Phillips, Pfefferkorn, and Peppers (1973). Lesquereux was hired by Owen and Worthen to work on the fossil floras and coal deposits of the area and he published special accounts between 1858 and 1883. He later included these data in his major monograph. Sellards (1902, 1903) and David White (between 1896 and 1912) worked in Illinois. Noe was professor for paleobotany in Chicago and published several papers on compression floras between 1922 and 1938. Janssen (between 1939 and 1946) was one of his students. Canright (1958, 1959) and Wood (1954 to 1963) studied the plant fossils of Indiana. Langford (1958, 1963) was an amateur collector who concentrated on the Mazon Creek flora and made many valuable collections. More recent and still-active workers are Pfefferkorn (since 1970), Jennings (since 1970), Leary (since 1972), and Gastaldo (since 1976).

Iowa and Nebraska

The Western Interior Coal Field covers parts of several states and paleobotanical work is spread over a large area. Condit and Miller (1951) reported a flora in excellent preservation from Iowa, and Eggert and Kryder (1969) published on a remarkable fructification. Reihman and Schabilion (1976a,b) reported on cuticles of foliage from coal balls. A single report deals with the Pennsylvanian compression floras from Nebraska (Pepperberg, 1910).

Kansas

Cridland, Morris, and Baxter (1963) published a revision and summary of the Pennsylvanian compression flora of Kansas. They found floras of Westphalian D and Stephanian age to be present. The Kansas section also continues without interruption into the Permian and many valuable Permian collections have been reported. In this transitional sequence, Pennsylvanian forms continue while Permian forms appear. New finds by Leisman (1971) and other workers when completely studied will aid in the resolution of this question.

Missouri

David White worked extensively in Missouri and published several papers (1892, 1893, 1897, 1898, 1915), including a monograph on the "Fossil Flora of the Lower Coal Measures of Missouri" (1899). Cridland (1967, 1968) and Cridland and Darrah (1969) reported on several plant fossils and Basson (1968) published a report on one flora.

Oklahoma

In Oklahoma, plant megafossils occur in uppermost Mississippian beds (White, 1937) and in Middle to Upper Pennsylvanian rocks (White, 1899; Wilson, 1963, 1972). There are numerous other reports of compression floras in the geologic literature. These are cited by Wilson (1960) who gives a complete bibliography.

Arkansas

The uppermost Mississippian beds with plant fossils from Oklahoma occur also in

Arkansas (White, 1936; Eggert and Taylor, 1971). Pennsylvanian plants were described by Moss (1850) and Lesquereux (1860), and White (1907) compared the floras known to him with those from the Kittanning Coals of the Appalachians.

Texas

Plant megafossils are not uncommon in the Upper Pennsylvanian of Texas, but only a few have been described or mentioned in the literature. Mamay (1972) described a new species of *Biscalitheca* from uppermost Pennsylvanian beds. Plant fossils in other areas have been mentioned in geologic publications, for instance, Sellards et al. (1932). Some paleobotanical research has been done recently, but it is still unpublished.

West

Oregon

Pennsylvanian plant megafossils occur in Oregon in the Spotted Ridge Formation. This area was a microcontinent or island are at some distance from the North American craton in Pennsylvanian time. Thus, it is not unusual to find forms that do not occur in other areas of North America. The age of the flora is Early Pennsylvanian as shown by the occurrence of *Mesocalamites* (sensu Hirmer) and *Pecopteris oregonensis* which belongs to the *Dactylotheca aspera* group. Read and Merriam (1940), Arnold (1953), and Mamay and Read (1956) reported on the flora and have provided complete descriptions.

Wyoming

Mamay (1975) reported a small florule from the Hoback Canyon area in Wyoming. The plants indicate an age near the Mississippian—Pennsylvanian boundary. Plant fossils have been mentioned from subsurface samples of Pennsylvanian rocks in Montana and the Dakotas and they can be expected in the surface outcrops of these beds.

Colorado

Compression floras occur in Colorado in the Lower Pennsylvanian (Read, 1934; Arnold, 1940; Chronic, 1958) and in other formations which are presumably higher in the section. Arnold (1941) and Elias (1942) both comment on the fact that *Walchia* and some other late Pennsylvanian forms, which they call"Permian," occur rather low in the section as judged by other criteria. They remark further that aridity indicators occur with hygrophile floras in alluvial fan deposits. The ancestral Rocky Mountains, which were forming during the Pennsylvanian, apparently influenced some floras, even though spots with water-logged soil were present to support the well-known hygrophilous flora of that time.

Utah

Pennsylvanian floras in Utah have been reported by Tidwell (1962, 1967, 1975; Tidwell, Medlyn, and Simper, 1974). The best described flora is that of the Manning Canyon Shale, which is Early Pennsylvanian in age. It contains *Crossopteris*, which is restricted to this area, and a few other rare forms such as *Noeggerathia*.

Nevada

Stigmaria stellata (= *S. wedingtonensis*), a form of Late Mississippian age, has been reported from two localities in Nevada (Rich, 1962; Pfefferkorn, 1972). Other plant fragments present in these sections indicate an early Namurian interval. At that time, islands formed in parts of Nevada and were covered with vegetation for a short time before marine sedimentation resumed.

Arizona

Blazey reported a fossil flora from the Mogollon Rim in Arizona (Blazey, 1974; Blazey and Canright, 1972; Canright and Blazey, 1974). It contains predominately Pennsylvanian forms and is characterized by the absence of lycopods and the presence of *Odontopteris, Taeniopteris,* and *Walchia*. Fragments of *Callipteris*, an indicator of Permian time, are also included.

New Mexico

New Mexico had a diverse topography in Pennsylvanian time and the deposition of the basal Pennsylvanian beds occurred at different times in different areas. Plant fossils are not uncommon in the terrestrial beds intercalated in the marine sequences. The oldest plant fossils that give biostratigraphic information indicate a Middle Pennsylvanian age although the Late Pennsylvanian and the Permian are also represented. Lists of plant fossils have been published in geologic publications (Northrop, 1961; Kelley and Northrop, 1975), and a specific study is now in preparation.

Mexico and South America

Silva (1970) reported a flora from the province of Tehuacan in Mexico. The figured plant fossils seem to indicate a late Westphalian D or Stephanian age, but further investigations are clearly needed.

Pennsylvanian floras are known in two areas of Venezuela (Benedetto and Odreman, 1977; Pfefferkorn, 1977) and one locality in Columbia (Gerth and Kräusel, 1931). The floras reported by Benedetto and Odreman (1977) are richer in species than the others, but all three publications demonstrate that the forms present belong to the Amerosinian Flora Realm. Gondwana forms are entirely missing. Considering that Gondwanaland collided with North America—Europe in the Carboniferous, it is clear that northern South America was where the Gulf of Mexico is now, therefore close to North America and thus in the tropical realm.

ACKNOWLEDGMENTS

We thank Sergius H. Mamay for his review. This study was supported by the Division of Earth Sciences, National Science Foundation, NSF grant EAR76—22562.

REFERENCES

Abbott, M. L. 1950. The Upper Freeport No. 7 coal parting compression flora of the Appalachian basin (abstract). *Am. Jour. Bot.* 37: 672.

Abbott, M. L. 1968. Lycopsid stems and roots and sphenopsid fructifications and stems from the Upper Freeport Coal of southeastern Ohio. *Palaeontographica Americana* 6: 5-49.

Andrews, E. B. 1875a. Descriptions of fossil plants from the Coal Measures of Ohio. *Ohio Geol. Sur. Rep.* 2 pt.2: 415-426.

Andrews, E. B. 1875b. Notice of new and interesting coal plants. *Am. Jour. Sci.* ser. 3, 10:462-466.

Arnold, C. A. 1930. A petrified Lepidophyte cone from the Pennsylvanian of Michigan. *Am. Jour. Bot.* 17: 1028-1032.

Arnold, C. A. 1934. A preliminary study of the fossil flora of the Michigan Coal Basin. *Contr. Mus. Paleontol. Univ.* Michigan 4: 177-204.

Arnold, C. A. 1940. *Lepidodendron johnsonii* sp. nov., from the Lower Pennsylvanian of central Colorado. *Contr. Mus. Paleontol. Univ.* Michigan 6 (2): 21-52.

Arnold, C. A. 1941. Some Paleozoic plants from central Colorado and their stratigraphic significance. *Contr. Mus. Paleontol. Univ.* Michigan 6 (4): 59-70.

Arnold, C. A. 1949. Fossil flora of the Michigan Coal Basin. *Contr. Mus. Paleontol. Univ.* Michigan 7: 131-269.

Arnold, C. A. 1953. Fossil plants of early Pennsylvanian type from central Oregon. *Palaeontographica B* 93: 61-68.

Arnold, C. A. 1966. Fossil plants in Michigan. *Michigan Botanist* 5:3-13.

Bassler, H. 1916. A cycadophyte from the North American coal measures. *Am. Jour. Sci. Ser.* 4, 42: 21-26.

Basson, P. W. 1968. The fossil flora of the Drywood Formation of southwestern Missouri. Univ. Missouri Stud. 44: 1-170.

Bell, W. A. 1938. Fossil flora of Sidney Coalfield, Nova Scotia. *Geol. Sur. Canada Mem. 215.* 334 pp.

Bell, W. A. 1944. Carboniferous rocks and fossil floras of northern Nova Scotia. *Geol. Sur. Canada Mem. 238,* 227 pp.

Bell, W. A. 1966. Carboniferous plants of eastern Canada. *Geol. Sur. Canada, Paper 66-11,* 76 pp.

Benedetto, G. and O. E. Odreman Rivas. 1977. Bioestratigrafia y paleoecologia de las unidades Permocarbonicas aflorantes en el area de Carache-Ague de Obispo, Estado Trujillo, Venezuela. *5th Cong. Geol. Venezol. Mem.* 1: 253-288.

Bertrand, P. 1934. Les flores houillères d'Amerique d'après les travaux de M. David White. *Soc. Geol. Nord. Ann.* (1933) 58: 231-254.

Bertrand, P. 1935. Nouvelles correlations stratigraphiques entre le Carbonifère des Etates-Units et celui de l'Europe occidentale d'après MM. Jongmans et Gothan. *Soc. Geol. Nord. Ann.* 60: 3-16.

Blazey, E. B. 1974. Fossil flora of the Mogollon Rim, central Arizona. *Palaeontographica B* 146: 1-20.

Blazey, E. B., and J. E. Canright. 1972. The fossil flora of the Mogollon Rim, Arizona, (Abs). *Am. Jour. Bot* 59 (Pt. 2): 659.

Bode, H. 1958. Die floristische Gliederung des Oberkarbons der Vereinigten Staaten von Nordamerika. *Zeitschr. Deutsch. Geol. Gesell.* 110: 217-259.

Bouroz, A. and J. Doubinger. 1977. Report on the Stephanian-Autunian boundary and on the contents of Upper Stephanian and Autuniar in their stratotypes. In *Symposium on Carboniferous Stratigraphy,* V.M. Holub, and R. H. Wagner eds. Prague; Geological Survey, pp. 147-169.

Bunbury, C. J. F. 1846a. On some remarkable fossil ferns from Frostburg, Maryland, collected by Mr. Lyell. *Quart. Jour. Geol. Soc. London* 2: 82-91.

Bunbury, C.J. F. 1846b. Notes on the fossil plants communicated by Mr. Dawson from Nova Scotia. *Quart. Jour. Geol. Soc. London* 2: 136-139.

Bunbury, C. J. F., 1847. On Fossil Plants form the coal formation of Cape Breton. *Quart. Jour. Geol. Soc. London* 3: 423-438.

Bunbury, C. J. F., 1852. Description of a peculiar fossil fern from the Sidney Coal field, Cape Breton. *Quart. Jour. Geol. Soc. London* 8: 31-35.

Canright, J.E. 1958. History of Paleobotany in Indiana. Proc, *Indiana Acad. Sci.* 67:268-273.

Canright, J. E. 1959. Fossil plants of Indiana. *Indiana Geol. Sur., Rep. Prog. 14,* 45 p.

Canright, J. E., and E. B. Blazey, 1974. A Lower Permian flora from Promontory Butte, central Arizona. *In Guidebook to Devonian, Permian and Triassic Plant Localities, East-Central Arizona,* S. R. Ash, ed. Paleobotanical Sec. Bot. Soc. America, pp. 57-62.

Chaloner, W. G., and W. S. Lacey 1973. The distribution of Late Palaeozoic Floras. *Spec. Pap. Palaeontol. No. 12* London pp. 271-289.

Chaloner, W. G. and S. V. Meyen. 1973. Carboniferous and Permian floras of the northern Continents. In *Atlas of Palaeobiogeography*A. Hallam, ed. pp. 169-186. Amsterdam: Elsevier.

Chronic, J. 1958. Pennsylvanian paleontology in Colorado. *Symposium on Pennsylvanian rocks of Colorado and adjacent areas.* Rocky Mountain

Assoc. Geologists. Denver, Color. pp. 13-16.

Churkin, M., and G. D. Eberlein. 1977. Ancient borderland terranes of the North American Cordillera. Correlation and microplate tectonics. *Geol. Soc. Am. Bull.* 88: 769-786.

Cist, Z. 1822. Letter. *Am. Jour. Sci.* 4: 1-7; 1825 9:165-166.

Condit, C., and A. K. Miller. 1951. Concretions from Iowa like those of Mazon Creek, Illinois. *Jour. Geol.* 59: 395-396.

Cridland, A. A. 1967. A Calamitean cone from the Drywood Formation (Pennsylvanian) of Missouri. (Abst). *Am. Jour. Bot.* 54: 652.

Cridland, A. A. 1968. *Alethopteris ambiqua* Lesquereux, a Pennsylvanian pteridosperm from Missouri, U.S.A. *Jour. Linn. Soc. London* 61: 107-111.

Cridland, A. A., and E. L. Darrah. 1968-1969 *Crossotheca urbani* sp. nov., a male fructification from the Pennsylvanian (Desmoinesian series, Verdigris Formation) of Missouri, U.S.A. *Palaeobotanist* 17: 93-96.

Cridland, A.A.; J. E. Morris; and R. W. Baxter. 1963. The Pennsylvanian plants of Kansas and their stratigraphic significance. *Palaeontographica B.* 112: 58-92.

Cross, A. T. 1958. Fossil flora of the Dunkard strata of eastern United States. *In* M.T. Sturgeon. The geology and mineral resources of Athens County, Ohio, *Ohio Div. Geol. Sur. Bull.* 57: 191-197.

Cross, A. T.; W. L. Smith; and T. Arkle, Jr. 1950. *Field Guide For Special Field Conference on Stratigraphy, Sedimentation and Nomenclature of Upper Pennsylvanian and Lower Permian Strata in the Northern Portion of the Dunkard Basin of Ohio, West Virginia and Pennsylvanian.* West Virginia Geol. Sur. 100 p.

Darrah, W. C. 1932. Recent paleobotanic investigations near Pittsburgh, Pennsylvania. *Proc. Pennsylvania Acad. Sci.* 6: 110-114.

Darrah, W. C. 1933. Stephanian in America.(Abs.). *Geol. Soc. Am. Proc. 1933*, p. 451.

Darrah, W. C. 1937a. American Carboniferous floras. *2nd Congr. Stratigr. Carbonifère, Compte Rendu* (Heerlen 1935) 1; 109-129.

Darrah, W. C. 1937b. Recent studies of American pteridosperms. *2nd Congr. Stratigr. Carbonifère, Compte Rendu (Heerlen 1935) 1:*131-137.

Darrah, W. C. 1970 (1969). A critical review of the Upper Pennsylvanian floras of Eastern United states, with notes on the Mazon Creek flora of Illinois. Gettysburg, Pa.: (Privately printed), 220 p.

Dawson, J. W. 1860. Recent researches in the Devonian and Carboniferous flora of British America. *Proc. Am. Assoc. Advanc. Sci.*, 1859, 13: 308-310.

Dawson, J. W. 1891. Carboniferous fossils from Newfoundland. *Geol. Soc. Am. Bull.* 2: 529-540.

Eggert, D. A., and R. W. Kryder. 1969. A new species of *Aulacotheca* (Pteridospermales) from the Middle Pennsylvanian of Iowa. *Palaeontology* 12: 414-419.

Eggert, D. A., and T. N. Taylor. 1971. *Telangiopsis* gen. nov., an Upper Mississippian pollen organ from Arkansas. *Bot. Gaz.* 132: 30-37.

Elias, M. K. 1942. *Walchia* associated with diagnostic early Pennsylvanian forms in central Colorado. (abs.) *Geol. Soc. Am. Bull.* 53 (12, Pt. 2.) 1800.

Englund, K. J., J. B. Roen; S. P. Schweinfurth; and H. H. Arndt. 1974. Description and preliminary report of the Pennsylvanian System stratotype study. (abs.) *Geol. Soc. Am. (Abstracts With Programs).* 6: 722.

Fontaine, W. F., and I. C. White. 1880. The Permian or Upper Carboniferous flora of West Virginia and southwestern Pennsylvania. *2nd Pennsylvania Geol. Surv. Rep. PP*, 143 p.

Gastaldo, R. A. 1976. A Middle Pennsylvanian nodule flora from Carterville, Illinois (Abs.) *Abs. Pap. Present. Meet. Bot. Soc. Am.*, Tulane University, New Orleans, 30 May—4 June 1976, p. 25.

Geoppert, H. R. 1836. Fossilen Farrenkraueter (systema filicum fossilium) *Nova Acta Leopoldina* 17, 486 p.

Gerth, H., and R. Kräusel. 1931. Beiträge zur Kenntnis des Carbons in Südamerika. II. Obercarbonische Pflanzenreste aus Columbien. *Neues Jahr. Mineral., Geol. Paläontol., Abh Abt. B., Beilage 65: 529-534.*

Gillespie, W. H., and J. A. Clendening. 1962. A lower Kittaning flora from northern West Virginia. *Proc. West Virginia Acad. Sci.* 34: 125-132.

Gillespie, W. H., and J. A. Clendening. 1966. West Virginia plant fossils, I. *Dolerotheca* and *Daubreeia*. *Proc. West Virginia Acad. Sci.* 38: 159-168.

Grew, E. S.; S. H. Mamay; and E. S. Barghoorn. 1970. Age of plant fossils from the Worcester Coal Mine, Worcester, Massachusetts. *Am. Journ. Sci.* 268: 113-126.

Harlan, R. 1835. Notice of fossil vegetable remains from the bituminous coal measures of Pennsylvania. *Trans. Pennsylvania Geol. Soc.* 1: 256-259.

Havlena, Vaclav. 1970. Einige Bemerkungen zur Phytogeographie and Geobotanik des Karbons und Perms. *6th Cong. Intern. Stratigr. Geol. Carbonifère, Compte Rendu Sheffield 1967.* 3: 901-911.

Hildreth, S. P., and S. G. Morton. 1836. Observations on the bituminous coal deposits of the valley of the Ohio, and accompanying rock strata, with notices of the fossil organic remains and the relics of vegetable and animal bodies, illustrated by a geological map, by numerous drawings of plants and shells and by views of interesting scenery. *Am. Jour. Sci.* 29: 1-154.

Anonymous. 1837. (Actually S. P. Hildreth.) Miscellaneous observations made during a tour in May 1835 to the Falls of the Cuyohoga, near Lake Erie; extracted from the diary of a naturalist. *Am. Jour Sci.* 31: 1-84.

Hitchcock, E. 1841. *Final Report on the Geology of Massachusetts.* Amherst, Mass. J. S. & C. Adams, Northampton, Mass: J. H. Butler. 831 p.

Janssen, R. E. 1939. Leaves and stems from fossil forests — A handbook of the paleobotanical collections in the Illinois State Museum. *Illinois State Mus. Pop. Sci. Ser.* 1, 190 p. (2nd printing, rev. 1957).

Janssen, R. E. 1946. Miniature fossil concretions of Mazon Creek. *Trans. Illinois State Acad. Sci.* (1945) 38: 83-84.

Jénnings, J. R. 1970. Preliminary report on fossil plants from the Chester Series (Upper Mississippian) of Illinois. *Trans. Illinois State Acad. Sci.* 63: 167-177.

Jongmans, W. J. 1937. Contribution to a comparison between the Carboniferous floras of the United States and western Europe. *2nd Congr. Stratigr. Carbonifère, Compte Rendu,* (Heerlen 1935), 1: 363-387.

Jongmans, W. J., and Walter Gothan. 1934. Florenfolge und vergleichende Stratigraphie des Karbons der östlichen Staaten Nord-Amerikas. Vergleich mit West-Europa. *Geol. Bur. Heerlen Jaarverslag over 1933,* pp. 17-44.

Kelly, V. C., and S. A. Northrop. 1975. Geology of Sandia Mountains and vicinity, New Mexico. *New Mexico Bur Mines Miner. Resour, Mem. 29,* 136 pp.

Lacoe, R. D. 1884. *Catalogue of the Palaeozoic fossil plants of North America.* (Extract) Pottsville Scientific Association, Pittston, Pa.: 15 p.

Lacoe, R. D. 1892. Supplement: fossil flora. *West Virginia Univ. Agric. Exper. Sta. Bull.* 24: 519-527.

Langford, G. 1958. The Wilmington coal flora from a Pennsylvanian deposit in Will County, Illinois. Downers Grove, Ill. Esconi Associates, 360 p.

Langford, G. 1963. The Wilmington coal fauna and additions to the Wilmington coal flora from a Pennsylvanian deposit in Will County, Illinois. Downers Grove, Ill: Esconi Associates, 280 p.

Leary, R. L. 1972 *Lacoea*, an unusual plant fossil. *Liv. Mus.* (Ill. State Mus. Springfield, Ill.) 34: 78-79,

Leisman, G. A. 1971. An upland flora of Upper Pennsylvanian near Hamilton, Kansas. *Am. Jour. Bot.* 58(5): 470 (abstract).

Lesquereux, L. 1854. New species of fossil plants from the anthracite and bituminous coal fields of Pennsylvania, collected and described by Leo Lesquereux. *Jour. Boston Soc. Nat. His.* 6: 413-431.

Lesquereux, L. 1858a. The fossil plants of the Coal Measures of the United States, with descriptions of the new species, in the cabinet of the Pottsville Scientific Association. (Extract) *Proc. Pottsville Sci. Assoc.* pp. 5-24.

Lesquereux, L. 1858b. On the order in the coal measures of Kentucky and Illinois and their relations to those of the Appalachian coal field. *Am. Jour. Sci.* ser. 2, 26: 110-112.

Lesquereux, L. 1860. Botanical and palaeontological report on the geological survey of Arkansas. *2nd Rep. Geol.* Arkansas, pp. 308-317.

Lesquereux, L. 1876. Partial list of coal plants from the Alabama fields and discussion of several coal seams. Alabama *Geol. Sur., Rep. Prog. 1875*, pp. 75-82.

Lesquereux, L. 1879-1884. Description of the coal flora of the Pennsylvanian and of the Carboniferous formation in Pennsylvania and throughout the United States. *2nd Geol. Sur. Pennsylvania*, 997 p. (Atlas 1879; vols. 1 and 2, 1880; vol. 2, 1884).

Lesquereux, L. 1883. Principles of Paleozoic Botany. *Indiana Dept. Geol. Nat. Hist. Annual Rep. 13* (1884), *Paleontology*,, pp. 6-231.

Lesquereux, L. 1884. The Carboniferous flora of Rhode Island. *Am. Naturalist* 18: 921-923.

Lyell, C. 1846. Coal field of Tuscaloosa, Alabama. *Am. Journ. Sci. Arts* Ser. 2 1: 371-376.

Lyell, C. 1846. Observation on the fossil plants of the coal field of Tuscaloosa, Alabama, with a description of some species by C. J. F. Bunbury. *Am. Jour. Sci.* Ser. 2, 2: 228-233.

Lyons, P. C. 1971. Correlation of the Pennsylvanian of New England and the Carboniferous of New Brunswick and Nova Scotia. (Abs). *Geol. Soc. Am. Abstracts with Programs* (Northeastern Section) 3 (1): 43-44.

Lyons, P. C., and W. C. Darrah. 1978. A late Middle Pennsylvanian flora of the Narragansett Basin, Massachusetts. *Geol. Soc. Am. Bull.* 89: 433-438.

Lyons, P.C.; B. Tiffney; and B. Cameron. 1976. Early Pennsylvanian age of the Norfolk basin, southeastern Massachusetts, based on plant megafossils. *Geol. Soc. Am., Mem. 146*, pp. 181-197.

Mamay, S. H. 1955. *Acrangiophyllum*, a new genus of Pennsylvanian Pteropsida based on fertile foliage. *Am. Jour. Bot.* 42: 177-183.

Mamay, S. H. 1972 *Biscalitheca suzanneana*, n. sp., from the uppermost Pennsylvanian of Texas. *Rev. Palaeobot. Palynol.* 14: 141-147.

Mamay, S. H. 1975. *In*"Stratigraphy and geologic history of the Amsden Formation (Mississippian and Pennsylvanian) of Wyoming," W.J. Sando; G. Mackenzie, Jr.' and J. T. Dutro, Jr.*U.S. Geol. Sur. Prof. Pap. 848-A*, 83 p.

Mamay, S. H., and C. B. Read. 1956. Additions to the flora of the Spotted Ridge Formation in central Oregon.

U.S. Geol. Sur. Prof. Pap. 274, pp. 211-226.

McKee, E. D., et. al. 1975. Paleotectonic investigations of the Pennsylvanian System in the United States. *U.S. Geol. Sur. Prof. Pap 853,* pt. I, 349 p; p. II, 192 p; p. III, charts and maps.

Melton, W. G. 1971. The Bear Gulch Fauna from central Montana. *Proc. North American Paleontol. Conven.* 1969, p I, pp 1202-1207.

Moss, T. F. 1850. Description of the new carpolite from Arkansas. *Proc. Acad. Nat. Sci. Philadelphia,* 5: 59.

Newberry, J. S. 1853. Fossil plants from the Ohio coal basin. *Ann. Sci.* (Cleveland) 1: 95-97.

Newberry, J. S. 1891. The genus *Sphenophyllum. Jour. Cincinnati Soc. Nat. Hist.* 13: 212-217.

Northrop, S. A. 1961 Mississippian and Pennsylvanian fossils of Albuquerque Country, New Mexico. *New Mexico Geol. Soc. 12th Field Conf. Guildebook to the Albuquerque Country,* pp. 105-112.

Noé, A. C. 1922. Fossil flora of Braidwood, Illinois, *Trans. Illinois State Acad. Sci.* 15: 396-397.

Noé, A. C., and R. E. Janssen. 1938. Identification key for Illinois plant fossils. *Trans. Illinois State Acad. Sci.* (1937), 30: 236-237.

Oleksyshyn, J. 1976. Fossil plants of Pennsylvanian age from northwestern Narragansett Basin. *Geol. Soc. Am. Mem 146,* p. 143-180.

Pepperberg, R. V. 1910. Preliminary notes on the Carboniferous flora of Nebraska. *Nebraska Geol. Suv.* 3 (Part. 11): 313-329.

Pfefferkorn, H. W. 1970. Use of plant megafossils in correlation of the Pennsylvanian of the Illinois Basin with the Upper Carboniferous of Europe. (Abs.) Geol. Soc. Am. Abstracts with Programs 2 7: p. 652.

Pfefferkorn, H. W. 1972. Distribution of *Stigmaria wedingtonensis* (Lycopsida) in the Chesterian (Upper Mississippian) of North America. *Am. Midland Naturalist* 88: 224-231.

Pfefferkorn, H. W. 1975. Biostratigraphy of the Pennsylvanian of the Illinois Basin (USA) based on plant megafossils. 8th Congr. *Internl. Carboniferous Stratigr. Geol.* Abs. (Moscow 1975) pp. 212-213.

Pfefferkorn, H. W. 1977. Plant megafossils in Venezuela and their use in geology. *5th Congr. Geol.* Venezolano *Memo. 1: 407-414.*

Phillips, T. L; H. W. Pfefferkorn; and R. A. Peppers. 1973. Development of paleobotany in the Illinois Basin. *Illinois State Geol. Sur. Circ. 480,* 86 p.

Read, C. B. 1934. A flora of Pottsville age from the Mosquito Range, Colorado. *U.S. Geol. Surv. Prof. Pap 185-D,* 79-96.

Read, C. B. 1944. In R.C. Moore, et al. Correlation of Pennsylvanian formations of North America. *Geol. Soc. Am. Bull.* 55: 657-706.

Read, C. B. 1947. Pennsylvanian floral zones and floral provinces. *Jour. Geol.* 55: 271-279.

Read, C. B. and S. H. Mamay. 1964. Upper Paleozoic floral zones and floral provinces of the United States. *U.S. Geol. Sur. Prof. Pap. 454K,* p. 1-35.

Read, C. B., and C. W. Merriam. 1940. A Pennsylvanian flora from central Oregon. *Am. Jour. Sci.* 238 (2) 107-111.

Reihman, M. A., and J. T. Schabilion. 1976a. Cuticles of two species of *Alethopteris. Am. Jour. Bot.* 63: 1039-1046.

Reihman, M. A., and J. T. Schabilion. 1976b. *Linopteris* and *Neuropteris* from Iowa coal balls. (Abs.) *Abs. Pap. Meet. Bot. Soc. Am.,* New Orleans 1976, p.30.

Remy, W. 1975. The floral changes at the Carboniferous-Permian boundary in Europe and North America. In "The age of the Dunkard," J. A. Barlow ed.

Proc. First I. C. White Memorial Symp. Morgantown, W. Va., 1972, pp. 305-352.

Remy, W., and R. Remy. 1977. *Die Floren des Erdaltertums.* Essen: Verlag Glückauf, 467 p.

Rich, M. 1962. Mississippian Stigmarian plant fossil from southern Nevada. *Jour. Paleontology, 36: 347-349.*

Round, E. M. 1921. Odontopteris genuina in Rhode Island. Bot. Gaz. 72: 397-403.

Round, E. M. 1927. Correlation of fossil floras in Henry County, Missouri, and the Narragansett Basin. *Bot. Gaz.* 83: 61-69.

Schopf, J. M. 1975. Pennsylvanian climate in the United States. In "Paleotectonic investigations of the Pennsylvanian System." E. C. McKee, E. J. Crosby, et al. *U.S. Geol. Sur. Prof. Pap. 853,* Pt II, p. 23-31.

Sellards, E. H. 1902. On the fertile fronds of *Crossotheca* and *Myriotheca,* and on the spores of other Carboniferous ferns from Mazon Creek, Illinois. *Am. Jour. Sci.* ser. 4, 14 (81, Art. 22): 195-202.

Sellards, E. H. 1903. *Codonotheca,* a new type of spore-bearing organ from the Coal Measures. *Am. Jour. Sci.* ser. 4, 16: 87-95.

Sellards, E. H.; W. S. Adkins; and F. B. Plummer. 1932. The geology of Texas, V. I, Stratigraphy. *Univ.* Texas *Bull. No. 3232,* 1007 p.

Silva Pineda, A. 1970. Plantas del Pensilvanico de la region de Tehuacán, Puebla. *Paleontol. Mexicana No. 29,* 47 p.

Stopes, M. C. 1914. The "Fern Ledges" Carboniferous flora of St. John, New Brunswick. *Geol. Sur. Canada, Mem. 41,* 142 p.

Tidwell, W. D. 1962. An early Pennsylvanian flora from the Manning Canyon Shale, Utah. *Brigham Young Univ. Geol. Stud.* 9: 83-102.

Tidwell, W. D. 1967. Flora of the Manning Canyon Shale, Part 1: A lowermost Pennsylvanian flora from the Manning Canyon Shale, Utah, and its stratigraphic significance. *Brigham Young Univ. Geol. Stud.* 14: 1-63.

Tidwell, W. D. 1975. Common fossil plants of western North America. Provo, Utah: *Brigham Young Univ. Press* 197 p.

Tidwell, W. D.; D. A. Medlyn and A. D. Simper. 1974. Flora of the Manning Canyon Shale, Part 2: Lepidodendrales. *Brigham Young Univ. Geol. Stud.* 21: 119-146.

Van Alstine, D. R., and J. de Boer. 1978. A new technique for constructing apparent polar wander paths and the revised Phanerozoic path for North America. *Geology* 6: 137-139.

Ward, L. F. 1892. Principles and methods of geologic correlation by means of fossil plants. *Am. Geologist* 9: 34-37.

Wegener, A. 1915. Die Entstehung der Kontinente und Ozeane. Braunschweig: Friedr. Vieweg & Sohn, 94 p. (Samml. Vieweg Heft 23).

White, David, 1892. A new Taeniopteroid fern and its allies. *Geol. Soc. Am. Bull.* 4: 119-132.

White, David, 1983. Flora of the outlying Carboniferous basins of southwestern Missouri. *U.S. Geol. Sur. Bull. 98,* 139 p.

White, David, 1896. Report on the fossil plants from the Hindostan Whetstone beds of Orange County, Indiana. In "The whetstone and grindstone rocks of Indiana." E. M. Kindle, *Indiana Dep. Geol. Nat. Resour. Annu. Rep. 20,* pp. 328-368.

White, David, 1897. Age of lower coals of Henry County. Missouri. *Geol. Soc. Am. Bull* (for 1896) 8: 287-304.

White, David, 1898. *Omphalophloios,* a new lepidodendroid type. *Geol. Soc. Am. Bull.* (for 1897) 9: 329-342.

White, David, 1899. Fossil flora of the Lower Coal Measures of Missouri. *U.S. Geol. Sur. Monogr. 37,* 467 p.

White, David. 1900. The stratigraphic succession of the fossil floras of the Pottsville formation in the southern anthracite coal field, Pennsylvania. *U.S. Geol. Sur. Annu. Rep. 20*, pp. 749-930.

White, David, 1901a. The Canadian species of the genus *Whittleseya* and their systematic relations. *Ottawa Naturalist* 15:98-110.

White, David. 1901b. Some palaeobotanical aspects of the Upper Palaeozoic in Nova Scotia. *Canadian Rec. Sci.* 8: 271-280.

White, David. 1902. Stratigraphy versus Paleontology in Nova Scotia. *Science* n.s. 16: 232-235.

White, David. 1904. The seeds of *Aneimites. Smithsonian Misc. Collect. 47* (Publ. no. 1548, vol. 2, p. 3, 1905), pp. 322-331.

White, David. 1905. The age of the Wise and Harlan Formations of southwestern Virginia. (Abs.) *Science* 22: 335-336.

White, David, 1907. Report on fossil plants from the coal measures of Arkansas. *U.S. Geol. Sur. Bull. 326*, pp. 24-31.

White, David, 1913. The fossil flora of West Virginia, West Virginia *Geol. Sur.* vol. V(A), pp. 390-453, 488-491.

White, David, 1915. *Notes on the fossil floras of the Pennsylvanian in Missouri.* Missouri Bureau of Geology and Mines, 2nd ser. vol. 13, p. 256-262.

White, David, 1936. Fossil flora of the Wedington Sandstone Member of the Fayetteville Shale. *U.S. Geol. Sur. Prof. Pap. 186-B*, pp. 13-41.

White, David. 1937. Fossil plants from the Stanley Shale and Jackford Sandstone in southeastern Oklahoma and western Arkansas. *U.S. Geol. Sur. Prof. Pap. 186-C*, pp. 43-67.

White, David, 1943. (Posthumously) Lower Pennsylvanian species of *Mariopteris*, *Eremopteris*, *Diplothmema* and *Aneimites* from the Appalachian region. *U.S. Geol. Sur. Prof. Pap. 197-C*, 140 p.

Wilson, L. R. 1960. Development of paleobotany in Oklahoma. *Oklahoma Geol. Notes* 20: 217-223.

Wilson, L. R. 1963. A geological history of Oklahoma's vegetation. *Shale Shaker* 13: 4-20.

Wilson, L. R. 1972. Fossil plants of the Seminole Formation (Pennsylvanian) in Tulsa County, Oklahoma. Tulsa's Physical Environment. *Tulsa Geol. Soc. Dig.* 37: 151-161.

Wood, J. M. 1963. The Stanley Cemetery flora (early Pennsylvanian) of Greene County, Indiana. *Bull. Indiana Dep. Conserv. and Geol. Sur.* 29: 1-73.

Wood, J. M. and Canright, J. E. 1954. The present status of paleobotany in Indiana with special reference to the fossils of Pennsylvanian age. *Proc. Indiana Acad. Sci.* 63: 87-91.

Zen, E., and S. H. Mamay. 1968. Middle Pennsylvanian plant fossils. Problematic occurrence in the Bronx. *Science* 161: 157-158.

ADDENDUM

The following paper appeared after this chapter was written: Gillespie, W. H., and H. W. Pfefferkorn. 1979. Distribution of commonly occurring plant megafossils in the proposed Pennsylvanian system stratotype. In *Proposed Pennsylvanian System Stratotype — Virginia and West Virginia*, K. J. England, H. H. Arndt and T.W. Henry (eds.), Falls Church, Virginia. Amercian Geological Institute, pp. 87-96.

4

PERMIAN AND TRIASSIC FLORAL BIOSTRATIGRAPHIC ZONES OF SOUTHERN LAND MASSES

James M. Schopf
Rosemary A. Askin

SUMMARY

Land plants and animals are of great importance in correlation of the predominantly terrestrial Permian and Triassic deposits of southern land masses. Throughout South America, southern Africa, India, Australia, and Antarctica, broad zonal arrangement of plant megafossils is recognized. Plant microfossils, though their botanical affinities are still little understood, are more widely distributed and allow finer resolution of biostratigraphic zones. The tetrapod vertebrate zonation is more clearly based on phyletic reasoning than are the more empirical plant zones, but vertebrate remains are scarce. Fossiliferous marine horizons are infrequent in most Gondwana basins, thus comparison with standard northern marine successions is difficult.

Gross compositional changes in plant megafossil and microfossil successions are useful for broad correlations. These include establishment of the glossopterids during the waning stages of the mostly Late Carboniferous glaciation; development of the *Glossopteris* flora through the Permian; its decline and replacement in the latest Permian-earliest Triassic by floras dominated by *Dicroidium*, and related corystosperms, with lycopsids important in the Early Triassic; and the decline of the *Dicroidium* flora at the end of the Triassic. The most precise correlations are based on range comparisons, though much basic research remains to be done to determine distribution and ranges of index species throughout Gondwana continents.

INTRODUCTION

The Permian and Triassic geologic record of southern land masses is notable for the remarkable continuity of fossil floras and faunas, and of some lithologic units, over the vast Gondwanaland area. There is, therefore, great potential for precise correlation between the southern continents. The southern Permian and Triassic are composed predominantly of continental sediments, and biostratigraphic zones based on plant megafossils and microfossils are therefore emphasized in this review. Important palynologic studies have recently been published that provide much improved stratigraphic control for many of the southern regions.Discussion is related to zones established for vertebrate fossils, which are less frequently preserved but, like the plants, closely reflect the terrestrial environment. Marine invertebrate fossils, where present, are important for age control, and are mentioned in this context.

The intention of this paper is to provide a brief summary, in particular for the northern hemisphere worker, of the Permian and Triassic floral biostratigraphic zones of southern, or Gondwana, land masses. Bearing this in mind, we present here an abridged review of plant megafossil and plant microfossil zonation in South America, southern Africa and Madagascar, India, Australia, and Antarctica.

In Gondwanaland, biostratigraphic zones based on tetrapod fossils seem most secure, although these fossils are not abundant or are lacking in many areas. Numerous zonation schemes based on plant megafossils and plant microfossils have been suggested but botanical evidence for their relationships is inadequate. Correlation with standard northern marine biostratigraphic zones is tenuous because marine deposits of Permian and Triassic age are poorly represented in southern regions.

Perhaps the greatest need in the correlation of different biostratigraphic schemes is for a better botanical understanding of the plant fossils. A greater emphasis on reproductive structures such as sporangia would aid in revealing the true alliances of palynomorphs, and also indicate which morphologic features of palynomorphs are intrinsic to each species and which are incidental intraspecific variations. Present difficulties and inadequacies in microfossil classification are exemplified by the taxonomic assignment of southern microfossil species to plant groups of the Euramerican area that are not represented in the megafossil record in Gondwanaland.

A vast wealth of data now exists on Permian and Triassic biostratigraphy. Various attempts at collating this information are now underway. The most notable of these are by Anderson and Anderson (1970), and J. M. Anderson (1973), who have compiled a comprehensive series of thirty-five charts with explanatory notes. These charts are intended as a preliminary review of the available material. They include maps showing the known global distribution of Permian and Triassic strata, stratigraphic columns, occurrences and ranges of plant megafossils, plant microfossils, vertebrate and invertebrate fossils, radiometric age data, and correlation charts for the strata.

GENERAL AGE ASSIGNMENTS

Nonmarine deposits predominate in Gondwanaland; and the land animals and vegetation are, therefore, of greatest importance for paleontologic zonation in this region. General global standards of correlation and zonation are based on marine biotas, however; and marine fossil horizons are infrequent in Gondwanaland. This is not the only source of difficulty. The most important and widespread marine zone in the Gondwana areas has an endemic fauna (*Eurydesma* assemblage) that is difficult to compare with faunas in the north. The *Eurydesma-Conularia* marine deposits are recognized in Australia, India, southern Africa, and South America. At the Mar del Plata Gondwana Symposium in 1967, a group of Australian paleontologists proposed that this zone be taken as the basal one in Permian, admitting that "this may be too high relative to the European and Russian type sections" (Banks, and fifteen other authors, 1969, p. 467). Although this marine *Eurydesma* zone has been accepted by some workers as Asselian (basal Permian), it occurs above the widespread upper Paleozoic glacial deposits, which had been regarded by others as basal or Early Permian. Another commonly accepted indicator of basal Permian age is the first appearance of *Gangamopteris* and *Glossopteris*. The stratigraphic position of the Carboniferous-Permian boundary in Gondwanaland is still not agreed upon. Whether one accepts the *Eurydesma* zone, first glossopterids, or the base or top of the glacials as the base of the Permian, it must be remembered that th *Eurydesma* fauna and *Glossopteris* flora are provincial and that correlative glacials are lacking in the north.

Age of Gondwana Glaciation

Du Toit (1927b, 1930a) believed the Gondwana glaciation commenced in the Lower Carboniferous, citing as evidence the *Cardiopteris-Rhacopteris* flora occurring in shales overlying the basal tillite in San Juan, Argentina. This conclusion was based on the now-incorrect assumption that the southern *"Rhacopteris"* (see Australia, below) indicates an Early Carboniferous age. Marine invertebrates between tillite beds at another San Juan locality were considered Late Carboniferous (Reed, 1927). More recently Closs (1969) reported Late Carboniferous goniatites in a marine intercalation between tillites in northeastern Uruguay.

Ages assigned to formations underlying the glacial deposits in Western Australia vary from latest Viséan to early Namurian as interpreted from brachiopods (Roberts, 1971), while conodonts and ostracodes indicate an early Namurian age (Jones, Campbell, and Roberts, 1973). In eastern Australia, beds underlying the glacials contain brachiopod faunas considered to be early Namurian to possibly as young as Westphalian (Roberts, Hunt, and Thompson, 1976).

The presently accepted age for beds overlying the glacial formations in Western Australia is Sakmarian. In their review of Australian Permian ammonoids, Glenister and Furnish (1961) assigned a Sakmarian age to the ammonoid faunas from these beds, namely the Nura Nura Member of the Canning Basin; the Callytharra Formation, Carnarvon Basin; and the Holmwood Shale, Perth Basin.

Palynological evidence from the Australian glacial beds suggests they may be contemporaneous with the northern Missourian-Virgilian, according to Kemp et al. (1977). Low-diversity palynomorph assemblages, characterized by large monosaccate pollen and cavate spores, were compared to assemblages with similar monosaccate pollen in Late Pennsylvanian (Desmoinesian-Virgilian) strata in North America and strata of similar age in Russia. Palynomorph assemblages from higher in the Australian glacial sequence include the first rare taeniate bisaccate pollen. Greater diversity in this type of pollen occurs in Virgilian and Wolfcampian deposits of North America. Of course, such long-ranging comparisons may be meaningless, since vastly different climatic conditions existed between North America and Australia, and few if any genera of plants identified as megafossils were common to the two regions.

The widespread Gondwana glacial deposits have long been held as an important lithostratigraphic datum. King (1958) suggested that glaciation may be of somewhat different ages in different parts of Gondwanaland. Kemp (1975), in a study of palynomorphs from the late Paleozoic glacial deposits, concluded however, that most of the deposits were essentially synchronous within limits of palynologic resolution. With reference to King's hypothesis, Kemp (1975) discussed the significance of plant microfossils in the Paganzo Basin, South America, in sediments regarded as equivalent to possible glacial strata. These resemble preglacial assemblages from Australia. Lavas with a radiometric age of 295 m.y. (Westphalian-Stephanian boundary; Thompson and Mitchell, 1972) overlie the microfossil-bearing beds. Until a more definite relationship of the microfossils with the glacial sequence can be established, the implications of this occurrence remain obscure .There is also no definitive evidence for the age of the basal glacial beds in Australia.

The Permian-Triassic Boundary

The boundary between the Permian and Triassic is another major stratigraphic datum important in Gondwana and global correlations. It too is a subject of controversy. As Tozer (1972) has indicated, there is no known locality where the latest Permian is conformably overlain by the earliest Triassic. One or more of the chronostratigraphic units always seems to be absent at the systemic boundary. In southern areas, inter-continental correlation of the Permian-Triassic boundary sequence has been most successfully accomplished by utilizing the vertebrate faunas that occur in the Ural region, in South Africa, South America, India, Antarctica, and Australia. The *Lystrosaurus* Zone is regarded as early Scythian (earliest Triassic), but is not present everywhere. The overlying *Cynognathus* Zone is accepted as late Scythian.

Cosgriff (1965, 1969) suggested that the *Lystrosaurus* Zone may be Permian. This conclusion was based on an occurrence of vertebrates considered younger than the *Lystrosaurus* Zone in the Blina and Kockatea Shales of Western Australia, and the Otoceratan (basal Triassic ammonoid zone; Spath, 1930) age assigned to an invertebrate fauna from the Kockatea Shale (Dickins and McTavish, 1963). Palynological evidence indicated that the Kockatea and Blina Shales are probably age equivalent (Balme, 1963).

More recently, *Lystrosaurus* has been identified from the Middle Volga region of the U.S.S.R., where it is regarded as Early Triassic (Kalandadze, 1975); new evidence and reassessment of the age of invertebrate faunas in Western Australia indicate that the Blina and Kockatea vertebrates occur in strata younger than Otoceratan; and the relationship of the vertebrates to the African vertebrate zones has been revised. Invertebrate fossils provide evidence that the Kockatea Shale spans an age of Griesbachian at the base of the formation, through Dienerian, to Smithian near the top (McTavish, 1973; McTavish and Dickins, 1974). Dolby and Balme (1976) noted the occurrence of age-equivalent plant microfossils (*Kraeuselisporites saeptatus* Zone) in the Kockatea and Blina Shales. The vertebrate faunas from the Blina Shale are from the upper part of the formation and the single amphibian skull from the Kockatea Shale is from the middle part of the formation in the Beagle Ridge (BMR 10) bore. Thus they occur in strata younger than the previously accepted Otoceratan age, or lower Griesbachian as defined by Tozer (1971). Cosgriff and Zawiskie (in press) now accept the *Lystrosaurus* Zone as Triassic.

The Western Australian amphibian vertebrates were originally thought to be allied with the *Cynognathus* Zone (Cosgriff, 1965), then were considered a transitional *Lystrosaurus-Cynognathus* assemblage (Cosgriff, 1969), and recently Camp and Banks (1978) again suggested affinity with the *Cynognathus* Zone. However, Cosgriff and Zawiskie (pers. comm.) now include these Blina and Kockatea vertebrates in the *Lystrosaurus* Zone. This interpretation is very significant for Gondwana biostratigraphy because of the association with marine invertebrate fossils. Palynomorphs, which could link marine and non-marine biostratigraphy, have not been found in the *Lystrosaurus*-bearing beds of South Africa, India, and Antarctica. If the *Lystrosaurus* Zone affinity of the Blina and Kockatea vertebrates is correctly interpreted, a Dienerian-early Smithian, and possible Griesbachian age is implied for the *Lystrosaurus* Zone.

The exact position of the Permian-Triassic boundary remains a controversial matter, even in the standard northern sequences; but, as Romer (1966a, p. 325) has said, "in reptilean evolution the Paleozoic-Mesozoic boundary is almost without meaning." In the plant record some major groups appeared and others died out during the latest Permian and earliest Triassic. Replacement of the "*Glossopteris* flora" by the "*Dicroidium* flora" occurred relatively rapidly, though sedimentary hiatuses in the Permian-Triassic boundary sequence greatly exaggerate the abruptness of the change. Transitional associations between the strongly contrasting Permian and Triassic floras occur in some basins and are particularly evident in the plant microfossil record. The nature of this change in Australia is described in detail by Balme (1969) and Balme and Helby (1973).

The marine beds with acritarch "swarms" in the Early Triassic succession in southern continents, as noted by Balme (1970) and Balme and Helby (1973), may prove to be one of the more reliable markers for relating Gondwana and Laurasian stratigraphy. They discussed the significance for biostratigraphy and paleogeography of the Scythian marine transgression, with its vast numbers of marine phytoplankton in such widely separated areas as Western Australia (Blina and Kockatea Shales), the Salt Range of Pakistan, Madagascar, western Arctic Canada, and northern U.S.S.R.

In southern areas, the acritarch swarms are associated with low diversity assemblages of certain genera of gymnospermic taeniate bisaccate pollen, and cavate spores of presumed lycopsid origin, believed to represent specialized plant communities of coastal lowlands (Balme, 1969). The rapid world-wide colonization of coastal lowlands by similar floras in the Scythian is striking, and is discussed by Balme (1970), and by Retallack (1975), who dealt with one particular lycopsid group — the Pleuromeiaceae. This plant group is one of the very few to be recognized in both northern and southern areas, and could therefore be very valuable in correlations.

Australian Invertebrate Age Control

Permian and Triassic zonal schemes based on marine invertebrates have been established in some southern regions. Those from Australia are particularly significant. Not only do they provide comprehensive coverage of many basins but several workers have attempted integration with the plant megafossil and microfossil zones, an inherently difficult task when both marine and terrestrial environments are involved. Zonal arrangements of invertebrate fossils has also been stressed in New Zealand studies, mainly by Waterhouse (e.g., 1967); but, as yet, plant microfossil data are scarce. Triassic plant megafossil zones have been recognized in the South Island of New Zealand by Retallack (1977).

Invertebrate faunas and zonal schemes are not discussed here. Relationship of the Australian Carboniferous and Permian plant microfossil zones to the invertebrate zones, especially those of Dickins, Malone, and Jensen (1964), Runnegar (1969), and Runnegar and McClung (1975) is summarized by Kemp et al. (1977) and shown here in Table 4.1. Provincial differences make it difficult to compare southern Permian faunas with those of the standard northern hemisphere succession, so much so that Runnegar and McClung (1975) suggested the use of an entirely separate southern "time scale" or chronostratigraphic framework. Scattered occurrences of Permian ammonoids in Australia (e.g. Glenister and Furnish, 1961) and Early Triassic conodonts in Western Australia (McTavish, 1973) have been most useful for correlation with the northern succession (Tables 4.1 and 4.2). The Australian conodonts are correlated by McTavish to zones proposed by Sweet (1970) for the Salt Range succession.

Radiometric Age Control

Radiometric age data are summarized by Anderson (1973), with additional

TABLE 4.1 *Correlation of Australian plant microfossil zones, marine invertebrate zones, and northern Upper Carboniferous and Permian chronostratigraphy (adapted from Kemp et al. [1977], Figure 12, with permission of Bureau of Mineral Resources, Australia).*

PALYNOLOGICAL UNITS Western Australia Kemp et al. (1977)	INTERNATIONAL AGES from Western Australian sequences	PALYNOLOGICAL UNITS eastern Australia Evans (1969); Kemp et al. (1977)	FAUNAL ZONES eastern Australia Dickins et al. (1964)	FAUNAL ZONES eastern Australia Runnegar (1969)	BRACHIOPOD ZONES Runnegar & McClung (1975)	INTERNATIONAL AGES Runnegar & McClung (1975); Roberts et al. (1976); Runnegar (1969)
P. reticulatus	Changhsingian?	P. reticulatus				
		Weylandites				
Unit VIII	late Guadalupian?					
					M. ovalis	
Unit VII	Guadalupian	Stage 5	Fauna IV	Fauna IV	M. isbelli	
					M. undulosa	
				Fauna III		Baigendzhinian
					M. brevis	
		Upper Stage 4b	Fauna III	Ulladulla fauna		
Unit VI	Leonardian to early Roadian	Upper Stage 4a			M. plana	
			Fauna II			
Unit V	Leonardian	Lower Stage 4		Fauna II	M. ovata	Aktastinian
			Fauna I			
Unit IV	Aktastinian to Leonardian				M. branxtonensis	
		Stage 3b				early Sakmarian - late Asselian
Unit III	Tastubian to Sterlitamakian			Allandale fauna	M. konincki	
		Stage 3a			M. elongata	
Unit II	No faunal evidence, possibly Virgilian on palynological grounds	Stage 2			T. campbelli	
Unit I	No faunal evidence. Probably at least as old as Missourian	Stage 1 or Potonieisporites				
Anabaculites yberti	early Namurian - ?	Anabaculites yberti			L. levis	Namurian-Westphalian

TABLE 4.2 *Correlation of Western Australian plant microfossil zones (Dolby and Balme, 1976), Salt Range conodont zones (Sweet, 1970), and Lower Triassic marine stages (Tozer, 1971). Correlations are based on McTavish (1973) and Dolby and Balme (1976). Conodont zones marked with an asterisk (*) are not reported from Western Australia but are included for completeness.*

PLANT MICROFOSSIL ZONES Western Australia Dolby & Balme (1976)	TRIASSIC STAGES	LOWER TRIASSIC CONODONT ZONES Salt Range, Pakistan Sweet (1970)
↑ Staurosaccites quadrifidus	↑ ANISIAN	
Tigrisporites playfordi	SPATHIAN	Neospathodus timorensis
		Neogondolella jubata
	SMITHIAN	Neospathodus waageni
		Neospathodus pakistanensis
Kraeuselisporites saeptatus	DIENERIAN	Neospathodus cristagalli*
		Neospathodus dieneri
		Neospathodus kummeli*
	GRIESBACHIAN	Neogondolella carinata*
		Anchignathodus typicalis* ↓

dates in Retallack (1977) and Webb and McNaughton (1978), and are not discussed here. Application of radiometric dates to the southern successions would greatly aid our interpretations of the paths and rates of floral migration and evolution, and thus would aid in distinguishing between real and apparent synchroneity of the floras.

FLORAL ZONES

The zonal concept most widely used in subdivision of Gondwana plant megafossil and microfossil successions is the assemblage-zone. This biostratigraphic zone is characterized by a natural association or assemblage of fossils. As defined by Hedberg (1976), the assemblage-zone is somewhat subjective - it does not have precisely defined limits. Murphy (1977) commented on the ambiguity and imprecison of assemblage-zones, noting that "characterization" is often substituted for "definition" of a unit. For greater precision and unambiguous definition, the boundaries of assemblage-zones should be delineated by the first appearances of diagnostic species, or one species. In practice, this is the case in definition of the lower limits of most of the Gondwana floral assemblage-zones.

The most obvious and most frequently cited assemblage-zones are the Permian "*Glossopteris* flora" and the Triassic "*Dicroidium* flora." Finer subdivision of these two broad plant megafossil zones is becoming possible as detailed studies clarify the prolific and confused nomenclature that characterizes much of Gondwana paleobotany, particularly within the genus *Glossopteris*. In a recent comprehensive study of Triassic floras of eastern Australasia, Retallack (1977) described three zones that included several plant associations within the "*Dicroidium* flora."

In contrast to the sometimes vague and undifferentiated plant megafossil zones, several palynostratigraphic zones can often be recognized in the same strata. Plant microfossils are of small size, produced in great quantities and widely dispersed, and composed of extremely resistant material. Thus they provide a better representation of the original flora than do leaf fossils, which partly explains the greater precision of zonal subdivision possible with the microfossils. Pollen and spores also exhibit greater morphologic diversity than does the more conservative leaf. Another factor contributing to precision in zonal subdivision may be that the microfossils have been more extensively studied with a strong emphasis on stratigraphic correlation for fossil fuel exploration.

A brief review of the principal literature on Gondwana Permian and Triassic plant megafossils and microfossils is given in the following sections. Carboniferous floras are mentioned where relevant to the periglacial vegetation. Depositional basins and geographic locations referred to in the text are shown in Figure 4.1.

Plant megafossil and microfossil zonal successions and their relations are shown in Tables 4.3, 4.4, and 4.5. Lack of space prevents inclusion of the lithostratigraphy for each basin, except for the Salt Range, and for the Karroo Basin and peninsular India where it has provided a traditional classification to which paleontologic studies are related. Table 4.6 gives the ranges of important groups and genera. Tables 4.3 to 4.6 are presented on the same vertical scale so that they may be easily related to each other. Only a broad indication of the age is given in Table 4.6, as inclusion of northern stages would imply a precision in age control throughout Gondwanaland that does not yet exist.

Reliable correlation of marine invertebrate faunas with the standard northern marine biostratigraphy is available throughout much of the western Australian Permian and Lower Triassic, and to a lesser extent in the eastern Australian Permian. Relationships between the faunal zones, northern chronostratigraphic units, and the palynologic successions are shown separately in Tables 4.1 and 4.2 Less certain comparison with northern stratigraphy can also be inferred from the ranges of "northern" plant microfossil species described from some southern basins; these age inferences are discussed by workers cited in Tables 4.3 to 4.5 and are not covered here.

These tables are generalized and the zonations are undoubtedly more complex than indicated. A boundary between biostratigraphic zones, though shown as a single straight line, may in reality represent a hiatus, or time-transgressive contact. It must also be noted that the boundaries of lithostratigraphic and biostratigraphic units do not necessarily coincide, although this is often the most practical way of presenting them.

Correlation of Gondwana floral zones relies on somewhat circular reasoning. Boundaries marked by first appearances of plant species over the vast Gondwana area may be diachronous in detail, but are accepted as essentially synchronous within the limits of the large-scale relative-time framework shown here.

Plant Megafossil Zones

The same broad changes in the plant megafossil succession can be observed throughout the Gondwana regions. Through the Permian they reflect climatic amelioration following glacial recession. *Gangamopteris-Glossopteris* vegetation characterizes the Gondwanaland Permian, with *Gangamopteris* dominating plant associations above the glacial beds. *Glossopteris* gradually

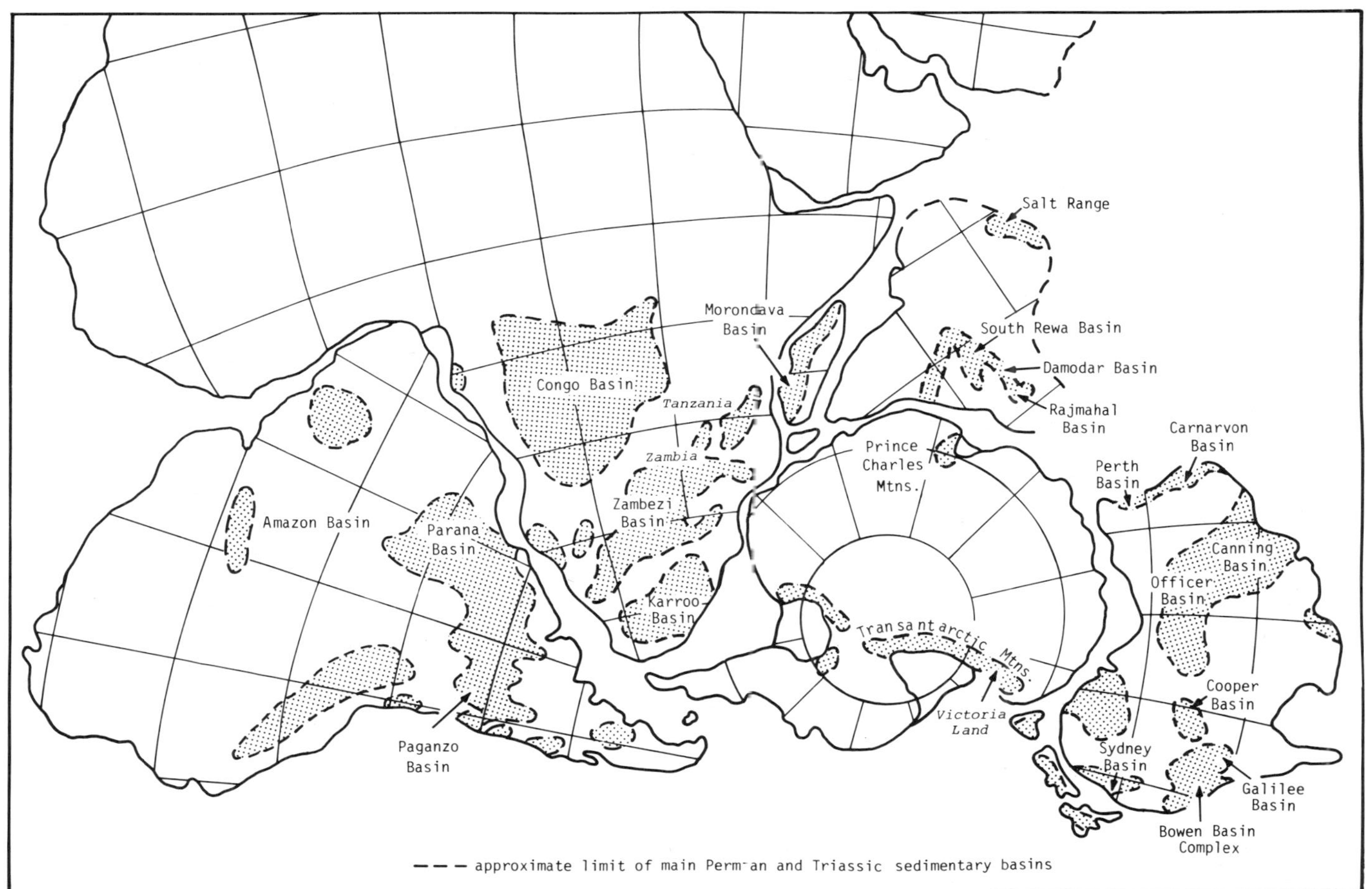

FIGURE 4.1 *Permian and Triassic sedimentary basins of Gondwanaland. The reconstruction is according to Smith and Briden (1977), and presents the present geography of New Zealand and West Antarctica rather than their Paleozoic-Mesozoic configurations.*

becomes more common higher in the stratigraphic succession until plant assemblages in Upper Permian strata consist predominantly of *Glossopteris* leaves.

A hiatus, of variable magnitude in different basins, occurs in all the southern successions at the level of the Permian-Triassic transition. The change from *Glossopteris* to *Dicroidium* flora thus appears very abrupt in many cases. In a few locations assemblages that represent transitional floras are found. These include occasional *Glossopteris* leaves, conifers, and pteridosperms such as *Lepidopteris* and *"Thinnfeldia" callipteroides*, which Retallack (1977) believes gave rise to *Dicroidium*. The corystosperm *Dicroidium* appeared in the earliest Triassic and, after the early Scythian, dominated Triassic assemblages throughout Gondwanaland, until it disappeared in the latest Triassic.

South America

Permian and Triassic plant megafossils have been widely studied in South America, although zonation of the floras has not been stressed. Menendez (1968, 1969) compiled a thorough bibliography of South American paleobotany and palynology. He also (1969) provided a general summary of over 150 occurrences of fossil floras in South America. Detailed references to older works of Frenguelli and others need not be cited here.

Archangelsky has contributed greatly to Argentine paleobotany for many years. He provided a general summary in 1971, and with Arrondo (1969, 1975), has been concerned with paleogeographic relationships. Broad subdivisions of the late Paleozoic Argentine floras were outlined by Archangelsky (1971) (Table 4.3, column 1), with the plant associations named after the rock units in which they occur. The glossopterids appeared in the Lubeckense, this zone being characterized by a mixed assemblage of *Gangamopteris*, *Glossopteris*, and "northern" elements. *Glossopteris* predominates in the Bonetense. Occurrences of plant megafossils in Brazil are summarized by Dolianiti (1952) and, more recently, by Rigby (1969, 1972). An outline of the South American succession is presented by Rocha-Campos (1973).

The Triassic of Argentina was subdivided by Stipanicic (1957, 1969) and Stipanicic and Bonetti (1969). Plant fossils occur only in their upper subdivision, "complex" IV (Ladinian-Norian), and represent the *Dicroidium* zone. They occur with vertebrate fossils, though correlation of the vertebrate zones is not agreed upon (Sill, 1969). There seems to be no record of the lowest Triassic *Lystrosaurus* Zone (Romer, 1966a, 1975).

Southern Africa

For many years the problems of correlation in southern Africa have been minimized by reference to the lithologic classification of the Karroo Basin of South Africa (Table 4.4, column 1). However, Van Eeden (1973) has suggested that important unconformities occur in the Karroo sequence,and Stapleton (1977) has emphasized a hiatus occurring below the Dwyka tillite. These and other works indicate that the succession of Karroo deposits is more complex than has often been presented, and now new formational names are often used instead of the traditional classification (e.g., Dingle, 1978). Comparison of Karroo strata in different areas is summarized by Anderson and Anderson (1970) and Anderson (1973).

The first reports of plant megafossils having correlational significance were by Seward (1897, 1903) and later by du Toit (1927a, 1930b, 1932a,b). Du Toit (1927a) recognized seven plant-bearing zones (A to G; Table 4.4, column 3) in the upper Karroo beds (Beaufort and Stromberg Series). Plumstead (1966, 1967) proposed a simple scheme for regional correlation, dividing the Karroo beds (Dwyka to Stormberg) into very broad indefinite floral zones (Table 4.4,

TABLE 4.3 *Correlation chart for plant megafossil and microfossil zones of South America and the Transantarctic Mountains, Antarctica; and the lithostratigraphy of the Salt Range, Parkistan, and peninsular India; a, Archangelsky (1971); b, Daemon and Quadros (1970); c, Kyle (1977).*

SOUTH AMERICA: Argentina — Plant megafossil zones a	SOUTH AMERICA: Paraná Basin, Brazil — Plant microfossil zones b	PAKISTAN: Salt Range — Lithostratigraphic units	INDIA: Peninsular basins — Lithostratigraphic units	ANTARCTICA: Transantarctic Mountains — Plant microfossil zones c
			Mahadeva	Alisporites zone: D
			Mahadeva	Alisporites zone: C
		Tredian Formation: Khatkiara Member	Mahadeva	Alisporites zone: B
		Tredian Formation: Landa Member	Mahadeva	Alisporites zone: B
		Mianwali Formation: Narmia Member	Mahadeva	Alisporites zone: A
		Mianwali Formation: Mittiwali Member	Panchet	
		Mianwali Formation: Kathwai Member	Panchet	
Bonetense	L	Chhidru Formation	Raniganj	
Bonetense	L	Wargal Limestone	Raniganj	
Bonetense	K	Wargal Limestone	Barren Measures	
Lubeckense: B	J	Amb Formation	Barakar	Protohaploxypinus zone
Lubeckense: B	I	Amb Formation	Barakar	Protohaploxypinus zone
Lubeckense: A	H		Karharbari	
Lubeckense: A	G		Talchir	Parasaccites zone
Trampeaderense				
Tupense				
1	2	3	4	5

TABLE 4.4 *Correlation chart for plant megafossil and microfossil zones, vertebrate zones, and lithostratigraphy of southern Africa and Madagascar; a, Kitching (1972); b, du Toit (1927a); c, Plumstead (1967); d, Hart (1969); e. Anderson (1977); f, Falcon (1975b); g, Utting (1978); h, Hart (1965); i, Manum and Tien (1973); j, Hoeg and Bose (1960); k. Bose and Kar (1966); l, Kar and Bose (1967); m, Goubin (1965).*

Karroo Basin, South Africa							
Lithostratigraphic Units		Vertebrate zones	Plant megafossil zones		Plant microfossil zones		
		a	b	c	d	e	
Stormberg	Cave Sandstone	(hatched)	G	Barren Zone VI	(hatched)	(hatched)	
Stormberg	Red Beds	(hatched)	F	Barren Zone VI	Goubin III	(hatched)	
Stormberg	Molteno	(hatched)	E	Dicroidium Zone V	Goubin III	(hatched)	
Stormberg	Molteno	(hatched)	D	Dicroidium Zone V	Goubin III	(hatched)	
Stormberg	Molteno	(hatched)	C	Dicroidium Zone V	Goubin II	(hatched)	
Beaufort	Upper	Cynognathus	B	Dicroidium Zone V	Goubin II	(hatched)	
Beaufort	Middle	Lystrosaurus	A	(hatched)	Goubin II	(hatched)	
Beaufort	Lower	Daptocephalus	A	Glossopteris Zone IV	Striatiti	Zone 7	
Beaufort	Lower	Cistecephalus	A	Glossopteris Zone IV	Striatiti	Zone 6	
Beaufort	Lower	Tapinocephalus	A	Glossopteris Zone IV	Striatiti	Zone 5	
Ecca	Upper	(hatched)	Glossopteris Zone IV		Zonati	Zone 4	e, d, c, b, a
Ecca	Middle	(hatched)	Mixed Glossopteris -Gangamopteris Zone III		Cingulati	Zone 3	d, c, b, a
Ecca	Lower	(hatched)	Mixed Glossopteris -Gangamopteris Zone III		Cingulati	Zone 2	d, c, b, a
Dwyka		(hatched)	Proto-Glossopteris Zone II		Cavati	Zone 1	
1 (hatched)		2 (hatched)	3 (hatched)		4 (hatched)	5 (hatched)	

SOUTHERN AFRICA — Plant microfossil zones: Zambezi Basin, Rhodesia (f)	Zambia (g)	Tanzania (h)	Tanzania (i)	Zaire (j, k, l)	MADAGASCAR Plant microfossil zones (m)
					Zone IV A
					Zone III B
					Zone III A
					Zone II B
					Zone II A
					Zone I C
					Zone I B
					Zone I A
Zone IV H					
Zone IV G					
Zone III F	Vittatina africana-Gondispora vrystaatensis Zone (AV)		Vesicaspora Zone	Assise à couches de houille	
		K2e$_2$		Assise des schistes noirs de la Lukuga	
Zone II E	Apiculatisporis levis-Vesicaspora potoniei Zone (LP)			Assise des schistes noirs de la Walikale	
Zone II D		K2e$_1$	Cordaitina Zone	Assises glaciaires et périglaciaires	
Zone I C	Plicatipollenites indicus-Cannanoropollis obscurus Zone (IO)				
Zone I B		K2c			
Zone I A					
6	7	8		9	10

TABLE 4.5 *Correlation chart for plant megafossil and microfossil zones of Australia; a, Segroves (1972); b, Dolby and Balme (1976); c, Balme (1964); d, Kemp et al. (1977); e, Evans (1969) f, Evans (1966); g, de Jersey (1975); h, Helby (1973); i, j, Retallack (1977, in press); k, Rigby (1973), l, Gould (1975).*

A U S T R A L I A Plant microfossil zones				
Perth Basin a b*	Carnarvon Basin b	Canning Basin c	Canning Basin d	eastern Australia d e
		Classopollis Microflora		
	Minutosaccus crenulatus Zone	Pteruchipollenites Microflora		
	Samaropollenites speciosus Zone			
	Staurosaccites quadrifidus Zone			
	Tigrisporites playfordi Zone			
Kraeuselisporites saeptatus Zone *	Kraeuselisporites saeptatus Zone	Taeniaesporites Microflora		
			Protohaploxypinus reticulatus Zone	Protohaploxypinus reticulatus Zone
				Weylandites Zone
		Striatites Microflora: Dulhuntyispora Assemblage	Unit VIII	Stage 5
Dulhuntyispora Assemblage			Unit VII	
Haplocystia Assemblage		Vittatina Assemblage		Upper Stage 4 b
			Unit VI	Upper Stage 4 a
Acanthotriletes Assemblage			Unit V	Lower Stage 4
			Unit IV	
Quadrisporites Assemblage		Nuskoisporites Assemblage	Unit III	Stage 3 b / a
Microbaculispora Assemblage			Unit II	Stage 2
			Unit I	Stage 1 (=Potonieisporites)
1	2	3	4 Anabaculites yberti Assemblage	5 Anabaculites yberti Assemblage

AUSTRALIA				
Plant microfossil zones				Plant megafossil zones
eastern Australia	southeastern Queensland	Sydney Basin		eastern Australia
f	g	h	i, j	k, l
Jl	Polycingulatisporites crenulatus Zone			Early Jurassic flora
Tr3a-d				Dicroidium flora
	Craterisporites rotundus Zone			
			Dicroidium odontopteroides Zone	
	Duplexisporites problematicus Microflora	Aratrisporites parvispinosus Zone		
		Aratrisporites tenuispinous Zone	Dicroidium zuberi Zone	
Tr2a-b		Protohaploxypinus samoilovichii	"Thinnfeldia" callipteroides Zone	"Thinnfeldia" callipteroides flora
Trlb		Lunatisporites pellucidus		
Trla		Protohaploxypinus reticulatus Zone		
P4		Dulhuntyispora Zone	Glossopteris and Gangamopteris flora	Glossopteris flora
			Botrychiopsis flora	Botrychiopsis (=Gondwanidium) flora
			Sphenopteridium flora	Pseudorhacopteris flora
6	7	8	9	10

column 3). A sequence from "Protoglossopterids," through a mixed *Gangamopteris-Glossopteris* zone to a *Glossopteris* zone was recognized. Four of du Toit's subdivisions (B to E) are grouped together in Plumstead's broad *Dicroidium* zone. Distribution of Triassic species in the Karroo Basin, mainly in the Molteno Formation, was given by Anderson and Anderson (1970) and Anderson (1974).

India

Permian plant megafossils of India have been extensively studied. Important early reviews giving a thorough appraisal of Indian paleobotany and related geological problems were presented by Sahni (1921, 1926, 1939). Later reviews and listings of plant megafossils include those by Surange (1966, 1975) and Lele (1976).

As in southern Africa, the Indian studies are referred to traditional lithostratigraphic units, or "stages" as they are generally known (Talchir, Karharbari, Barakar, Barren Measures, and Raniganj; Table 4.3, column 4). The floral zones do not depart from these lithostratigraphic units. *Gangamopteris* is most abundantly represented in the Talchir glacial stage and the Karharbari stage. Glossopterid diversification continued through the Permian, and *Glossopteris* predominates in the Raniganj stage.

The Triassic floras are less well known and their relationships are controversial. The Panchet, Nidpur, and Parsora floras all include *Glossopteris* and *Dicroidium*. *Lystrosaurus* fauna remains occur in the upper Panchet in the Raniganj coalfield, but are not associated with plant remains, including palynomorphs. In their review of Indian Triassic plant megafossils and microfossils, Roy Chowdhury et al. (1975) describe four successive zones corresponding to the lower Panchet Formation, upper Panchet Formation, Nidpur Beds, and Parsora Formation, although they suggest somewhat younger ages for the Nidpur and Parsora floras than are often accepted. Problems still remain in determining precise stratigraphic relationships of the various fossil localities.

Australia

Conventional paleontologic studies of upper Paleozoic plant megafossils were made by Walkom (1922, 1928, 1938, 1941, 1944). Zonation was not stressed and age determinations were rendered in a less formalized fashion. The partly glacial Kuttung sequence with *Rhacopteris* was then regarded as Lower Carboniferous (Walkom, 1944, 1949) on the basis of the northern *Rhacopteris* distribution. However, Westphalian invertebrates have since been reported in these beds (Campbell, 1962; Campbell and McKellar, 1969). In an apparent attempt to relieve misgivings about placing *Rhacopteris* in the Upper Carboniferous, contrary to its usual designation, Lower Carboniferous, Rigby (1973) has proposed a generic separation of the Australian fossils. He considers that fine lineations between the veins in the pinnae are a basis for discrimination and assigns the fossils previously identified with *Rhacopteris ovata* (McCoy) Walkom to his genus *Pseudorhacopteris*. This is in keeping with a postulated period of independent phylesis for the Australian plants following the more ancient introduction of a northern parent stock. It would also seem to account for the significant difference in age.

Rigby (1973) recognized a *Pseudorhacopteris* zone and a younger *Gondwanidium* zone in beds that Campbell regarded as Upper Carboniferous. Rigby's taxonomic assignments have not all been upheld by Morris (1975), Gould (1975), and Retallack (in press), but regardless of the lack of taxonomic unanimity these zones (Table 4.5, columns 9 and 10) have biostratigraphic value in eastern Australia. Retallack (in press), recognizes three distinctive floras in

TABLE 4.6 *Range chart for important plant groups and genera in the southern land masses during Late Carboniferous to Early Jurassic time.*

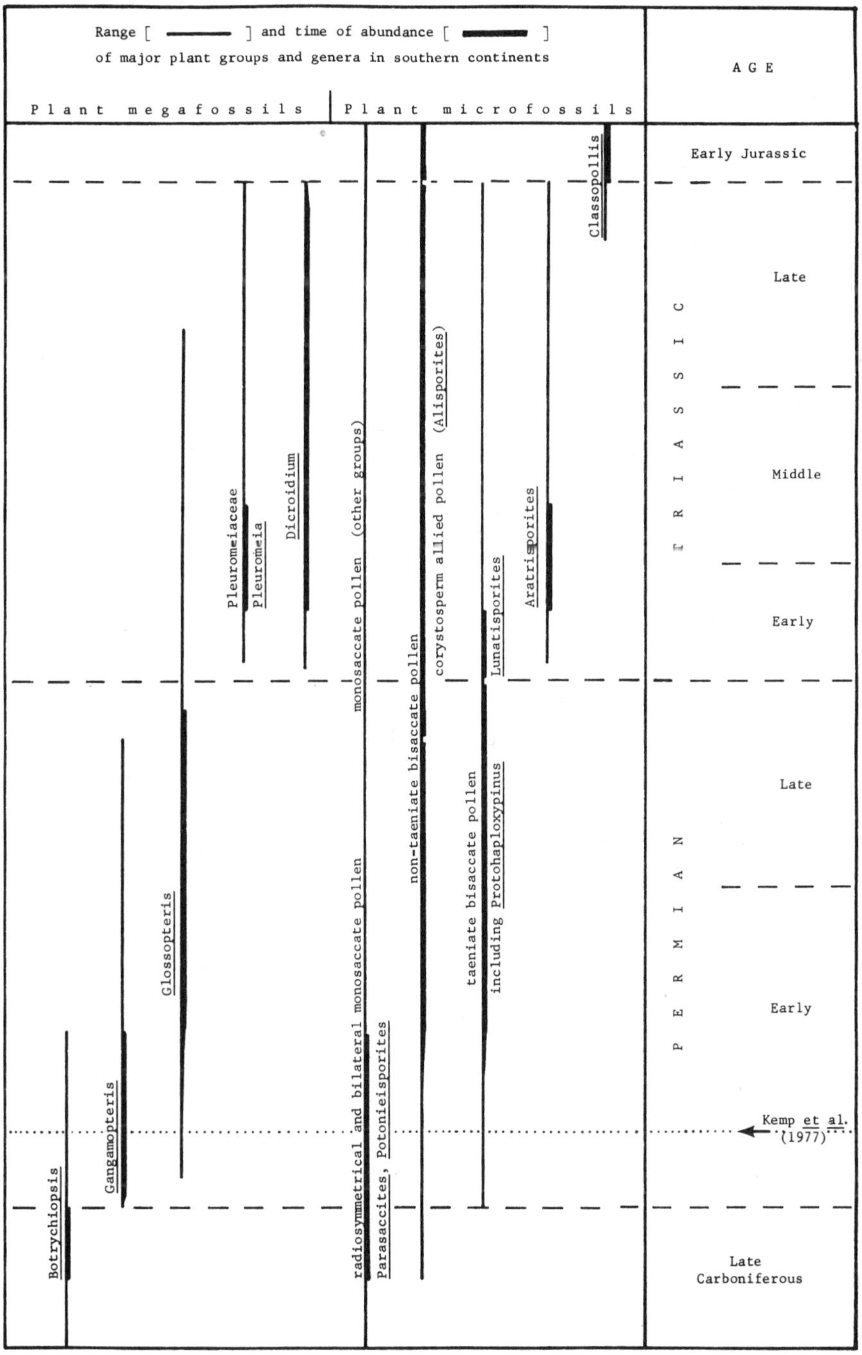

Range [———] and time of abundance [▬▬▬]
of major plant groups and genera in southern continents

Plant megafossils	Plant microfossils	AGE
Botrychiopsis Gangamopteris Glossopteris Pleuromeiaceae Pleuromeia Dicroidium	radiosymmetrical and bilateral monosaccate pollen Parasaccites, Potonieisporites monosaccate pollen (other groups) non-taeniate bisaccate pollen corystosperm allied pollen (Alisporites) taeniate bisaccate pollen including Protohaploxypinus Lunatisporites Aratrisporites Classopollis	Early Jurassic TRIASSIC: Late, Middle, Early PERMIAN: Late, Early Kemp et al. (1977) Late Carboniferous

the Upper Carboniferous succession, and notes the relationship of the megafossil zones with Helby's (1969; in Kemp et al., 1977) palynological zones (Table 4.5, column 5) and with the onset and amelioration of glacial climate. The impoverished *Botrychiopsis* flora (=*Potonieisporites* palynological Assemblage), contrasts sharply with the underlying luxuriant *Sphenopteridium* flora (=*Anabaculites yberti* Assemblage), and is thought to represent a periglacial tundra vegetation (Kemp et al., 1977; Retallack, in press).

Rigby (1966) also described Permian megafossils from the Perth and Collie Basins in Western Australia. Studies of plant megafossils have continued in Australia ever since the initial controversy about the age of the Indian and Australian *Glossopteris* floras, but lack of distinctive floral zones within the Permian has limited their stratigraphic use. In his review of Australian pre-Tertiary floras, Gould (1975) included the Permian floras above the *Botrychiopsis* flora as an undifferentiated *Glossopteris* zone. Retallack (in press) describes *Gangamopteris* floras in the post-glacial strata, with a transition to *Glossopteris* - dominated floras that is so gradual that biostratigraphic subdivision is difficult.

Greater subdivision of the Triassic is possible. Townrow (1956, 1960, 1966) has described a series of species of *Lepidopteris* that probably have great stratigraphic value. He notes the occurrence of *Lepidopteris martinsii* in the Upper Permian of Germany, and *L. stormbergensis* in the Lower and Middle Triassic of southern Africa and Madagascar. *L. madagascariensis* is characteristic of the Middle and Upper Triassic. In the northern hemisphere *L. stuttgardiensis* is present in beds equivalent to the lower Keuper (Karnian), and *L. ottonis* is characteristic of the Rhaetian. There is some overlap, as might reasonably be expected, but the series is noteworthy as it suggests that these southern plants derived their ancestry from the north. *Lepidopteris* may be a derivative of the *Alethopteris-Callipteris* lineage of the northern Carboniferous and Permian that moved southward during the Early Triassic (Schopf, 1973).

A two-fold division of the Triassic outlined by Gould (1975) includes the *"Thinnfeldia" callipteroides* flora, which spans the Permian-Triassic boundary, and the *Dicroidium* flora (Table 4.5, column 10). Gould noted progressive changes in the *Dicroidium* flora of eastern Australia and the restricted ranges of some species.

In a study that encompassed plant megafossil data, environmental interpretation including consideration of soil type, and marine invertebrate and radiometric dating control, Retallack (1977) reconstructed the Triassic vegetation of eastern Australia and New Zealand. He described the nature and distribution of twelve fossil plant associations, and proposed a series of four oppel-zones (Hedberg, 1976) defined by evolutionary lineages of *Dicroidium* and related pteridosperms along with the distribution of other species. The three zones of the Sydney Basin are shown in table 4.5 (column 9). The youngest zone, the *Yabeiella* Oppel-zone, is recognized in various Queensland and South American localities. Retallack outlined the distribution of these four zones throughout Gondwanaland.

Probably the most useful Triassic plant fossil for biostratigraphy is a lycopod. *Pleuromeia* and its allies are characteristic of the German Bunter, occur extensively in the Soviet Union, and also show an extensive distribution in Australia. The distribution of these plants has been recently given in some detail by Retallack (1975). He suggests that *Pleuromeia* was a salt-marsh or mangrove plant that inhabited coastal areas and estuaries. Its vegetative parts and particularly its heterospores, assignable to several distinguishable taxa, are highly distinctive and widely distributed. Surely the spores and vegetative structures allied with *Pleuromeia* will provide a basis for reliable correlations of Triassic deposits. Its greatest value may be in reconstructing Triassic paleogeography.

Antarctica

Permian and Triassic megafossils were first reported from Antarctica in 1914 by Seward. Since then, Plumstead (1962, 1975) and Schopf (1962, 1965, 1968, 1976) have made extensive studies of Antarctic collections. Rigby and Schopf (1969) gave a comprehensive listing of Permian plant fossils.

The *Glossopteris* floral succession is not easily subdivided. A vague zonal distribution, as seen elsewhere in Gondwanaland, is recognizable however. *Gangamopteris* is common immediately above the tillite (Plumstead, 1975; Kyle, 1976), and *Glossopteris* dominates collections higher up the section from the coal measures of the central Transantarctic Mountains. The later Permian assemblages seem to include glossopterid leaf types that have a more open type of venation with larger areoles.

Dicroidium-zone fossils are well represented higher in the section throughout the Transantarctic Mountains. Retallack (1977) includes these in his Middle Triassic *Dicroidium odontopteroides* Oppel-zone, but plant microfossil evidence (Kyle, 1977; Kyle and Fasola, 1978) indicates that the majority of these occurrences (in upper Lashly Formation, southern Victoria Land; lower Falla Formation, Beardmore Glacier area) are actually equivalent to his Upper Triassic *Yabeiella* Oppel-zone.

Plant Microfossil Zones

Throughout the southern land masses the same broad changes can be observed in the plant microfossil succession.

Palynomorph assemblages associated with the glacial deposits are characterized by large monosaccate pollen *(Parasaccites* and *Potonieisporites)*. The younger glacials and immediately post-glacial sediments include the first rare occurrences of taeniate bisaccate pollen corresponding with the appearance of glossopterids. Higher in the Permian succession relative abundances of monosaccate pollen decline and taeniate bisaccate pollen (including *Protohaploxypinus)* increase. Upper Permian strata are characterized by the predominance of taeniate and nontaeniate bisaccate pollen. Various workers (e.g., Pant and Nautiyal, 1960) have extracted *Protohaploxypinus* pollen from glossopterid microsporophylls, and the high proportion of these pollen is analogous to the abundance of *Glossopteris* in the megafossil record.

The plant microfossil record shows transitional assemblages in latest Permian strata and into the Triassic. Basal Triassic marine sediments contain an overwhelming abundance of small spinose and smooth acritarchs. These beds and terrestrial strata of equivalent age are characterized by low diversity assemblages including trilete cavate lycopod spores *(Lundbladispora, Densoisporites,* and *Kraeuselisporites)* and taeniate bisaccate gymnosperm pollen *(Lunatisporites)*. Immediately succeeding assemblages in many areas are characterized by the monolete cavate spore *Aratrisporites* corresponding to the spread of the coastal lycopod group Pleuromeiaceae (Retallack, 1975). Above the earliest Triassic, nontaeniate bisaccate pollen *(Alisporites)* predominate, analogous to the predominance of *Dicroidium* in megafossil associations. *Alisporites* pollen have been extracted from the corystosperm microsporophyll *Pteruchus* associated with *Dicroidium* foliage (Townrow, 1962).

The Triassic-Jurassic boundary is traditionally delineated by the appearance of *Classopollis*, and although *Classopollis* is now known to range down into the Triassic, large numbers of this characteristic form are still used to mark the transition into the Jurassic.

South America

Systematic studies of plant microfossils have been made in various areas of South America, mainly in Argentina and the Paraná basin of southern Brazil and adjacent countries, by Menendez (1969) and several other workers. The only zonations recognized are those by Daemon and his coworkers. In the Upper Carboniferous and Permian successions of the Paraná basin, Daemon (1966) outlined a succession of three palynological zones, including six subzones. Later, Daemon and Quadros (1970) described six zones (G-L) (Table 4.3, column 2) with numerous subdivisions, which they regarded as Upper Pennsylvanian to Upper Permian. Broad subdivisions were recognized in the Silurian to Tertiary succession of the Amazon Basin by Daemon and Contreiras (1971), including four zones in the Upper Carboniferous and Permian. In her discussion of Gondwana glacial palynology, Kemp (1975) recognized assemblages equivalent to eastern Australian zones in the Paraná and Paganzo Basins. These were Stage 3, Stage 2, possibly Stage 1, and a pre-Stage 1 assemblage, which can probably be equated with the *Anabaculites yberti* Zone.

Southern Africa

Plant microfossil zonations have been proposed for the Upper Carboniferous and Permian succession by Hart (1965), Christopher and Hart (1971), and Manum and Tien (1973) in Tanzania; Høeg and Bose (1960), Bose and Kar (1966), and Kar and Bose (1967) in Zaire; Utting (1976, 1978) in Zambia; Falcon (1973, 1975a,b, 1976) in Rhodesia; Hart (1969) and Anderson (1977) in South Africa. In Gabon, which is further northwest than generally accepted "Karroo" outcrops, Jardine (1974) has described three Permian assemblage-zones.

Hart described three zones based on microfossils in the lower coal measures of Tanzania (Hart, 1960, 1965; Christopher and Hart, 1971). He later continued these studies in South Africa (Hart, 1967, 1969, 1971) and, in combination with work by Goubin (1965) in Madagascar, recognized six microfossil zones in the Permian and Triassic of southern Africa (Table 4.4, column 4). This work was a precursor to the later and somewhat more detailed zonations of the Permian sections by Falcon and Anderson.

Recent studies by Falcon are based on the Matabola borehole of the mid-Zambezi Basin of central Rhodesia. An earlier paper (Falcon, 1973) included generic identifications and a general summary of the palynologic composition of each lithostratigraphic unit. Falcon (1975b) defined eight biostratigraphic subdivisions (Table 4.4, column 6), and also (1976) applied numerical analysis to the palynologic successions. The major zones determined by numerical methods appear to conform to the zones she had previously identified by visual methods.

The most extensive study of plant microfossils from South Africa has been recently published by Anderson (1977). In the northern Karroo Basin, he recognized twenty-one biostratigraphic units (seven assemblage-zones and eighteen subzones; Table 4.4, column 5) whose boundaries have been taken to coincide with the lithostratigraphic units or, in the case of the three youngest zones, with vertebrate zones (Table 4.4, column 2). Palynomorphs in the southern Karroo Basin are poorly preserved and here eight assemblage-zones are recognized. The scheme is influenced by defining zone limits by lithology, instead of fossil occurrences, which makes comparison with other palynological zonations more difficult. Anderson provides an excellent photographic coverage illustrating his concept of each taxon, although he has adopted some slightly unorthodox methods in assigning priority to names (Anderson, 1977, p. 20).

Anderson did not find palynomorphs in strata of the *Lystrosaurus* Zone. In reference to the Permian-Triassic boundary problem, it is worth noting here the microfossils from three samples described by Stapleton (1978) from the uppermost Beaufort Group (i.e., younger than *Lystrosaurus* Zone) and lower Molteno Formation. These were regarded as a transitional Permian-Triassic assemblage, and they occur in beds contemporaneous with and younger than the *Cynognathus* Zone of generally accepted late Early Triassic age. Comparisons to "northern" species, as made by Stapleton, must, however, be viewed with caution (Foster, 1978), as discussed later. Furthermore, at least one form described as "known only from the Upper Permian" is recognized throughout the southern Triassic succession.

Madagascar

Following an earlier summary of the palynostratigraphy by Jekhowsky and Goubin (1964), Goubin (1965) outlined a zonal succession from the Morondava Basin of Madagascar. The Upper Permian to Upper Jurassic succession was divided into twelve zones included in four main subdivisions (Table 4.4, column 10) based on frequency variations of many diagnostic species. These species were selected because they were "peculiarly abundant" and the defined zones thus are believed to be quite independent of lithofacies and environment.

India

A large amount of descriptive work on the Indian Permian plant microfossils has been carried out, chiefly by Bharadwaj and his colleagues at the Sahni Institute (Bharadwaj, 1962, 1969, 1971, 1972, 1974; with Tiwari, 1964; with Srivastava, 1969). As with the plant megafossils, most of these microfossils are referred to the major lithostratigraphic or "stage" subdivisions (Table 4.3, col. 4). The palynological succession is similiar to that observed in other Gondwana basins, showing a trend from the monosaccate dominated Talchir assemblages to the bisaccate (taeniate and nontaeniate) dominated Raniganj assemblages.

In addition to the major stage subdivisions, a great variety of palynological zonal schemes have been proposed by different authors, usually with different terminology for each coalfield or sampling area without much attempt at correlation between the various zonal schemes. Most of the minor subdivisions are based on generic relative abundances and probably reflect local facies changes. Tiwari (1974) reviewed and synthesized the many zonal schemes recognized throughout India in the Barakar stage. Bharadwaj (1972) briefly summarized the Permian and Triassic palynological succession of India.

In comparison with the Permian, relatively little work has been carried out on Indian Triassic palynology. In his review of the Indian Mesozoic succession, Bharadwaj (1969) recognized two major subdivisions in the Triassic: the Early Triassic *Decisporis* Mioflora, and the *Alisporites* complex Mioflora, which includes three assemblages. A zonation of microfossils was also described by Maheshwari (1975). Maheshwari and Banerji (1975) described a transitional Permian-Triassic assemblage from the Maitur Formation (lower Panchet), below the occurrence of *Lystrosaurus* and *Dicroidium*.

Salt Range, Pakistan

An extensive study of Permian to Middle Triassic pollen and spores from the Salt and Surghar Ranges of Pakistan (Table 4.3, column 3) was made by Balme (1970). Sarjeant (1970, 1973) described acritarchs from the same area.

Considering that the samples available to Balme represented the section very unequally, Balme's study is a model in many respects. The Salt Range microfossils are critically located both geographically and stratigraphically. Assemblage-zones are most appropriate for use within the same basin or province of sedimentation, and Balme refrained from comparison with assemblage-zones that were well established for some Australian basins. Range comparisons were made, though his conclusions were cautious. Balme pointed out the necessity of using well-defined, unusually distinctive species for such comparisons.

Australia

The Australian Permian and Triassic palynological successions are the most studied and best understood in the southern continents. Balme (1964) established the first biostratigraphic zonation (Table 4.5, column 3) for the Late Devonian to Early Cretaceous plant microfossil succession in Australia. His zonation was based mainly on data from Western Australia and, where possible, was extended to eastern Australian basins. In Western Australia more recent microfossil zonations have been established in Permian strata of the Perth Basin by Segroves (1972) (Table 4.5, column 1) and in the Canning Basin by Balme (in Kemp et al., 1977) (Table 4.5, column 4). Evans (1969) proposed five zones based on eastern Australian successions; these have since been extended, subdivided, and refined by Helby (1969, 1973), Paten (1969), Norvich (1971, 1974), Price (1976, and Kemp et al. (1977) (Table 4.5, column 5). An excellent review of Australian Carboniferous and Permian palynostratigraphy has been presented by Kemp and others (1977). This review synthesizes all available data, including some previously unpublished work, and discusses the composition, age, and distribution of zones throughout Australia.

A sequence of Early to Late Triassic zones is recognized in the Carnarvon Basin (and in other Western Australian basins) by Dolby and Balme (1976) (Table 4.5, column 2). In eastern Australia a series of uppermost Permian and Triassic palynological "units" was proposed by Evans (1966) (Table 4.5, column 6). Helby (1973) outlined a sequence of six zones in the Late Permian to Middle Triassic of the Sydney Basin (Table 4.5, column 8). De Jersey and his coworkers have made detailed and comprehensive studies of Triassic and Jurassic microfossils of Queensland over many years, and these results were synthesized by de Jersey (1975) when he proposed four zones (plus one subzone) in the Middle Triassic to Early Jurassic of southeastern Queensland (Table 4.5, column 7).

Antarctica

Kyle (1977) proposed a preliminary zonation for the Permian and Triassic plant microfossil succession in southern Victoria Land. Two broad zones separated by barren strata were recognized in the Early Permian. Four more clearly defined subdivisions were proposed in the Triassic (Table 4.3, column 5). This zonal succession has since been extended further south in the central Transantarctic Mountains (Kyle and Schopf, in press). Additional assemblages comparable to the Late Permian Stage 5 of eastern Australia were recognized from the Prince Charles Mountains (Balme and Playford, 1967; Kemp, 1973) and the central Transantarctic Mountains (Kyle and Schopf, in press). No microfossils have been extracted from the *Lystrosaurus*-bearing beds.

DISCUSSION

Changes in land plant and animal assemblages reflect the changing terrestrial environment through Permian and Triassic time. The successive fossil associations are recognized as biostratigraphic zones. Although it is not clear how much of a fossil association is a true reflection of the original community, or to what extent it is affected by differential preservation or methods of study, the empirically defined zones still provide our best means of comparing successions in southern land areas.

Plant microfossils are widely distributed and have become increasingly important for biostratigraphy. Zonations based on palynomorphs have greatly extended and supplemented the biostratigraphy that had previously depended largely on tetrapod vertebrate fossils, which are much more sporadic in occurrence. Plant megafossils occupy a somewhat intermediate position for purposes of biozonation, since they also vary greatly in occurrence but are more common than vertebrate remains. They generally lack the advantage of a clear phyletic basis of interpretation that is best exemplified by tetrapods. As more becomes known about floral reproductive structures, however, it is becoming possible to develop a biozonation keyed to phyletic sequence. The fossil reproductive structures are also vital to our understanding of the botanical affinities of the palynomorphs, and thus to the integration of plant megafossil and microfossil zones.

Tables 4.4 and 4.5 show the correlations between zones established in different basins of Australia and southern Africa. Many of these correlations are based upon those suggested by the authors cited in the tables. Intercontinental correlations are obtained by comparing Tables 4.3 to 4.5; Table 4.6 summarizes the ranges and times of abundance of plant groups or genera used in making broad correlations. Ranges of diagnostic species are also important in comparing plant zones. The main large-scale events indicated in Table 4.6 and utilized in the correlations are briefly outlined below.

The older glacials in Australia include low-diversity assemblages that contrast sharply with the underlying luxuriant preglacial *Pseudorhacopteris (Sphenopteridium)* flora. The paucity of species in these Upper Carboniferous glacial assemblages (monosaccate-dominated Stage 1 microfossil zone, *Botrychiopsis* megafossil zone) was believed to reflect a periglacial flora (Kemp et. al., 1977), likened by Retallack (in press) to a tundra vegetation. Possible Stage 1 microfossil assemblages in Brazil were reported by Kemp (1975), and Archangelsky (1971) has described an equivalent megafossil association (Trampeaderense) from Argentina. Throughout Gondwanaland the advent of the glossopterids occurs in the uppermost glacial and immediately postglacial deposits. Leaves included in the genus *Gangamopteris* are the first to appear, with *Glossopteris* appearing soon after. The palynological expression of this event is the appearance (e.g., in Stage 2) of the first rare taeniate bisaccatepollen. First occurrences of these distinctive microfossils form a valuable biostratigraphic marker throughout the southern land areas.

The plant megafossil succession through the Permian is characterized by an increase in the dominance of *Glossopteris* and a decline of *Gangamopteris*. In the microfossil succession large monosaccate pollen common in the glacials become less important. Bisaccate pollen, including taeniate and non-taeniate forms, increase and predominate in Upper Permian strata. Relative proportions of monosaccate and bisaccate pollen have been used in making broad comparisons throughout the Gondwanaland Permian. Ranges of diagnostic species are important in establishing more precise zonations and correlations. However, not all species useful as index fossils on one continent are widespread geographically. The index spore *Dulhuntyispora dulhuntyi,* which defines the base of microfossil zones Stage 5 and Unit VII in Australia, has only one reported occurrence outside this continent: from the northern Karroo Basin where it is restricted to Zone 4d (Anderson, 1977). Ranges of other diagnostic species, however, are useful in tracing equivalents of these Australian zones. The species *Bascanisporites undosus*, *Praecolpatites sinuosus*, *Didecitriletes ericianus*, *Indospora clara*, and *Guttulapollenites hannonicus*, for example, have been useful in making comparisons of Australian Stage 5 (Unit VII) assemblages with those in Antarctica, peninsular India and the Salt Range, Madagascar, and southern Africa.

A noticeable change in composition of the floras, which is particularly evident in the plant microfossil record, occurred in latest Permian assemblages. By earliest Triassic time new plant groups were established and the glossopterids formed only a minor residual element of the flora in a few areas. Transitional latest Permian assemblages including new "Triassic" species have been reported from Australia (e.g., lower "*Thinnfeldia*" *callipterioides* plant megafossil zone, Gould, 1975, and Retallack, 1977; *Protohaploxypinus reticulatus* microfossil zone, Kemp et al., 1977), and the upper Chhidru Formation of the Salt Range, Pakistan (Balme, 1970). In South Africa plant microfossils from the vertebrate *Daptocephalus* Zone, which underlies the *Lystrosaurus* Zone, are of Permian aspect and lack the new species characteristic of assemblages mentioned above (Zone 7, Anderson, 1977).

Extensive marine transgression occurred at the base of the Triassic with a resulting phytoplankton bloom attested to by vast numbers of acritarchs in marine strata in widely separated parts of the world. Low-diversity coastal-plant associations of lycopods and certain groups of gymnosperms were also widespread (Balme, 1969, 1970; Balme and Helby, 1973; Retallack, 1975). Plant microfossils of this association occur with probable *Lystrosaurus* Zone vertebrates in Western Australia and are of Griesbachian—Smithian age (*Kraeuselisporites saeptatus* Zone, Dolby and Balme, 1976). The *Lystrosaurus* Zone is a useful datum for indicating earliest Triassic age, as is the marine transgressive sequence with acritarch swarms and associated coastal plant assemblage. Many problems still remain concerning ranges of species and integration of zonal schemes in the Permian-Triassic boundary sequence. For example, in India the *Lystrosaurus* fauna occurs in the upper Panchet. The lower Panchet contains transitional Permian-Triassic palynomorph assemblages (Maheshwari and Banerji, 1975) including the pollen species *Lunatisporites pellucidus*. In Australia the appearance of *L. pellucidus* marks the top of the Permian *Protohaploxypinus reticulatus* Zone (Kemp et al., 1977) and the base of the Triassic *Kraeuselisporites saeptatus* and *Lunatisporites pellucidus* Zones (Dolby and Balme, 1976; Helby, 1973). It also appears at the base of the Triassic in the Salt Range, Pakistan, in the Kathwai Member of the Mianwali Formation (Balme, 1970). In Madagascar, however, it occurs in the upper Permian Zone I of the lower Sakamena, though in extremely small numbers, then becomes abundant in Zone IIA of Earlv Triassic age (Goubin, 1965).

Lycopods formed an important component of the younger Scythian floras and *Dicroidium* became established during that time. *Dicroidium* and related pteridosperms dominate the Triassic floras throughout most of Gondwanaland. Recognition of distinctive palynomorphs with restricted ranges has aided zonation of the Australian Triassic and comparison of a similar succession in the Transantarctic Mountains. Retallack (1977) made an important contribution to Gondwana Triassic biostratigraphy with the recognition of four plant megafossil zones in eastern Australia and elsewhere throughout Gondwanaland.

The main floral events outlined above have been accepted as essentially synchronous within the limits of paleontologic resolution, this premise forming the basis of correlations. The possibility of time-transgressive zones reflecting migrations across Gondwanaland must be borne in mind however. With the paleontologic data we have at present, it is not possible to determine to what extent diachronous events are represented in the Gondwana geologic record. The most obvious possible diachronous event is the late Paleozoic glaciation. Paleomagnetic data showing the shift in relative pole position eastward across Gondwanaland support King's (1958) model of glaciation beginning in South America, with the glacial center moving eastward toward Antarctica and Australia. This suggests that deglaciation occurred first in South America and was followed by establishment of *Gangamopteris* and then *Glossopteris* associations, which appeared at progressively later times eastward across Gondwanaland. Kemp (1975) noted the possibility of older glacial floras in South America; however, contrary to what might be expected in King's model, microfossils in the upper glacial beds in South America resemble postglacial assemblages (Stage 3) from Australia. More data is required to test King's hypothesis, and even then the conclusions may be ambiguous as the floral succession presumably reflects climatic amelioration after glaciation. This problem may only be solved by obtaining radiometric dates from rocks with direct unambiguous association with the glacial deposits.

Dolby and Balme's (1976) study of Western Australian Triassic microfossils is important to our understanding of Gondwana plant geography. They draw attention to the marked difference between western and eastern Australian assemblages during the Middle and Late Triassic. The strong floral provincialism recognized in Gondwanaland at that time contrasts with the more uniform floras prior to the Middle Triassic. The two discrete floras were considered in relation to the reconstructed Gondwana paleogeography. It was suggested that one represents high-latitude vegetation (the typical *Alisporites* assemblages of southwestern and eastern Australia, Antarctica, southern Madagascar, and southern Madagascar, and southern South America), and the other, which includes many "northern" species, a narrow belt of midlatitude (30°-35°S) temperate rain-forest vegetation (northwestern Australia and northern Madagascar).

The presence of palynomorph species characteristic of European and North American Permian and Triassic sequences is particularly evident in Gondwana basins bordering the Tethys region (Salt Range, India, Madagascar, Western Australia) and also down into southern Africa and South America. These "northern" species have been used for comparison with the northern hemisphere succession. Correlations using species apparently common to two or more floristic provinces are questionable because the plant megafossil record, though admittedly less complete, does not support such wide distribution of most plant groups. There is a difference of opinion on whether the presence of a few genera of plant megafossils in both northern and southern areas is the result of migration from one province into another, or the result of homoplasy or parallel evolution. Chaloner and Lacey (1973) and Lacey (1975) believe that a combination of both migration and homoplasy has played a part in the occurrence of mixed floras. Foster (1978) discussed the morphologic variability and distribution of certain species of palynomorphs presumed common to Permian strata of the Euramerican province and eastern Australia. He advised caution in the use of "shared" species for correlation, and suggested that many, in particularly the Upper Permian taeniate bisaccates, may be the result of parallel evolution.

From the preceding review it is obvious that broad intercontinental correlations are possible, and in some cases on a detailed scale. Much basic research remains to be done in many regions before meaningful detailed comparisons are possible throughout the Permian and Triassic of southern land areas.

ACKNOWLEDGMENTS

Dr. Robin J. Helby reviewed an early draft of the manuscript. The paper has since been revised after the untimely death of Dr. James M. Schopf in September 1978. Drs. Walter C. Sweet, Stephen R. Jacobson, William J. Zinsmeister, and Mr. John M. Zawiskie provided helpful discussion on particular aspects, and Drs. Greg J. Retallack, David H. Elliot, and Paul F. Ciesielski critically reviewed later drafts of the manuscript. We extend our grateful thanks to all those who helped in its preparation.

The paper was prepared as a continuation of work under an initial National Science Foundation grant AG-82 to the U.S. Geological Survey, and a recent National Science Foundation grant DPP 76-83030.

REFERENCES

Anderson, H. M. 1974. A brief review of the flora of the Molteno "Formation" (Triassic), South Africa. *Palaeontol. Africana* 17:1-10.

Anderson, H. M., and J. M. Anderson 1970. A preliminary review of the biostratigraphy of the uppermost Permian, Triassic and lowermost Jurassic of Gondwanaland. *Palaeontol. Africana* 13 (suppl.): 1-22.

Anderson, J. M. 1973. The biostratigraphy of the Permian and Triassic. Pt. 2—A preliminary review of the distribution of Permian and Triassic strata in time and space. *Palaeontol. Africana* 16: 59-83.

Anderson, J. M. 1977. The biostratigraphy of the Permian and Triassic. Pt. 3—A review of Gondwana Permian palynology with particular reference to the northern Karroo Basin, South Africa. *Mem. Bot. Surv. South Africa, No. 41*, pp. 1-67.

Archangelsky, S. 1971. Las Tafofloras del sistema Paganzo en la Republica Argentina. *Ann. Acad. Brasil Cienc.*, 43(suppl.): 67-88.

Archangelsky, S., and O. G. Arrondo, 1969. The Permian Taphofloras of Argentina with some considerations about the presence of "northern" elements and their possible significance. *Gondwana Stratigraphy*, IUGS Symposium, Buenos Aires (Mar del Plata), October 1967: UNESCO. pp. 197-212.

Archangelsky, S., and O.G. Arrondo, 1975. Paleogeografia y plantas fosiles en el Permico inferior Austrosudamericano. *Actas 1st Cong. Argentino Paleontol. Bioestratigr.*, (Tucumán, Argentina, 1975) 1: 479-496.

Balme, B. E. 1963. Plant microfossils from the Lower Triassic of Western Australia. *Palaeontology* 6(1): 12-40.

Balme, B. E. 1964. The palynological record of Australian pre-Tertiary floras. *Ancient Pacific Floras, L. M., Cranwell*, ed., Univ. Hawaii Press, pp. 48-80.

Balme, B. E. 1969. The Permian-Triassic boundary in Australia. *Geol. Soc. Australia Spec. Publ. no. 2*, pp. 99-112.

Balme, B. E. 1970. Palynology of Permian and Triassic strata in the Salt Range and Surghar Range, West Pakistan. In Kummel and Teichert, 1970, pp. 305-453.

Balme, B. E., and R. J. Helby, 1973. Floral modifications at the Permian-Triassic boundary in Australia. In Logan and Hills, 1973, pp. 433-444.

Balme, B. E., and G. Playford, 1967.

Late Permian plant microfossils from the Prince Charles Mountains, Antarctica. *Rev. micropaleontol.* 10(3): 179-192.

Banks, M. R. et al 1969. Correlation charts for the Carboniferous, Permian, Triassic and Jurassic Systems in Australia. *Gondwana Stratigraphy,* IUGS Symposium, Buenos Aires (Mar del Plata), October 1967: UNESCO, pp. 467-470.

Bharadwaj, D. C. 1962. The miospore genera in the coals of Raniganj Stage (Upper Permian), India. *Palaeobotanist 9(1-2): 68-106 (1960).*

Bharadwaj, D. C. 1969. Palynological succession through the Mesozoic era in India. *Jour. Palynology.* 5(2):85-94.

Bharadwaj, D. C. 1971. Lower Gondwana microfloristics. *Sem. Paleopalynol. Indian Stratigr., Proc.,* pp. 41-50.

Bharadwaj, D. C. 1972. Palynological subdivisions of the Gondwana sequence in India. *2nd Gondwana Symp. Proc. and Pap.* (South Africa 1970): IUGS Commission on Stratigraphy, pp. 531-536.

Bharadwaj, D. C. 1974. Palynological subdivisions of Damuda Series. In *Aspects and Appraisal of Indian Paleobotany.* K. R. Surange, R. N. Lakhanpal and D. C. Bharadwaj, (eds.). Lucknow: Birbal Sahni Inst. of Paleobotany, pp. 392-396.

Bharadwaj, D. C. and S. C. Srivastava 1969. Some new miospores from Barakar Stage, Lower Gondwana, India. *Palaeobotanist* 17(2): 220-229 (1968).

Bharadwaj, D. C., and R. S. Tiwari, 1964. The correlation of coalseams in Korba Coalfield, Lower Gondwanas, India. *5th Congr. Internat. Stratigr. Geol. Carbonifere, Compte Rendu* (Paris 1963) pp. 1131-1143.

Bose, M. N., and R. K. Kar 1966. Palaeozoic sporae dispersae from Congo. Pt. I. Kindu-Kalima and Walikale regions. *Mus. Roy. Afrique Cent. Tervuren. Belg. Ann. Ser. IN-8, Sci. Geol. no. 53,* pp. 1-168.

Camp, C. L., and M. R. Banks. 1978. A proterosuchian reptile from the Early Triassic of Tasmania. *Alcheringa* 2(1): 143-158.

Campbell, K. S. W. 1962. Marine fossils from the Carboniferous glacial rocks of New South Wales. *Jour. Paleontology* 36(1): 38-52.

Campbell, K.S. W. (ed). 1975. *Gondwana Geology.* Papers of the 3rd Gondwana Symp. (Canberra 1973): Australian National Univ. Press, pp. 99-108, 125-172, 397-413, 425-441, 469-473.

Campbell, K. S. W, and R. G. McKellar 1969. Eastern Australian Carboniverous invertebrate sequence and affinities. In *Stratigraphy and Palaeontology: Essays in Honour of Dorothy Hill,* K. S. W. Campbell, ed. Canberra: Australian National Univ. Press, pp. 77-119.

Chaloner, W. G., and W. S. Lacey. 1973. The distribution of Late Paleozoic floras. In *Organisms and Continents Through Time,* N. F. Hughes, ed. London: Palaeontological Association, pp. 271-289.

Christopher, R. A., and G. F. Hart. 1971. A statistical model in palynology. *Geoscience and Man* 3:49-56.

Closs, D. 1969. Intercalation of goniatites in the Gondwanic glacial beds of Uruguay. *Gondwana Stratigraphy,* IUGS Symposium, Buenos Aires (Mar del Plata), October 1967: UNESCO, pp. 197-212.

Cosgriff, J. W. 1965. A new genus of *Temnospondyli* from the Triassic of Western Australia. *Jour. Roy. Soc. Western Australia* 48(3): 65-90.

Cosgriff, J. W. 1969. *Blinasaurus,* a brachyopid genus from Western Australia and New South Wales. *Jour. Roy. Soc. Western Australia* 52(1): 65-68.

Cosgriff, J. W., and Zawiskie, J. M. In

press. A new species of Rhytiodosteidae from the *Lystrosaurus* Zone and a review of the Rhytidosteoidea. *Palaeontol. Africana.*

Daemon, R. F. 1966. Ensaio sobre a Distribuicao e Zoneamento dos Esporomorfos do Paleozoico Superior da Bacia do Paraná. Bol. Téc. Petrobrás (Rio de Janiero) 9(2): 211-218.

Daemon, R. F., and C. J. A. Contreiras, 1971. Zoneamento palinologico da Bacia do Amazona. *An. XXV Congr. Bras. Geol., Soc. Bras. Geol.* (Sao Paulo 1971) 3:81-88.

Daemon, R. F., and L. P. Quadros, 1970. Bioestratigrafia do Neopaleozoico da Bacia do Paraná. *An. XXIV Congr. Bras. Geol., Soc. Bras. Geol.* (Brasilia 1970), pp. 359-412.

de Jersey, N.J. 1975. Miospore zones in the Lower Mesozoic of southeastern Queensland. In Campbell, 1975, pp. 159-172.

Dickins, J. M., and R. A. McTavish 1963. Lower Triassic marine fossils from the Beagle Ridge (BMR 10) Bore, Perth Basin, Western Australia. *Jour. Geol. Soc. Aust.* 10(1): 123-140.

Dickins, J. M.; E. J. Malone; and A. R. Jensen, 1964. Subdivision and correlation of the Permian Middle Bowen Beds, Queensland. *Bur. Min. Resourc., Aust. Geol. Geophys. Rept.* 70: 1-12.

Dingle, R. V. 1978. South Africa. In *The Phanerozoic geology of the world*, vol. II. *The Mesozoic, A*, M. Moullade and A. E. M. Nairn, (eds). New York: Elsevier, pp. 401-434.

Dolby, J. H., and B. E. Balme. Triassic palynology of the Carnarvon Basin, Western Australia. *Rev. Palaeobot. Palynol.* 22: 105-168.

Dolianiti, E. 1952. La flore fossile du Gondwana au Brasil d'apres sa position stratigraphique. *19th Intern. Geol. Congr., Symposium sur les series de Gondwana* (Algiers 1952) pp. 285-301.

Du Toit, A.L. 1927a. The fossil flora of the Upper Karroo Beds. *South African Mus. Ann.* 22(2): 289-420.

Du Toit, A. L. 1927b. A geological comparison of South America with South Africa. *Carnegie Inst. Wash., Pub. No. 381*, pp. 28-41. (also pp. 129-150).

Du Toit, A. L. 1930a. A brief review of the Dwyka glaciation of South Africa. *15th Intern. Geol. Congr., Compte Rendu (South Africa 1929)* 2:90-102.

Du Toit, A. L. 1930b. A short review of the Karroo fossil flora. *15th Intern. Geol. Congr., Compte Rendu* (South Africa 1929) 2:239-251.

Du Toit, A. L. 1932a. Some fossil plants from the Gondwana beds of Uganda. *South African Mus. Ann.* 29(4): 395-406.

Du Toit, A. L. 1932b. Some fossil plants from the Karroo system of South Africa. *South African Mus. Ann.* 28(4): 369-393.

Evans. P. R. 1966. Mesozoic stratigraphic palynology in Australia. *Australian Oil Gas Jour.* 12: 58-63.

Evans, P. R. 1969. Upper Carboniferous and Permian palynological stages and their distribution in eastern Australia. *Gondwana Stratigraphy.* IUGS Symposium, Buenos Aires (Mar del Plata), October 1967: UNESCO, pp. 41-54.

Falcon, R. M. S. 1973. Palynology of the Lower Karroo succession in the Middle Zambezi Basin. In "The Palaeontology of Rhodesia," G. Bond, *Rhod. Geol. Survey Bull.*, 70:43-71.

Falcon, R. M.S. 1975a. Application of palynology in sub-dividing the coal-bearing formations of the Karroo sequence in Southern Africa. *South African Jour. Sci.* 71(11): 336-344.

Falcon, R. M. S. 1975b. Palynostratigraphy of the Lower Karroo sequence in the Central Sebungwe District, Mid-Zambezi Basin, Rhodesia. *Palaeontol. Africana* 18:1-29.

Falcon, R. M. S. 1976. Numerical

methods in the definition of palynological assemblage zones in the Lower Karroo (Gondwana) of Rhodesia. *Palaeontol. Africana* 19:1-20.

Foster, C. B. 1978. Aspects of palynological correlation between the Late Permian of the Euramerican and Gondwanan (eastern Australian) provinces. *Queensland Govt. Min. Jour.* 79(921): 376-379.

Glenister, B. F., and W. M. Furnish, 1961. The Permian ammonoids of Australia. *Jour. Paleontology* 35(4): 673-736.

Goubin, N. 1965. Description et repartition des principaux pollenites permiens, triasiques et jurassiques des Sondages du Bassin de Morondava (Madagascar). *Inst. Francais Pétrol. Ann. Combust. Liq. Rev.* 20(10):1415-1461.

Gould, R. E. 1975. The succession of Australian pre-Tertiary megafossil floras. *Bot. Rev.* 41:453-483.

Hart, G. F. 1960. Microfloral investigation of the Lower Coal Measures (K2); Ketewaka-Mchuchuma Coalfield, Tanganyika. *Geol. Surv. Tanganyika, Bull.* 30:1-18.

Hart, G. F. 1965. Microflora from the Ketewaka-Mchuchuma Coalfield, Tanganyika. *Geol. Surv. Tanganyika, Bull.* 36:1-27 (1963).

Hart, G. F. 1967. Micropalaeontology of the Karroo deposits in South and Central Africa. In *IUGS Reviews* prepared for the first symposium of Gondwana stratigraphy, Haarlem, Netherlands: IUGS secretariat, 1967, pp. 161-172.

Hart, G. F. 1969. The stratigraphic subdivision and equivalents of the Karroo sequence as suggested by palynology. *Gondwana Stratigraphy.* IUGS Symposium, Buenos Aires (Mar del Plata), October 1967: UNESCO, pp. 23-35.

Hart, G. F. 1971. The Gondwana Permian palynofloras. *An Acad. brasil Cienc.* 43(supl.):145-175.

Hedberg, H. D. (ed.) 1976. *International Stratigraphic guide*, by the International Subcommission on Stratigraphic Classification of IUGS Commission on Stratigraphy. New York: Wiley, 200 p.

Helby, R. J. 1969. The Carboniferous-Permian boundary in eastern Australia: an interpretation on the basis of palynological information. *Spec. Publ., Geol. Soc. Austral.* 2: 69-72.

Helby, R. J. 1973. Review of Late Permian and Triassic palynology of New South Wales. *Spec. Publ. Geol. Soc. Austral.* 4:141-155.

Høeg, O. A., and M. N. Bose, 1960. The Glossopteris flora of the Belgian Congo. *Ann. Mus. Congo Belge Ser. 8, Scien. geol.* 32:1-106.

Jardiné, S. 1974. Microflores des formations du Gabon attribuées au Karroo. *Rev. Palaeobot. Palynol.* 17:75-112.

Jekhowsky, B. de, and N. Goubin, 1964. Subsurface palynology in Madagascar: a stratigraphic sketch of the Permian, Triassic and Jurassic of the Morondava Basin. In "Palynology in Oil Exploration," A. T. Cross, (ed.) Symposium, Research Committee SEPM, San Francisco, March 1962. *Soc. Econ. Paleontol. Mineral. Spec. Publ. 11:116-130.*

Jones, P. J.,; K. S. W. Campbell; and J. *Roberts 1973.* Correlation chart for the Carboniferous System of Australia. *Bur. Min. Resour., Austral. Geol. Geophys. Bull.* 156A: 1-40.

Kalandadze, N. N. 1975. The first discovery of *Lystrosaurus* in the European regions of the USSR. *Paleont. Zhur.* 4: 140-152.

Kar, R. K., and M. N. Bose. 1967. Palaeozoic sporae dispersae from Congo, Pt. III, Assise des schistes noirs de la Lukuga. *Mus. Roy. Afr. Cent.* (Tervuren, Belgique), *Ann. ser. IN-8º, Sci. Geol.* no. 54, pp. 3-59.

Kemp, E. M. 1973. Permian flora from Beaver Lake area, Prince Charles

Mountains, Antarctica. 1. Palynological examination of samples. *Bur. Min. Resour., Austral. Geol. Geophys. Bull.* 126:7-12.

Kemp, E. M. 1975. The palynology of Late Paleozoic glacial deposits of Gondwanaland. In Campbell, 1975, pp. 397-413.

Kemp, E.M.; B. E. Balme; R. J. Helby; R. A. Kyle; G. Playford; and P. L. Price. 1977. Carboniferous and Permian palynostratigraphy in Australia and Antarctica: a review. *Bur. Min. Resour., Austral. Geol. Geophys. Jour.* 2:177-208.

King, L. C. 1958. Basic palaeogeography of Gondwanaland during the late Paleozoic and Mesozoic eras. *Quart. Jour. Geol. Soc. London,* 114:47-70.

Kitching, J. W. 1972. A short review of the Beaufort zoning in South Africa. *2nd Gondwana Symp. Proc. and Papers* (South Africa, 1970): IUGS Commission on Stratigraphy, pp. 309-312.

Kummel, B., and C. Teichert (eds.). 1970. Stratigraphic Boundry Problems: Permian and Triassic of West Pakistan. *Univ. Kansas Dept. Geol. Spec. Publ.* 4:207-453.

Kyle, R. A. 1976. Palaeobotanical studies of the Permian and Triassic Victoria Group (Beacon Supergroup) of south Victoria Land, Antarctica. Ph.D. thesis, Victoria University of Wellington, New Zealand. 306 p.

Kyle, R. A. 1977. Palynostratigraphy of the Victoria Group of south Victoria Land, Antarctica. *New Zealand Jour. Geol. Geophys.* 20(6): 1081-1102.

Kyle, R. A, and A. Fasola 1978. Triassic palynology of the Beardmore Glacier area of Antarctica. *Palinologia* 1: 313-319.

Kyle, R. A., and J. M. Schopf. In press. Permian and Triassic palynostratigraphy of the Victoria Group, Transantarctic Mountains. In *Antarctic Geosciences,* C. Craddock, (ed.) *3rd Symp. Antarctic Geol. Geophys.* (Madison, Wisconsin, 1977).

Lacey, W. S. 1975. Some problems of "mixed" floras in the Permian of Gondwanaland. In Campbell, 1975, pp. 125-134.

Lele, K. M. 1976. Late Palaeozoic and Triassic floras of India and their relation to the floras of northern and southern hemispheres. *Palaeobotanist* 23(2):89-115 (1974).

Logan, A., and L. V. Hills (eds.) 1973. The Permian and Triassic systems and their mutual boundary. *Canad. Soc. Pet. Geol., Mem.* 2:35-73, 379-424, 433-444.

Maheshwari, H. K. 1975. Triassic miofloras south of the Tethys. *Palaeobotanist* 22(3): 236-244 (1973).

Maheshwari, H. K., and J. Banerji. 1975. Lower Triassic palynomorphs from the Maitur Formation, West Bengal, India. *Palaeontographica* 152B: 149-190.

Manum, S. V., and N. D. Tien. 1973. Palyno-stratigraphy of the Ketewaka coalfield (Lower Permian), Tanzania. *Rev. Palaeobot. Palynol* 16:213-227.

McTavish, R. A. 1973. Triassic conodont faunas from Western Australia. *N. Jahrb. Geol. Paläontol. Abh.* 143(3):275-303.

McTavish, R. A., and J. M. Dickins. 1974. The age of the Kockatea Shale (Lower Triassic), Perth Basin— a reassessment. *Geol. Soc. Austral. Jour.* 21(2):195-201.

Menendez, C. A. 1968. Bibliografia Paleobotanica de America del Sur, Revista del Museo Argentino de Ciencias Naturales "Bernardino Rivadavia" Instituto Nacional de Investigacion de las Ciencias Naturales. *Paleontologia* 1(6):1-229.

Menendez, C. A. 1969. Die fossilen floren Sudamerikas. In *Biogeography and Ecology in South America,* vol. 2, *Monographiae Biologicas (VanOyeed),*

E. J. Fittkau et al., eds. The Hague: W. Junk, vol. 19, pp. 519-561.

Morris, N. 1975. The *Rhacopteris* flora in New South Wales. In Campbell, 1975, pp. 99-108.

Murphy, M. A. 1977. On time—stratigraphic units. *Jour. Paleontology* 51(2):213-219.

Norvick, M. 1971. Some palynological observations on Amerada Thunderbolt No. 1 Well, Galilee Basin, Queensland. Unpublished Records Bur. Min. Resour., Austral. Geol. Geophys., 1971/53.

Norvick, M. 1974. Permian and Late Carboniferous palynostratigraphy of the Galilee Basin, Queensland. Unpublished Records Bur. Min. Resour., Austral. Geol. Geophys., 1974/141.

Pant, D.D. and D. D. Nautiyal. 1960. Some seeds and sporangia of *Glossopteris* flora from Raniganj coalfield, India. *Palaeontographica* 107B:1-64.

Paten, R. J. 1969. Palynologic contributions to petroleum exploration in the Permian formations of the Cooper Basin, Australia. *Austral. Pet. Explor. Assoc. Jour.* 9:79-87.

Plumstead, E. P. 1962. Fossil floras of Antarctica (with an appendix on fossil wood by R. Kräusel). In *Trans-Antarctic Expedition 1955-1958, Sci. Repts. 9, Geol.* London: T.A.E. Committee, 154p.

Plumstead, E. P. 1966. Recent paleobotanical advances and problems in Africa. In *Symposium on Floristics and Stratigraphy of Gondwanaland, 1964.* Lucknow: Birbal Sahni Institute of Palaeobotany, pp. 1-12.

Plumstead, E. P. 1967. A review of contributions to the knowledge of Gondwana mega-plants and floras of Africa published since 1950. In *IUGS Reviews* prepared for the first symposium of Gondwana stratigraphy, 1967. Haarlem, Netherlands: IUGS secretariat, pp. 139-148.

Plumstead, E.P. 1975. A new assemblage of plant fossils from Milorgfjella, Dronning Maud Land. *British Antarctic Surv., Sci. Rep. No. 83,* Cambridge, 30p.

Price, P. L. 1976. Permian palynology of the Bowen Basin. In *A guide to the geology of the Bowen and Surat Basins in Queensland,* A. R. Jensen; N. F. Exon; J. C. Anderson; and W. H. Koppe. *25th Intern. Geol. Cong. Excursion Guide 3C.*

Reed, F. R. C. 1927. Upper Carboniferous fossils from Argentina; Appendix. In Du Toit, 1927b, pp.129-150.

Retallack, G. J. 1975. The life and times of a Triassic lycopod. *Alcheringa* 1(1):3-29.

Retallack, G. J. 1977. Reconstructing Triassic vegetation of eastern Australasia: a new approach for the biostratigraphy of Gondwanaland. *Alcheringa* 1(3-4):247-277, + microfiche suppl. G1-J17.

Retallack, G. J. In press. Late Carboniferous to middle Triassic megafossil floras from the Sydney Basin. In "A guide to the Sydney Basin," C. Herbert, and R. J. Helby, *Geol. Surv. New South Wales. Rept.*

Rigby, J. F. 1966. The Lower Gondwana floras of the Perth and Collie Basins, Western Australia. *Palaeontographica* 118B (4-6):113-152.

Rigby, J. F. 1969. A re-evaluation of the Pre-Gondwana Carboniferous flora. *An Acad. brasil. Cienc.* 41(3):393-413.

Rigby, J. F. 1972. The distribution of Lower Gondwana plants in the Parana Basin of Brazil. *2nd Gondwana Symp. Proc. and papers,* (South Africa, 1970): IUGS Commission on Stratigraphy, pp. 575-584.

Rigby, J. R. 1973. *Gondwanidium* and other similar upper Palaeozoic genera, and their stratigraphic significance: *Publ. Geol. Surv. Queensland, Palaeont. Pap. No. 24,* pp. 1-10.

Rigby, J. F., and J. M. Schopf. 1969. Stratigraphic implications of Antarctic paleobotanical studies. *Gondwana Stratigraphy*, IUGS Symposium, Buenos Aires (Mar del Plata), October 1967: UNESCO, pp. 91-106.

Roberts, J. 1971. Devonian and Carboniferous brachiopods from the Bonaparte Gulf Basin, Northwestern Australia. *Bur. Min. Resour., Austral. Geol. Geophys. Bull.* 122: 1-319.

Roberts, J.; J. W. Hunt; and D. M. Thompson, 1976. Late Carboniferous marine invertebrate zones of eastern Australia. *Alcheringa* 1(2):197-225.

Rocha-Campos, A. C. 1973. Upper Paleozoic and Lower Mesozoic paleogeography, and paleoclimatological and tectonic events in South America. In Logan and Hills, 1973, pp. 398-424.

Romer, A. S, 1966a. The Chanares (Argentina) Triassic reptile fauna. 1. Introduction. *Mus. Comp. Zool. Breviora no. 247,* pp. 1-14.

Romer, A. S. 1966b. *Vertebrate Paleontology,* 3rd ed. Chicago: Univ. of Chicago Press, 468 p.

Romer, A. S. 1975. Intercontinental correlations of Triassic Gondwana vertebrate faunas. In Campbell, 1975, pp. 469-473.

Roy Chowdhury, M. K.; M. V. A. Sastry; S. C. Shah; G. Singh; and S. C. Ghosh, 1975. Triassic floras in India. In Campbell, 1975, pp. 149-159.

Runnegar, B. 1969. The Permian faunal succession in eastern Australia. *Spec. Publ. Geol. Soc. Austral.* 2:73-98.

Runnegar, B., and G. McClung 1975. A Permian time scale for Gondwanaland. In Campbell, 1975, pp. 425-441.

Sahni, B. 1921. The present position of Indian palaeobotany (Presidential Address). Proc. 8th Indian Sci. Congr., Sec. Bot., *Proc. Asiatic Soc. Bengal.* N.S., 17(4):152-175.

Sahni, B. 1926. The southern fossil floras: a study in the plant-geography of the past (Presidential Address). *Proc. 13th Indian Sci. Congr.,* (Bombay 1926), pp. 229-254.

Sahni, B. 1939. Present advances in Indian palaeobotany. *Proc. 25th Indian Sci. Congr.* (Calcutta 1938), pt. 2, sec. 5 (Botany), pp. 133-176.

Sarjeant, W. A. S. 1970. Acritarchs and tasmanitids from the Chhidru Formation, Uppermost Permian of West Pakistan. In Kummel and Teichert, 1970, pp. 277-304.

Sarjeant, W. A. S. 1973. Acritarchs and tasmanitids from the Mianwali and Tredian Formations (Triassic) of the Salt and Surghar Ranges, West Pakistan. In Logan and Hills, 1973, pp. 35-73.

Schopf, J. M. 1962. A preliminary report on plant remains and coal of the sedimentary section in the Central Range of the Horlick Mountains, Antarctica. *Ohio State Univ., Inst. Polar Studies Rept. 2,* pp. 1-61.

Schopf. J. M. 1965. Anatomy of the axis in *Vertebraria.* In *Geology and Paleontology of the Antarctic,* Am. Geophys. Union, pub. no. 1299, J. B. Hadly, ed. Antarctic Research Ser., no. 6, pp. 217-228.

Schopf, J. M. 1968. Studies in Antarctic Paleobotany. *Antarc. Jour. U. S.,* 3(5):176-177.

Schopf, J. M., 1973. Contrasting plant assemblages from Permian and Triassic deposits in southern continents. In Logan and Hills, 1973, pp. 379-397.

Schopf, J. M. 1976. Morphologic interpretation of fertile structures in glossopterid gymnosperms. *Rev. Palaeobot. Palynol.* 21:25-64.

Segroves, K. L. 1972. The sequence of palynological assemblages in the Permian of the Perth Basin, Western Australia. In *2nd Gondwana Symp. Proc. and papers* (South Africa 1970): IUGS Commission on Stratigraphy, pp. 511-529.

Seward, A. C., 1897. The *Glossopteris*

flora: an extinct flora of a Southern Hemisphere continent. *Sci. Progress* (London), n.s., 1 (2): 178-201.

Seward, A. C. 1903. Fossil floras of Cape Colony. *South African Mus. Ann.* 4(1):1-122.

Seward, A. C. 1914. Antarctic fossil plants. In "British Antarctic ("Terra Nova") Expedition, 1910." *Nat. Hist. Rept. British Museum (Natural History), Geology* 1(1):1-49.

Sill, W. D. 1969. The tetrapod-bearing continental Triassic sediments of South America. *Am. Jour. Sci.* 267:805-821.

Spath, L. F. 1930. The Eotriassic invertebrate fauna of East Greenland. *Medd. Grønland* 82(1):5-90.

Stapleton, R.P. 1977. Carboniferous unconformity in southern Africa. *Nature* 268:222-223.

Stapleton, R. P. 1978. Microflora from a possible Permo-Triassic transition in South Africa. *Rev. Palaeobot. Palynol.* 25:253-258.

Stipanicic, P. N. 1957. El sistema Triasico en la Argentina. In *El Mesozoico del hemisferio occidental y sus correlaciones mundiales*, A. G. Rojas et al. (ed. com.) Congreso Geologico Internacional XXa Sesion (Ciudad de Mexico 1956) Sec. II, pp. 73-111.

Stipanicic, P.N. 1969. Las sucesiones Triasicas Argentinas. In *Gondwana Stratigraphy, IUGS* Symposium, Buenos Aires (Mar del Plata), October 1967: UNESCO, pp. 1121-1139.

Stipanicic, P.N., and M.I.R. Bonetti, 1969. Consideraciones sobre la cronologia de los terrenos Triasicos Argentinos. In *Gondwana Stratigraphy*, IUGS Symposium, Buenos Aires (Mar del Plata), October 1967: UNESCO, pp. 1081-1119.

Surange, K. R. 1966. Distribution of *Glossopteris* flora in the Lower Gondwana Formations of India. In *Symposium on Floristics and Stratigraphy of Gondwanaland.* Palaeobot. Soc. Spec. Sess: 1964, Lucknow, India: Birbal Sahni Inst. Palaeobot. pp. 55-68.

Surange, K. R. 1975. Indian lower Gondwana floras: a review. In Campbell, 1975, pp. 135-147.

Sweet, W. C. 1970. Uppermost Permian and Lower Triassic conodonts of the Salt Range and Trans-Indus Ranges, W. Pakistan. In Kummel and Teichert, 1970, pp. 207-275.

Thompson, R., and J. G. Mitchell. 1972. Palaeomagnetic and radiometric evidence for the age of the lower boundary of the Kiaman magnetic interval in South America. *Geophys. Jour. Roy. Astron. Soc.* 27:207-214.

Tiwari, R. S. 1974. Inter-relationships of palynofloras in the Barakar Stage (Lower Gondwana), India. *Geophytology* 4(2): 111-129.

Townrow, J. A. 1956. The genus *Lepidopteris* and its southern hemisphere species. *Avh. Norske Vidensk-Akad. Oslo, Mat.-Naturvidensk Kl.* 1956 no. 2, 28p.

Townrow, J. A. 1960. The Peltaspermaceae, a pteridosperm family of Permian and Triassic age. *Palaeontology* 3(3):333-361.

Townrow, J. A. 1962. On *Pteruchus*, a microsporophyll of the Corystospermaceae. *Bull. Brit. mus. (Nat. Hist.), Geol.* 6(2):287-320.

Townrow, J. A. 1966. On *Lepidopteris madagascariensis* Carpentier (Peltaspermaceae) *Jour. Proc. Roy. Soc. New South Wales*, 98:203-214.

Tozer, E. T. 1971. Triassic time and ammonoids: problems and proposals. *Canadian Jour. Earth Sci.* 8:989-1031.

Tozer, E. T. 1972. The earliest marine Triassic rocks: their definition, ammonoid fauna, distribution and relationship to underlying formations. *Bull. Canadian Petroleum Geol.* 20(4):643-650.

Utting, J. 1976. Pollen and spore

assemblages in the Luwumbu coal formation (Lower Karroo) of the North Luangwa valley, Zambia, and their biostratigrapic significance. *Rev. Palaeobot. Palynol.* 21:295-315.

Utting, J. 1978. Lower Karroo pollen and spore assemblages from the coal measures and underlying sediments of the Siankondobo coalfield, Mid-Zambezi valley, Zambia. *Palynology* 2:53-68.

Van Eeden, O. R. 1973. The correlation of the subdivisions of the Karroo system. *Trans. Geol. Soc. S. Africa,* 76 (3):201-206.

Walkom, A. B. 1922. Palaeozoic floras of Queensland. Part 1. The flora of the Lower and Upper Bowen Series. *Queensland Geol. Surv., Publ. no. 270,* 64p.

Walkom, A. B. 1928. Notes on some additions to the *Glossopteris* flora in New South Wales. *Proc. Linnean Soc. New South Wales,* 53(5): 555-564.

Walkom, A. B. 1938. A brief review of the relationships of the Carboniferous and Permian floras of Australia. In W. J. Jongmans, ed. *2nd Congress Carboniferous Stratig.* Heerlen 1935, 3:1335-1342.

Walkom, A. B., 1941. On a new species of *Annularia* from New South Wales. *Rec. Austral. Mus.* 21(1):43-44.

Walkom, A. B. 1944. The succession of Carboniferous and Permian floras in Australia. *Jour. Roy Soc. New South Wales* 78(2):4-13.

Walkom, A. 1949. Gondwanaland: a problem of palaeogeography (Presidential Address). *Australian and New Zealand Assoc. Adv. Sci. Rept.,* 27th meeting, Hobart, Jan. 1949, pp. 1-13.

Waterhouse, J. B. 1967. Proposal of Series and Stages for the Permian in New Zealand. *Trans. Roy. Soc. New Zealand, Geol.* 5:161-180.

Webb, J. A., and N.J. McNaughton. 1978. Isotopic age of the Sugars Basalt. *Queensland Govt. Mining Jour.* 79(925):591-595.

5

UPPER TRIASSIC FLORAL ZONES OF NORTH AMERICA

Sidney R. Ash

SUMMARY

Three floral zones based on plant megafossils have been recognized in the Upper Triassic rocks of North America. Although the zones have not been definitely recognized elsewhere they can be correlated with the stages of the Triassic in Germany by using pollen and spores. The oldest is the zone of *Eoginkgoites*. It is middle Carnian (early Late Triassic) and occurs in the lowest beds of the Chinle Formation in Utah, Arizona, and New Mexico; the Popo Agie Formation in Wyoming; and in the lower part of the Newark Supergroup in Pennsylvania and North Carolina. Above the zone of *Eoginkgoites* is the zone of *Dinophyton*, which is late Carnian (middle Late Triassic). It occurs in the lower and middle part of the Chinle Formation in Utah; Arizona, and western New Mexico; the Dockum Group in eastern New Mexico and adjacent areas in Texas; the Dolores Formation in Colorado; and the middle part of the Newark Supergroup in Pennsylvania. The youngest and as yet unnamed zone occurs in the upper part of Newark Supergroup in Connecticut and possibly in the Santa Clara Formation in northwestern Mexico. The zone appears to be Rhaeto-Liassic (latest Triassic-earliest Jurassic) and the upper part may be equivalent to the zone of *Thaumatopteris* in Greenland and western Europe.

INTRODUCTION

The Upper Triassic plant megafossils of North America first became known to science in 1823 when Dr. Edward Hitchcock read a paper before the American Geological Society in which he noted their occurrence in Connecticut (Hitchcock, 1823). Since then they have been described from many localities in North America. In this paper their stratigraphic distribution is summarized, three floral zones based on these fossils are tentatively differentiated, and the zones are correlated by means of plant microfossils with the Triassic stages in Germany.

It is difficult to estimate how many species comprise the entire Upper Triassic flora of North America because some of the individual floras, particularly those in the eastern United States and Mexico have not been revised for many years and certain plant names may need to be changed or may even have priority over names in general use. Furthermore, recent work by

Delevoryas, Hope, and myself has demonstrated that these strata continue to yield previously unknown taxa. At present, I estimate that about 120-140 well-characterized species have been described from these strata. In addition, there are probably 30 to 40 poorly or incompletely known species, which, if validated, would raise the total number of species.

Until recently no one seems to have attempted to differentiate any floral zones in the Upper Triassic of North America. Perhaps this is because little attention has been paid to the stratigraphic distribution of these plants. Also, until the large-pollen and spore floras in some of the Upper Triassic deposits were studied by Traverse, Cornet, Dunney and others, it was not practical to correlate the North American floral zones with the Triassic stages in Germany.

GEOLOGIC SETTING

Continental deposits of Late Triassic and possibly Early Jurassic age that are known to contain fossil plants are exposed in five regions in North America as shown on Figure 5.1. In the eastern part of the continent they are confined to some narrow fault basins which extend along the Atlantic coast from North Carolina to Nova Scotia. In the western United States they are much more widely distributed and are exposed over large areas in the Colorado Plateau region of New Mexico, Utah, and Colorado; in eastern New Mexico and adjacent areas in the panhandle of Texas; and in Wyoming. Smaller exposures occur also in South Dakota, Montana, Colorado, Oklahoma, Kansas, and Nevada. Continental deposits of Late Triassic age are exposed at a number of places in the state of Sonora in northwestern Mexico and possibly elsewhere in that country. (Maldonado-Koerdell, in Reeside et al., 1957). Several of these deposits reportedly contain plant fossils, but only those in Sonora are considered in this paper.

The continental Late Triassic deposits in North America consist principally of red beds and in places of small amounts of coal, limestone, and basalt. Typically they occur in discontinuous, intertonguing beds that cannot be traced any great distance. General reports on some of them are included with the correlation charts prepared by the Triassic subcommittee of the Committee on Stratigraphy of the National Research Council (Reeside et al., 1957) and with the Paleotectonic maps of the Triassic System pubished by the U.S. Geological Survey. (McKee et al., 1959). The continental Upper Triassic of the Rocky Mountains has been summarized more recently (Mallory et al., 1972) and a detailed report on the Upper Triassic rocks of the Colorado Plateau region has been published (Stewart et al., 1972). A similar report has been published (Reinemund, 1955) on the Deep River Coal Field of North Carolina.

In the following paragraphs, the stratigraphy of the continental Upper Triassic rocks is summarized by area.

WESTERN NORTH AMERICA

Western United States

Dockum Group

In the panhandle of Texas and adjacent areas in eastern New Mexico, Oklahoma, Kansas, and southeastern Colorado, Upper Triassic strata are assigned to the Dockum Group. These strata consist largely of variegated and red claystone and mudstone together with minor amounts of gray sandstone and conglomerate. In this region the Upper Triassic rocks have a maximum thickness of a little over 650m. but they usually are much thinner (see McKee et al., 1959 pl. 5.). Typically, the strata are nearly flatlying. The Dockum has been divided into about a dozen formations but many of them are discontinuous and cannot be

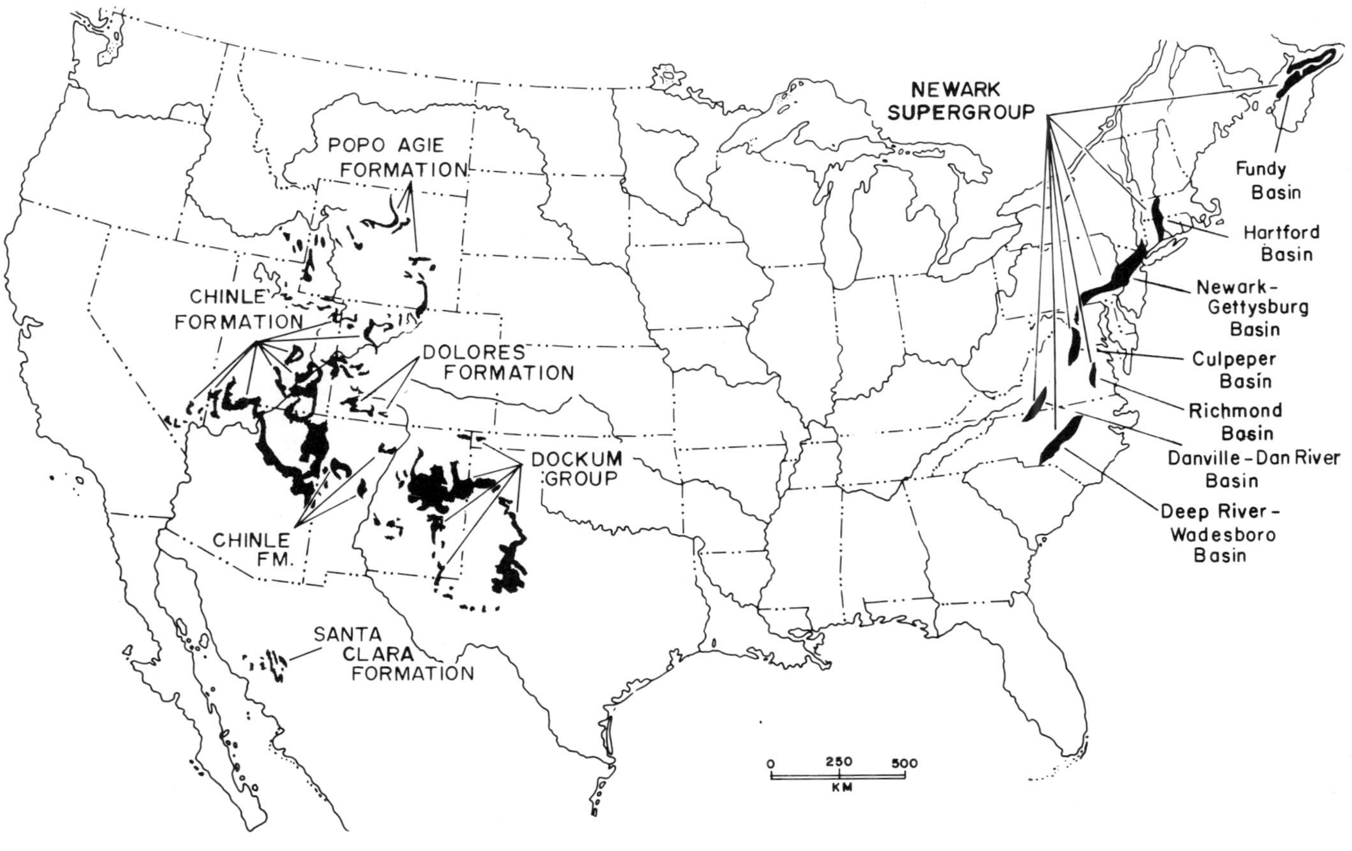

FIGURE 5.1 *Principal exposures of continental deposits of Late Triassic and Early Jurassic age in North America. (Adapted from the "Geologic Map of North America," compiled by North American Geologic Map Committee, E. N. Goddard, Chairman, U.S.G.S., 1965).*

recognized over the entire region. Fossil plants are usually represented by compressions and cuticles. The Dockum appears to be entirely of Late Triassic age.

Chinle Formation

The Upper Triassic strata of continental origin in most of the Colorado Plateau region of the western United States are referred to the Chinle Formation. This formation which has recently been studied in detail (Stewart et al., 1972), ranges up to about 550 m in thickness. It consists predominatly of variegated and red claystone and siltstone but it also contains several thin, widespread beds of drab sandstone, conglomerate, and silty limestone. Typically, these strata are nearly flat-lying, but near uplifts they are often steeply dipping. The Chinle is divided into several members but none are continuous throughout the area. The Chinle Formation is more or less equivalent to the Dockum Group although, as shown elsewhere in this report, the basal part of the Chinle is probably slightly older than the basal part of the Dockum. Most of the leaves in the formation are represented by compressions that contain the cuticle. The Chinle appears to be entirely of Late Triassic age.

Dolores Formation

A 160-190 m thick sequence of red-brown sandstone and siltstone in southwestern Colorado, which is correlative with the Chinle Formation in adjacent areas, is usually called the Dolores Formation. Locally the unit also contains a thin bed of light-colored sandstone and conglomerate. Plant fossils in the formation consist of impressions and casts. Sparse paleontological evidence and stratigraphic correlation suggest that the Dolores is Late Triassic in age (Stewart et al., 1972).

Popo Agie Formation.

The Popo Agie Formation consists of about 35 m of sandstone and siltstone in the Wind River basin of Wyoming. The sandstone beds are mottled pale red and gray while the siltstone beds are banded purple and ocher. This formation is thought to be of Late Triassic age because of its stratigraphic position and is usually correlated with the lower part of the Chinle Formation on the basis of sparse vertebrate fossils (Gregory, 1957). All the fossil plants known from the formation are impressions.

Mexico

Santa Clara Formation.

The Santa Clara Formation is exposed in Sierra de San Javier about 200 km southeast of Hermosillo, Sonora, Mexico. It is about 400 m thick and consists of alternating beds of sandstone, claystone, shale, and coal (De Cserna, 1961). Plant fossils are common in some of the shales, particularly in the carbonaceous shales associated with the coal beds. They are however, represented only by impressions.

Eastern North America

Newark Supergroup

The strata in the fault basins of eastern North America are assigned to the Newark Supergroup although different formational names generally are used in each of the basins (see Reeside et al., 1957). These strata consist principally of red arkosic sandstone, conglomerate, and mudstone, together with smaller amounts of black shale, coal, and drab claystone. In some of the basins, lava flows also occur in the Newark Supergroup. Typically, the strata dip either westerly at about 10^0 -20^0 or easterly at about the same angle. The thickness of the Newark varies from about 1,000 m in the Richmond Basin to nearly 10,000 m in the Newark-Gettysburg Basin. Many of the plant fossils are represented by impressions, but cuticles have been obtained from some of them. Until recently, it was generally thought the Newark was entirely of Late Triassic age (see McLaughlin, in Reeside et al. 1957). Now, however, the work reported by Cornet, Traverse, and McDonald (1973), Cornet and Traverse (1975), and Cornet (1977a), indicates that the upper part of the Newark in the Newark-Gettysburg and the Hartford basins is of Liassic age.

FLORAL ZONES

Three zones based on plant megafossils have been differentiated in the Late Triassic deposits of North America (see Tables 5.1 and 5.2). The oldest zone recognized is here termed the zone of *Eoginkgoites*. Above it is the zone of *Dinophyton* and above it is a presently unnamed zone that may extend into the Lower Jurassic.

Zone of Eoginkgoites

The zone of *Eoginkgoites* characterizes the basal beds of the Chinle Formation in the southwestern United States (Table 5.1) and the lower but not lowest beds of the Newark Supergroup in the eastern United States (Table 5.2). It also may occur in the Popo Agie Formation in Wyoming but this is not certain. The flora does not seem to be as large as that in the overlying zone of *Dinophyton* but it also has not been as intensively studied. Palynological studies indicate (Cornet, 1977a,b) that this zone is middle Carnian (early Late Triassic).

Plant megafossils especially characteristic of this zone include the ferns *Phlebopteris utensis* Arnold (1956) and *Pekinopsis auriculata* Hope and Patterson (1970); the cycad *Leptocycas gracilis* Delevoryas and Hope (1971); the bennettitaleans *Ischnophyton iconicum* Delevoryas and Hope (1976); *Eoginkgoites* Bock (1952) and *Zamites* n. sp.; and the small round plant structures called "red disks." Floras of the zone often contain taxa that occur also in the overlying zone. Such taxa include *Cynepteris lasiophora* Ash (1969), *Wingatea plumosa* (Daugherty, 1941), *Clathropteris walkeri* Daugherty (1941), and *Zamites powellii* Fontaine (in Fontaine and Knowlton, 1890). Represen-

TABLE 5.1 *Upper Triassic and Lower Jurassic floral zones of the southwestern United States and northern Mexico.*

Floral Zone	Southwestern Colorado	Southeastern Utah Northeastern Arizona Northwestern New Mexico	Eastern New Mexico And Texas	Northern Mexico (Sonora)	Stage
Upper				Santa Clara Formation	?Rhaeto—Liassic
Dinophyton	Dolores Formation	lower part of the Petrified Forest Member and the Monitor Butte Member of the Chinle Formation	Trujillo, Tecovas, and Santa Rosa Formations of the Dockum Group		late Carnian, ?Norian
Eoginkgoites		Shinarump and Temple Mountain Members of the Chinle Formation			middle Carnian

TABLE 5.2 *Upper Triassic and Lower Jurassic floral zones of the eastern United States.*

Floral Zone	North Carolina (Deep River Basin)	Virginia (Richmond Basin)	Pennsylvania (Newark—Gettysburg Basin)	Connecticut (Hartford Basin)	Stage
Unnamed				Shuttle Meadow Formation of the Newark Supergroup	?Rhaeto—Liassic
Dinophyton			New Oxford Formation of the Newark Supergroup		late Carnian, ?Norian
Eoginkgoites	Cumnock and Pekin Formations of the Newark Supergroup	lower part of the Newark Supergroup	Stockton Formation of the Newark Supergroup		middle Carnian

tative members of the zone are shown in Figure 5.2.

Chinle Formation

In the southwestern United States, the zone of *Eoginkgoites* has been recognized at two localities in the Temple Butte Member of the Chinle in southeastern Utah (Ash, 1975). Forms represented in the member at these localities include:

Araucarioxylon sp.
Podozamites sp.
Brachyphyllum spp.
Zamites powellii

The zone is represented also at six localities in the Shinarump Member of the Chinle Formation in the same general area in Utah (Ash, 1975). At these localities this member contains a fairly large and diverse flora which includes:

Selaginella?
Equisetites sp.
Phlebopteris utensis
Phlebopteris sp.
Cynepteris lasiophora
Cladophlebis sp.
Araucarioxylon sp.
Podozamites sp.
Brachyphyllum spp.
Pagiophyllum spp.
Williamsonia sp.
Zamites powellii
Zamites n. sp.
Eoginkgoites davidsonii Ash
"red disks"

The zone of *Eoginkgoites* occurs at two localities in northeastern Arizona, where both localities are in the Shinarump Member of the Chinle Formation (Ash, 1977). The fossils collected from these two localities include:

Equisetites sp.
Cladophlebis sp.
Phlebopteris smithii (Daugherty) Arnold, 1947
Cynepteris lasiophora
Pelourdea sp.
Nilssoniopteris n. sp.
Ctenophyllum braunianum Goeppert
Zamites powellii
Eoginkgoites davidsonii

A small flora in the basal beds of the Chinle Formation at a locality near Thoreau, New Mexico, is probably in the

FIGURE 5.2 *Representative plant fossils from the Zone of* Eoginkgoites *(a, b, Compsostrobus neotericus, x 2/3; c,* Leptocycas gracilis, *x 1/3; d,* Eoginkgoites *sectoralis, x l; e, Ischnophyton iconicum, x 1/3; f, Pekinopsis auriculata, ca x l; g, Zamites n. sp., ca x 1/3).*

zone of *Eoginkgoites*. This flora, which has not been described or studied in detail contains:

Araucarioxylon sp.
Pagiophyllum spp.
Brachyphyllum spp.
"red disks"

Popo Agie Formation

The Popo Agie Formation of the Chugwater Group contains poorly preserved impressions of fossil plants at a number of localities in the Wind River basin of central Wyoming (Berry, 1924). The flora is tentatively correlated with the zone of *Eoginkgoites* because the Popo Agie has been correlated by means of vertebrate fossils with the lower part of the Chinle Formation and Dockum Group by Colbert and Gregory (in Reeside et. al., 1957). Also, the flora includes the remains of large linear pinnae that compare with pinnae recently described from the Chinle Formation as *Eoginkgoites davidsoni* (Ash, 1977). The following species occur in the formation:

Equisetum sp.
Pterophyllum browni Berry
Pterophyllum?
Zamites sp.
Eoginkgoites?
Podozamites?

Newark Supergroup

In the eastern United States, the zone of *Eoginkgoites* is characteristic of the Pekin Formation of the Newark Supergroup in North Carolina and of strata that are probably in the upper part of the Stockton Formation of the Newark Supergroup in Pennsylvania. It may eventually be found in the Cumnock Formation of the Newark Supergroup in North Carolina and in the lower strata of the Newark Supergroup in Virginia.

The most completely and best known of these floras is that from the Pekin Formation. It occurs in the Deep River Basin about ten miles northwest of Sanford, North Carolina and is presently being described by Hope and Delevoryas. Currently, these worker, (Hope, 1975; Hope and Patterson, 1970; Delevoryas and Hope; 1971, 1975, 1976) recognize the following forms in the flora:

Neocalamites virginiensis (Fontaine) Berry
N. knowltonii Berry
Danaeopsis plana Fontaine
Cladophlebis borenensis Hope
C. daughertyi Ash
Phlebopteris smithii
Wingatea plumosa
Clathropteris walkeri
Cynepteis lasiophora
C. oblonga (Emmons) Hope
Pekinopsis auriculata
Leptocycas gracilis
Ischnophyton iconicum
Zamites powellii
Otozamites hespera Wieland
Eoginkgoites sp.
Compsostrobus neotericus Delevoryas and Hope
Voltzia andrewsii Delevoryas and Hope
Pelourdea sp.
Phoenocopsis sp.

Many years ago Ebenezer Emmons (1856, 1957) described a small flora from the Cumnock Formatin of the Newark Supergroup in the Deep River Basin about nine miles northwest of Sanford, North Carolina; this flora was subsequently redescribed by Fontaine (1883, see Ward 1900) and Bock (1969). Although no examples of *Eoginkgoites* are known from this flora, it is nevertheless tentatively placed in the zone of *Eoginkgoites* because the pollen and spores in the formation indicate that it is older than units containing the zone of *Dinophyton* (Cornet, 1977a). This flora needs to be thoroughly reevaluated before it will be of much stratigraphic value, but published data (Bock, 1969; Emmons, 1856, 1857; Fontaine, 1883; Ward, 1900) show that it includes one or more species of the ferns *Cynepteris*, *Phlebopteris*, and *Todites;* the cycadophytes *Pterophyllum*, *Anomozamites*, and *Ctenophyllum;* and many conifers such as *Pagiophyllum* and *Podozamites*. Large leaves similar to *Macrotaeniopteris* also are commmon in the flora. Particularly interesting is the occurrence in the flora of the leaf *Sagenopteris* (Fontaine, 1883), a leaf that is practically unknown elsewhere in the Triassic of North America. The flora does compare with the flora in the underlying Pekin Formation and with other floras of the zone of *Eoginkgoites*.

A flora characteristic of the zone of *Eoginkgoites* in the upper part of the Stockton Formation of the Newark Supergroup in the Newark-Gettysburg Basin near Carversville, Pennsylvania (Willard et al., 1959, pl. 2) has been described by Wherry in 1916 (Willard et al., 1959), A.P. Brown (1911), and Bock (1952, 1969). Those authors recognize the following forms in this flora: *Neocalamites* sp., *Taeniopteris*, *Zamites velderi* Brown, *Z. powellii*, *Otozamites* sp., *Eoginkgoites sectoralis* Bock, *E. gigantea* Bock, *Pterophyllum*, *Nilssonia*, *Pagiophyllum*, *Podozamites*, and *Cycadospadix* sp. Notable by their absence are the ferns.

Fontaine (1883) and Bock (1969) described a large flora from several localities in the lower part of the Newark Supergroup in the Richmond Basin of Virginia which is apparently in the zone of *Eoginkgoites*, although that genus has not been recognized in the flora. Cornet (1977a) indicates, however, that the pollen and spores in these rocks show that the flora is somewhat older than the Pekin Formation in North Carolina, which contains the zone of *Eoginkgoites*. The flora described by Fonatine and Bock contains many ferns, including *Cynepteris*, *Todites*, *Mertensides*, *Clathropteris*, and *Cladophlebis;* some leaves resembling *Ctenophyllum*, *Pterophyllum*, *Baiera*, *Podozamites;* and a pinnate leaf that has been called *Sphenozamites*. Many of these fossils are the remains of very large leaves. For example, the pinnae of *Clathropteris* are up to 30 cm in length and 20 cm wide. Apparently, they represent leaves that were originally as much as 70 cm in diameter. The pinnae of *Sphenozamites* are 12 cm long and 30 cm wide and the entire leaves probably were as much as 1 m long and 60 cm wide.

Zone of Dinophyton

Floras of the zone of *Dinophyton* are widespread and numerous in the United States. In the southwestern United States they occur in the lower but not lowest part of the Chinle Formation, throughout the Dockum Group, and in the Dolores Formation (Table 5.1). Floras of the zone also are found in the middle part of the Newark Supergroup in the eastern United States (Table 5.2). The total flora of this zone is quite large; it probably would include more than a hundred species if the many taxa erected by Bock (1969) could be validated. Palynological studies by Cornet (1977a,b) indicate that this floral zone is late Carnian (middle Late Triassic).

The zone is characterized by the occurrence of *Dinophyton* Ash (1970) and the absence of *Eoginkgoites*. Other noteworthy forms include *Selaginella anasazii* Ash (1972); the fern *Todites fragilis* Daugherty (1941); the fern-like foliage called *Cladophlebis daughertyi* Ash (1969) and *Marcouia neuropteroides* Daug. (1941); the bennettitalean *Nilssoniopteris ciniza* Ash (1978); and the enigmatic forms, *Dechellyia gormani* Ash (1972) and *Sanmiguelia lewisii* Brown (1956). Representative members of the zone are shown in Figure 5.3.

Chinle Formation

The classic locality in the southwestern United States for plant fossils of the zone of *Dinophyton*, and for

FIGURE 5.3 *Representative plant fossils from the Zone of* Dinophyton *(a,* Todites fragilis, *x 1/2; b,* Phlebopteris smithii, *x 2; c,* Marcouia neuropteroides, *x 1; d, "pinwheel" and leafy shoot of* Dinophyton spinosus, *x 2; e,* Cladophlebis yazzi, *x 2; f,* Nilssoniopteris ciniza, *x 1/2; g,* Pagiophyllum navajoensis, *x 5; h,* P. readiana, *x 5; i,* P. duttonia, *x 5; j,* Williamsonia nizhonia, *x 1; k,* Dechellyia gormanii, *x 2; 1,* Zamites powellii, *x 1; m,* Sanmiguelia lewisi, *x 1/2).*

Upper Triassic plant fossils as well, is in Petrified Forest National Park in eastern Arizona. As summarized elsewhere (Ash, 1974), plant fossils have been known to science in that area since 1851; and the first large-scale collecting of Upper Triassic leaves anywhere in the Southwest was accomplished in the Park during the 1930's. In this area, the plant fossils occur in the lower part of the Petrified Forest Member of the Chinle and in the Sonsela Sandstone Bed which divides the member into upper and lower parts. The flora as listed by Ash (1974) is as follows:

Isoetites circularis (Emmons) Brown
Chinlea campii (Daugherty) Miller
Lycostrobus chinleana Daugherty
Neocalamites virginiensis (Fontaine) Berry
Equisetites sp.
Todites fragilis
Cynepteris lasiophora
Wingatea plumosa
Phlebopteris smithii
Clathroperis walkeri
Cladophlebis daughertyi
C. yazzia Ash
Itopsidema vancleavii
Zamites powellii
Nilssoniopteris sp.
Lyssoxylon grigsbyi Daugherty
Bairera arizonica Daugherty
Dadoxylon chaneyi Daugherty
Samaropsis puerca Daugherty
Araucarioxylon arizonicum Knowlton
Araucariorhiza joae Daugherty
Woodworthia arizonica Jeffrey
Pagiophyllum simpsonii Ash
Brachyphyllum hegewaldia Ash
Schilderia adamanica Daugherty
Sphenopteris arizonica Daugherty
Marcouia neuropteroides
Carpolithus chinleana
Dinophyton spinosus Ash

A large flora is known from several localities in the Monitor Butte Member in western New Mexico, especially near Fort Wingate. That flora has been only partially described (Ash, 1967, 1968, 1969, 1970, 1972b, Gould, 1971). It includes the following forms:

Neocalamites sp.
Equisetites sp.
Todites fragilis
Cynepteris lasiophora
Wingatea plumosa
Phlebopteris smithii
Clathropteris walkeri
Cladophlebis daughertyi
Marcouia neuropterioides
Pelourdea poleoensis
Araucarioxylon arizonicum
Pagiophyllum navajoensis Ash
P. zuniana Ash
P. duttonia Ash
P. readiana Ash
Brachyphyllum sp.
Zamites powelii
Nilssoniopteris ciniza
Lyssoxylon grigsbyi
Williamsonia nizhonia Ash
?Dechellyia gormanii
Dinophyton spinosus

A small but significant flora is known from the basal beds of the Monitor Butte Member of the Chinle Formation in northeastern Arizona (Ash, 1972c). It is thought that this flora is in the zone of *Dinophyton* because it contains a few poorly preserved fossils that probably are the remains of the leafy shoot and "pinwheels" of that taxon. One of the other fossils *(Dechellyia gormanii)* from the locality has also been tentatively identified from the zone near Fort Wingate, New Mexico. The flora as reported (Ash, 1972c) consists of:

Selaginella anasazia
Dechellyia gormanii Ash
Masculostrobus clathratus Ash

Dinophyton spinosus
?Pelourdea sp.

The coniferous cone *M. clathratus* is noteworthy because it contains pollen grains similar to those previously identified from the Chinle Formation as both *Equisetosporites* and *Ephedra*. This cone probably belongs to the same plant that bore the foliage and seeds described from the same locality as *D. gormanii*.

Dolores Formation

Only a few plant fossils are known from the Dolores Formation of southwestern Colorado (Arnold, 1965, Brown, R., 1956, Tidwell et al. 1977). The flora probably correlates with the zone of *Dinophyton* although that taxon has not been recognized in it. However, *Sanmiguelia lewisii*, which is known from the zone in the Trujillo Formation of the Dockum Group in Texas, does occur in the flora. The following plant megafossils are known from the Dolores Formation or equivalent strata in the area:

Neocalamites sp.
Brachyphjyllum sp.
*Pelourdea poleoensis (*Daugherty) Arnold
Sanmiguelia lewisii

At least some of the structures Becker (1972) reported in association with *Sanmiguelia* in this formation are probably inorganic. For example, the "partially dichotomous pinnate branching structures" reported by him (p. 183, pl. 38, fig. 1, pl. 39, fig. 1) compare with the structures described from the Newark Supergroup in Connecticut by Newberry (1888, pp. 82-84, pl. 21, figs. 1, 2) as a seaweed and called *Dendrophycus triassicus*. More recently, such structures have been interpreted as rill marks (Lull, 1915) and as dendritic surge marks by High and Picard (1968). The latter authors reported their occurrence along modern stream banks. The other structures reported by Becker are too poorly characterized to identify with any degree of certainty at this time, but they also probably are of inorganic origin.

Dockum Group

The zone of *Dinophyton* is characteristic of the Dockum Group of eastern New Mexico and adjacent areas in Texas. The flora known from the Dockum is small in comparison to that of the Chinle Formation in the Colorado Plateau region. This is probably directly related to the relatively small amount of paleobotanical investigations that have been conducted in the Dockum Group. Limited paleobotanical investigations have been conducted in the Dockum by Torrey (1923), Daugherty (1941) and Ash (1972c, 1976).

Near Santa Rosa, New Mexico, the Santa Rosa Sandstone of the Dockum Group contains the following forms typical of the zone of *Dinophyton:*

Cynepteris lasiophora
Zamites powellii
Pelourdea poleoensis
Dinophyton spinosus
Samaropsis sp.

In an adjacent area, the Chinle Formation of the Dockum Group contains *Zamites powellii* and petrified wood similar to *Araucarioxylon arizonicum*.

I have collected fossils typical of the zone of *Dinophyton* from several localities in the Tecovas Formation of the Dockum near Crosbyton, Texas. They include:

Zamites powellii
Pagiophyllum simpsonii
Pagiophyllum n. spp.
Brachyphyllum n. spp.
Dinophyton spinosus

The zone of *Dinophyton* occurs also in the Trujillo Formation of the Dockum Group and I have collected the following species characteristic of the zone from several localities:

Phlobopteris smithii
Clathropteris walkeri
Zamites powellii
Pelourdea poleoensis
Araucarioxylon airzonicum
Woodworthia arizonica
Pagiophyllum simpsonii
Pagiophyllum n. spp.
Brachyphyllum n. spp.
Sanmiguelia lewisii

Newark Supergroup

In the eastern United States a zone of *Dinophyton* flora occurs in the New Oxford Formation of the Gettysburg Basin southwest of York Haven, Pennsylvania (Table 5.2). This flora, which was first studied by Wanner and Fontaine (Ward, 1900), needs to be revised. The original authors identified the following forms in the flora: *Cladophlebis*, *Phlebopteris*, *Cynepteris*, *Todites*, *Pseudodaenaeopsis*, *Macrotaeniopteris*, *Pterophyllum*, *Anomozamites*, *Ctenophyllum*, *Zamites*, and *Sphenozamites*. More recently, Cornet (1977b) has recognized the leafy shoot and "pinwheel" structure of *Dinophyton* in the flora, together with the leafy shoots of *Pagiophyllum simpsonii* and *Pagiophyllum diffusum* (Emmons) Cornet and several seed cones identified as *Glyptolepis*. An unidentified cone studied by Cornet (1977b) contains the pollen *Patinasporites densus*.

Upper Zone

The floras of the uppermost beds in the continental Triassic in North America are poorly known, but several lines of evidence suggest that they probably represent a third floral zone. Until more is known about these floras it seems appropriate to merely record the zone but not to name it. Floras of this unnamed zone occur in the upper part of the Newark Supergroup in the eastern United States and in the Santa Clara Formation in Sonora, Mexico (Figure 5.4). As noted below, the zone may extend into the Early Jurassic.

Newark Supergroup

The Shuttle Meadow Formation of the Hartford Basin of Connecticut contains a small flora originally described by Newberry (1888). More recently, Cornet et al (1973) have reported the occurrence of several additional plant megafossils in the flora of this formation. It is now known to include *Baiera müensteriana* Unger, *Pagiophyllum brevifolium* Newberry, *Otozamites latior* Saporta, *O. brevifolius* Braun, *Cycadinocarpus chapini* Newberry, and *Clathropteris meniscoides* Brongniart. The strata also contain pollen and spores of Rhaeto-Liassic age (Cornet and Traverse, 1975), so the observed flora may be correlated with the zone of *Thaumatoperis* proposed by Harris (1937).

The Portland Formation in Connecticut and Massachusetts contains a small flora that may also belong to the same unnamed zone as the Shuttle Meadow flora or it may be still younger. This flora was recorded by Cornet et al. (1973), Cornet and Traverse (1975), and Cornet (1977a), who reported that it includes the plant megafossils *Equisetites*, *Podozamites*, and *Hirmerella* and is associated with a large palynoflora of Jurassic age.

Santa Clara Formation

The Santa Clara Formation in Sonora, Mexico, contains a large impression flora that has never been adequately studied. Approximately 50 species have been identified from the formation, but only 19 have been described (Newberry, 1876, Silva Pineda, 1961). They include:

Asterocarpus platyrachis Fontaine
Thaumatopteris sp.
Mertensides bullatus (Bunbury) Fontaine
Clathophlebis roesserti (Presl) Saporta
Alethopteris whitneyi Newberry

Ctenophyllum braunianum Schimper
Taeniopteris magnifolia Rogers
T. auriculata (Fontaine) Berry
Pterophyllum fragile Newberry
P. affine Nathorst
Zamites truncatus Zeiller

The flora of the Santa Clara Formation also contains some leaves resembling *Glossopteris* (Figure 5.4) and similar to leaves from the Jurassic of southern Mexico, which have been named *Mexiglossa* Delevoryas and Person (1975). None of the species recognized in this flora have been identified in the Chinle flora but many occur in the Newark flora. Also some (e.g., *Clathophlebis roesserti* and *Mertensides bullatus)* occur elsewhere in Rhaeto-Liassic strata (see Silva-Pineda, 1961). The absence of "typical" Chinle forms such as *Dinophyton* and *Zamites powellii* in the Santa Clara flora and the presence of some Rhaeto-Liassic species, including *Thaumatopteris,* suggests that the flora may represent a zone younger than the zone of *Dinophyton.* Possibly, the upper part of the zone may be equivalent to the early Liassic zone of *Thaumatopteris* proposed by Harris (1937).

CONCLUSION

Analysis of plant megafossils in the continental Upper Triassic rocks of North America shows that three biostratigraphic zones can be differentiated in these strata and can be correlated in turn with the Triassic stages in Germany by using associated pollen and spores. The oldest of these zones, the zone of *Eoginkgoites,* is of middle Carnian (early Late Triassic) age. Floras attributed to this zone have been identified at several localities in the eastern United States and at a few places in the southwestern part of the country. These floras are not as well known as the floras of the overlying zone, but characteristic taxa appear to include several species of *Eoginkgoites* and *Zamites* and the small round plant structures called "red disks." The next youngest zone, the zone of *Dinophyton,* is late Carnian and possibly Norian (middle Late Triassic) in age. Floras attributed to this zone are widely distributed in the United States, particularly in the southwest where they have been intensively studied during the past decade. Fossils characteristic of this zone include *Dinophyton, Nilssoniopteris, Dechellyia, Sanmiguelia,* and many conifers. Above the zone of *Dinophyton* is an unnamed upper zone that probably is of Rhaeto-Liassic (latest Triassic-earliest Jurassic) age. Floras of this upper zone are very poorly known and have been identified at only a few localities in the northeastern United States and in northwestern Mexico. They contain several ferns, such as *Thaumatopteris* and *Todites,* and cycadophytes, such as *Otozamites, Zamites,* and *Pterophyllum.* If these determinations prove correct, the upper part of the zone would correlate with the early Liassic zone of *Thaumatopteris* in Greenland and western Europe.

ACKNOWLEDGMENTS

Much of my own research reported here was supported by a grant from the Earth Sciences Section, National Science Foundation (Grant GA-25620) and by grants from the Geological Society of America, the Society of the Sigma Xi, and Weber State College. This support is acknowledged with thanks. I am grateful to Dr. R. C. Hope, Campbell College, Buie's Creek, North Carolina for furnishing some of the photographs in Figure 5.2. Throughout the paper I have taken the liberty of correcting the orthographic errors made by myself and others as authorized by article 73 of the Code of Botanical Nomenclature.

FIGURE 5.4 *Representative plant fossils from the upper zone, Santa Clara Formation, Sonora, Mexico (a,* Taeniopteris *sp.; b,* Todites *sp.; c,* Cladophlebis *sp.; d,* Ptilophyllum *sp.; e,* Otozamites *sp.; f,* Zamites *sp.; g,* Pelourea *sp.; h,* Pterophyllum *sp.; i, j,* Glossopteris-*like leaf; all x 1).*

REFERENCES

Arnold, C. A. 1947. *An Introduction to Paleobotany*. New York: McGraw-Hill, 433 p.

Arnold, C. A. 1956. Fossil ferns of the Matoniaceae from North America. *Paleontol. Soc. India Jour.* 1:118-121.

Arnold, C. A. 1965. *Cordiates*-type foliage associated with palm-like plants from the Upper Triassic of southwestern Colorado. *Jour. Indian Bot. Soc.* 42A: 4-9.

Ash, S. R. 1967 The Chinle (Upper Triassic) megaflora of the Zuni Mountains New Mexico. *New Mexico Geol. Soc. Guidebook, 18th Field Conf.*, pp. 125-131.

Ash, S. R. 1968. A new species of *Williamsonia* from the Upper Triassic Chinle Formation of New Mexico. *Jour. Linnean Soc. London*, 61:113-120.

Ash, S. R. 1969. Ferns from the Chinle Formation (Upper Triassic) in the Fort Wingate area, New Mexico. *U.S. Geol. Surv., Prof. Paper 613-D*, pp. D1-D52.

Ash, S. R. 1970. *Dinophyton*, a problematical new plant genus from the Upper Triassic of southwestern United States. *Palaeontology* 13:646-663.

Ash, S. R. 1972a. Late Triassic plants from the Chinle Formation in northeastern Arizona. *Palaeontology* 15:598-618.

Ash, S. R. 1972b. *Marcouia*, gen. nov., a problematical plant from the Late Triassic of the southwestern U.S.A. *Palaeontology* 15:424-429.

Ash, S. R. 1972c. Upper Triassic Dockum flora of eastern New Mexico and Texas. *New Mexico Geol. Soc. Guidebook, 23rd Field Conf.*, pp. 124-128.

Ash, S. R. 1974. The Upper Triassic Chinle flora of Petrified Forest National Park, Arizona. In *Guidebook to Devonian, Permian and Triassic Plant Localities, East-Central Arizona*, S. R. Ash, ed. Paleobot. Sec., Bot. Soc. Am., pp. 43-50.

Ash, S. R. 1975. The Chinle (Upper Triassic) flora of southeastern Utah. *Four Corners Geol. Soc. Guidebook, 8th Field Conf.*, Canyonlands, pp. 143-147.

Ash, S. R. 1976. Occurrence of the controversial plant fossil *Sanmiguelia* in the Upper Triassic of Texas. *Jour. Paleontoloty* 50:799-804.

Ash, S. R. 1977. An unusual bennettitalean leaf from the Upper Triassic of the southwestern United States. *Palaeontology* 20:641-659.

Ash, S. R. 1978. Plant megafossils. In *Geology, paleontology and paleoecology of a Late Triassic lake in western New Mexico*, S. R. Ash, ed. Brigham Young Univ. Geol. studies, vol. 25, pt. 2, ch. 4.

Becker, H. F. 1972. *Sanmiguelia*, an enigma compounded. *Palaeontographica* 138 (pt. 3): 181-185.

Berry, E. W. 1924. Fossil plants and Unios in the red beds of Wyoming. *Jour. Geol.* 32:488-497.

Bock, 1952. New eastern Triassic Ginkgos. *Bull. Wagner Free Inst. Sci. Philadelphia*. 27:9-14.

Bock, 1969. The American Triassic flora and global distribution. *Geol. Center Research Ser.*, vols. 3 and 4.

Brown, A. P. 1911. New cycads and conifers from the Trias of Pennsylvania. *Proc. Acad. Nat. Sci. Philadelphia*. 63:17-21.

Brown, R. 1956. Palmlike plants from the Dolores Formation (Triassic) southwestern Colorado. *U.S. Geol. Sur. Prof. Paper 274-H*, pp. 205-209.

Cornet, B. 1977a. The palynostratigraphy and age of the Newark Supergroup. Ph.D. thesis, Penn. State Univ., 505 p.

Cornet, B. 1977b. Preliminary investigations of two Late Triassic conifers from York County, Pennsylvania.

Geobotany pp. 169-172. Plenum Pub. Co. N.Y.

Cornet, B., and A. Traverse 1975. Palynological contributions to the chronology and stratigraphy of the Hartford Basin in Connecticut and Massachusettes. *Geosci. Man.* 11:1-33.

Cornet, B., A. Traverse; and N. G. McDonald. Fossil spores, pollen and fishes from Connecticut indicate Early Jurassic age for part of the Newark Group. *Science* 182: 1243-1247.

Daugherty, L. h. 1941. The Upper Triassic flora of Arizona. *Carnegie Inst. Wash. Pub. 526.* 108p.

DeCserna, G. A. 1961. Estratigrafia del Triasico Superior de la parte central del Estado de Sonora. Part 1 of "Paleontologia del Triasico Superior de Sonora" *Univ. Nac. Auton. Mexico Inst. Geol. Paleontol. Mexican no. 11,* 18p.

Delevoryas, T., and R. C. Hope. 1971. A new Triassic cycad and its phyletic implications. *Postilla,* no. 150, 22p.

Delevoryas, T., and R. C. Hope. 1975. *Voltzia andrewsii* n. sp., an Upper Triassic seed cone from North Carolina, U.S.A. *Rev. Palaeobot. Palynol.* 20:67-74.

Delevoryas, T., and R. C. Hope. 1976, More evidence for a slender growth habit in Mesozoic cycadophytes. *Rev. Palaeobot. Palynol.* 21: 93-100.

Delevoryas, T., and C. P. Person. 1975. *Mexiglossa varia* gen. et sp. nov., a new genus of glossopteroid leaves from the Jurassic of Oaxaca, Mexico. *Palaeontographica, B* 154:114-120.

Emmons, E. 1856. *Geological Report on the Midland Counties of North Carolina.* New York: Putnam, 347p.

Emmons, E. 1857. *American Geology,* pt. 5. Albany: Sprague, 152p.

Fontaine, W. M. 1883. Contributions to the knowledge of the older Mesozoic flora of Virginia. *U.S. Geol. Surv. Monogr. 6,* 144p.

Fontaine, W. M., and F. H. Knowlton. 1890. Notes on Triassic plants from New Mexico. *Proc. U.S. Natl. Mus.* 13:281-285.

Gould, R. E. 1971. *Lyssoxylon grigsbyi,* a cycad trunk from the Upper Triassic of Arizona and New Mexico. *Am. Jour. Bot.* 58:239-248.

Gregory, J. T. 1957. Significance of fossil vertebrates for correlation of Late Triassic continental deposits of North America. In El Mesozoico del Hemisferio Occidental y sus correlaciones mundailes." *20th Internl. Geol. Cong.* (Mexico City 1956) sec. 2, pp. 7-26.

Harris, T. M. 1937. The fossil flora of Scoresby Sound, east Greenland, Pt. 5. Stratigraphic relations of the plant beds. *Medd. Grønland,* 112 (pt. 2): 1-114.

High, L. R., and D. M. Picard, 1968. Dendritic surge marks ("Dendrophycus") along modern stream banks. *Univ. Wyoming, Contrib. Geol.* 7:1-16.

Hitchcock, E. 1823. Geology and mineralogy, and scenery of the regions contiguous to the River Connecticut, with a geological map and drawings of organic remains. *Am. Jour. Sci.* 6:1-86, 201-236; 7:1-30.

Hope, R. C. 1975. A paleobotanical analysis of the Sanford Triassic basin, North Carolina. Ph. D. thesis, Univ. South Carolina, 74p.

Hope, R. C., and O. F. Patterson 1970. *Pekinopteris auriculata:* A new plant from the North Carolina Triassic. *Jour. Paleontology* 44:1137-1139.

Lull, R. S. 1915. Triassic life of the Connecticut Valley. *Conn. State Geol. Nat. Hist. Surv. Bull.* 24:126.

McKee, E. D., et al. 1959. Paleotectonic maps of the Triassic System. U.S. *Geol. Surv., Misc. Geol. Inv. Map. I-300,* 33p.

Mallory, W. W., et al. (eds.). 1972. *Geologic Atlas of the Rocky Mountain*

Region. Denver: Rocky Mountain Assoc. Geol., 331p.

Newberry, J. S. 1876. Geological report. In *Report of the Exploring Expedition from Santa Fe, New Mexico...*, J. N. Macomb, Washington: U.S. Army Eng. Dept., pp. 9-118.

Newberry, J. S. 1888. Fossil fishes and fossil plants of the Triassic rocks of New Jersey and the Connecticut Valley. *U.S. Geol. Surv. Monogr. 14,* 152p.

Reeside, J. B., Jr. (chm). 1957. Correlation of the Triassic formations of North America exclusive of Canada, with a section on the correlation of continental Triassic sediments by vertebrate fossils by E. H. Colbert and J. T. Gregory. *Geol. Soc. Am. Bull.* 68:1451-1513.

Reinemund, J. A. 1955. Geology of the Deep River Coal Field, North Carolina. *U.S. Geol. Surv. Prof. Paper 246,* 159p.

Silva Pineda, A. 1961. Flora fosil de la Formacion Santa Clara (Carnico) del Estado de Sonora. Part 2 of "Paleontologia del Triasico Superio de Sonora" *Univ. Nac. Auton. Mexico, Inst. Geol. Paleontol. Mexicana no. 11,* 36p.

Stewart, J. H., et al. 1972. Stratigrpahy and origin of the Chinle Formation and related Upper Triassic strata in the Colorado Plateau Region. *U.S. Geol. Surv. Prof. Paper 690,* 336p.

Tidwell, W. D., et al. 1977. Additional information concerning the controversial Triassic plant *Sanmiguelia. Palaeontographica, B* 163:143-151.

Torrey, R. E. 1923. The comparative anatomy and phylogeny of the coniferales. Pt. 3, Mesozoic and Tertiary coniferous woods. *Boston Soc. Nat. Hist. Mem.* vol. 6.

Ward, L. H. 1900. The Older Mesozoic, In. "Status of the Mesozoic floras of the United States." *U.S. Geol. Surv. 20th Annu. Rept.*, pt 2, pp.213-748.

Wherry, E. T. 1916. Two new fossil plants from the Triassic of Pennsylvania. *Proc. U.S. Nat. Mus.* 51:327-329.

Willard, B., et al. 1959. Geology and mineral resources of Bucks County, Pennsylvania. *Penn. Topogr. Geol. Surv. Bull. C9,* 243p.

Addendum

Since this paper was submitted for publication, I have examined the large collection made by Dr. Reinhart Weber and his students from the Santa Clara Formation in Sonora, Mexico. It is apparent from their work and my examination that at least a part of the Santa Clara flora is equivalent to the zone of *Eoginkgoites* and does not correlate with my unnamed upper zone. A summary of their work appeared recently. (Weber, R. 1980. New Observations on late Triassic flora of Santa Clara Formation, Sonora, Mexico, abstr. for Intern. Paleobotanical Cong., Reading, Eng. p. 61).

6

THE IMPORTANCE OF DEPOSITIONAL SORTING TO THE BIOSTRATIGRAPHY OF PLANT MEGAFOSSILS

Robert A. Spicer

SUMMARY

Both fluid and biological sorting of potential plant megafossils during transport and depositional processes produce a pattern that has previously been overlooked. This pattern is so complex that casual observation is insufficient to interpret it in terms of the source vegetation. The treatment of fluid sorting on a theoretical basis is also extremely complex; therefore, laboratory experiments, and examples of deposition in modern environments, must be used to illustrate problems that could arise when attempts are made to reconstruct the source vegetation using detailed stratigraphic analysis.

The pattern of plant remains relative to the entombing sediment, when deposited in fluviolacustrine deltaic environments under conditions of steady deposition, may be used to determine spatial species separation in the source vegetation. Flood deposits, however, yield a statistically more valid sample of the source vegetation as a whole although spatial information may be lost. Studies of the matrix grain size and sedimentary structures associated with plant fossil assemblages can yield information that is valuable in correcting some of the biases inherent in depositional sorting.

INTRODUCTION

Substantial sorting of plant remains takes place during transport from the growing site to the place of deposition; and, as a result, allochthonous plant fossil assemblages yield a highly biased sample of the source vegetation. Until recently, there have been few attempts to examine sorting phenomena or the relationship between the deposited assemblage and the source communities (Chaney, 1924; Fegruson, 1971; McQueen, 1969; Birks, 1973).

The more or less two-dimensional nature of most plant debris (leaves, fronds, and some fruits and seeds), together with the time-dependent fluctuations in submerged density due to progressive waterlogging, fragmentation, and breakdown by microorganisms, effectively preclude the use of any theoretical fluid-dymanics equations to predict the behavior of dispersed plant organs during aqueous transport. Transport by wind is no less complicated; and, because physical evidence of wind strength and direction is rarely preserved in the stratigraphic column, the effects may be even more difficult to interpret. In view of these problems, a more empirical approach to the subject of sorting is presented here, where examples of both field and laboratory studies of modern depositional environments are cited.

WIND TRANSPORT

The selective transport of leaves, fruits, and seeds by wind is obvious even

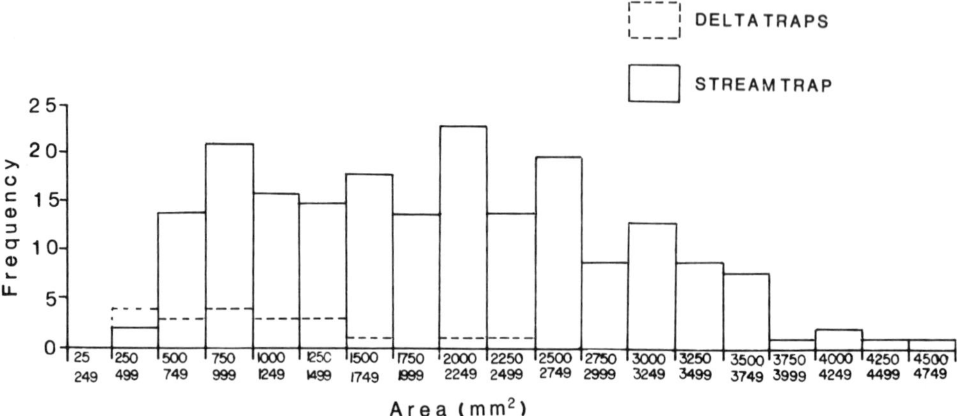

FIGURE 6.1 *Size frequency histogram for* Fagus sylvatica *leaves recovered from both stream and delta traps. The stream trap was in a part of the stream that flowed past a number of mature trees. These trees formed a closed canopy over the stream, and the sample may be considered to be representative of litter on the woodland floor. Using the Fisher-Behrens test for two-sample comparisons with unequal variance (Campbell, 1967), the null hypothesis that the two populations were equal as regards their means had to be rejected because the observed* d *= 6.94 exceeds the published value at the 1 percent level.*

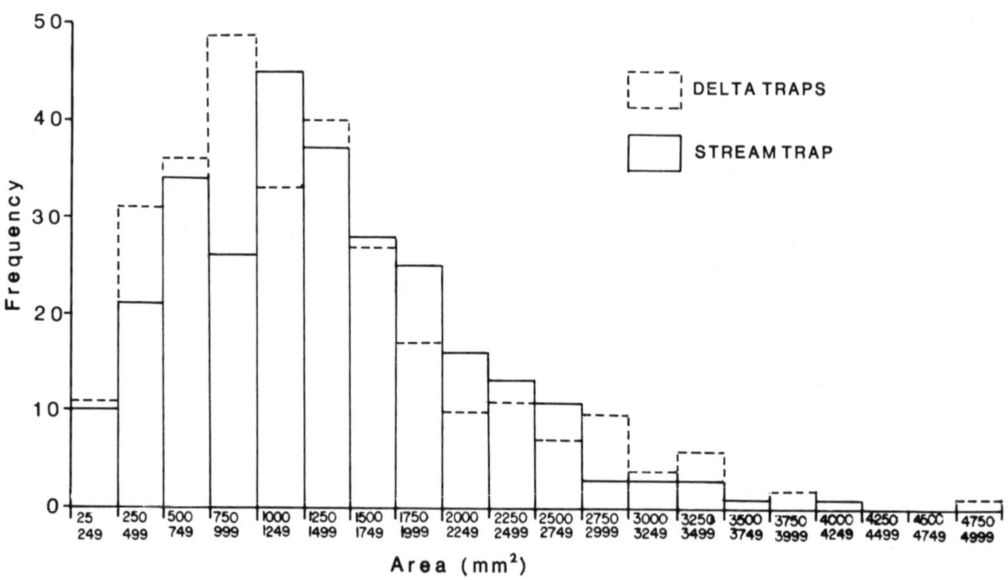

FIGURE 6.2 *Size frequency histogram for* Alnus glutinosa *leaves recovered from both stream and delta traps. The Fisher-Behrens test showed no significant difference in the means (*d *= 0.56).*

to a casual observer; and the special morphological characteristics of various fruits and seeds that facilitate wind dispersal are commonly studied in elementary botany classes. Laboratory experiments on both artificial and natural leaves (Spicer and Ferguson, unpub. data) show that, as expected, leaves with a small weight per unit area have a lower fall velocity in still air than do heavier leaves and thus are likely to be carried further by wind. Typically, these are large shade leaves.

Quantitative field sampling, however, shows that leaves blown directly on to an open lake surface from distantly growing trees are predominantly the small, dense "sun" leaves (Spicer, 1975). Figure 6.1 illustrates the difference in size distribution between wind transported *Fagus sylvatica* L. leaves recovered from delta traps (lake surface) and leaves of the same species recovered from traps situated on the surface of a woodland stream directly beneath a stand of mature *F. sylvatica* trees. This selectivity in favor of small leaves is not demonstrated in Figure 6.2, in which the sizes of *Alnus glutinosa* (L.) Gaertn. leaves recovered from similar traps are presented. In the Figure 6.1 experiment the source trees were some distance away from the delta traps, whereas in the second experiment the source trees were growing along the lake margin as well as along the stream.

The predominance of small sun leaves in the lake samples results from the fact that only the "sun" leaves are exposed to relatively high wind velocities and therefore have any chance of being blown significant distances. Trees growing around the lake margin can contribute leaves to the lake sediments by direct leaf fall and minimal lateral transport. The leaf population derived from such trees will closely resemble that found on the forest floor and, as such, will consist of both "sun" and "shade" leaves. Lacustrine sediments deposited in small basins are consequently likely to contain a population of fossil leaves that reflects the full size range of leaves, at abscission, in the original forest. Conversely, with deposits laid down in basins where the minimum lake width is large compared with the canopy height, this is not so. With increasing distance from the shore the basin will become relatively enriched with sun leaves from both the locally and distantly growing source taxa.

If the local community is restricted in area or leaf productivity, then its contribution to the fossil assemblage may become diluted, an effect analogous to that described by Chaloner (1958), Neves (1958) and Chaloner and Muir (1968) for spores.

The preceding argument applies only to a forest with a complete (closed) canopy. In savanna vegetation more of the individual crowns will be exposed to high winds; consequently, the deposited leaf population would seem likely to reflect more accurately the size distribution of the total leaf population. This supposition, however, remains to be tested.

The distance any given leaf will be transported by wind will partly depend on the height at which it grew. Understory plants in a closed-canopy forest will be sparsely represented in the fossil deposit because transport of abscissed organs by wind will be minimal. The filtering effect of neighboring plants will also ensure that only riparian understory plants have any significant chance of entering a depositional environment under normal circumstances. If the stream or lake is wide enough that the canopy is broken, then the increased light available may produce riparian vegetation that bears little resemblance to the bulk of the forest and, therefore, a biased fossil record of the regional vegetation.

WATER TRANSPORT

The transport of plant debris by water is in many ways more complex than transport by wind, inasmuch as dispersed organs are more susceptible to progressive destruction by microorganisms and mechanical fragmentation. It has been demonstrated that significant mechanical fragmentation of leaves only occurs after microbial preconditioning; fresh leaf material is very robust (Ferguson, 1971; Spicer, 1975). Similarly, microbial colonization predisposes leaves to invertebrate attack (Kaushik, 1969; Petersen and Cummins, 1974; Kaushik and Hynes, 1971). Therefore, those factors that affect microbial colonization also affect the likelihood of a species entering the fossil record. Unfortunately, such factors are commonly of a chemical rather than a physical nature and are rarely preserved.

Fragmentation, whatever the cause, is primarily a water-related phenomenon, and by studying the separate distributions of both whole and fragmented leaves, much can be learned about the source vegetation.

Spicer (1975) examined the distribution of deposited leaves in an existing modern fluviolacustrine environment. Systematic sampling across a single bedding plane within deltaic sediments, followed by multivariate statistical analyses of the resultant data, revealed that those species growing furthest away from the depositional site were represented in fragmented leaf form; those species with both distant and local distributions existed as whole leaves and as fragments; and those growing locally existed as more or less whole leaves. Moreover, the fragments were associated with deltaic stream distributaries and thus the pattern of leaf distributions across a single bedding plane could be related to the spatial distribution of the source taxa.

In consolidated sediments it is rare that such a bedding plane can be completely exposed for examination; but, fortunately, the vertical patterns of plant deposition can also be useful for reconstructing paleovegetation.

By means of laboratory experiments, Jopling (1960, 1963, 1965a,b) was able to show that the shape of the vertial profile of a delta front depends on the distribution of sediment grain size, the velocity of the in-

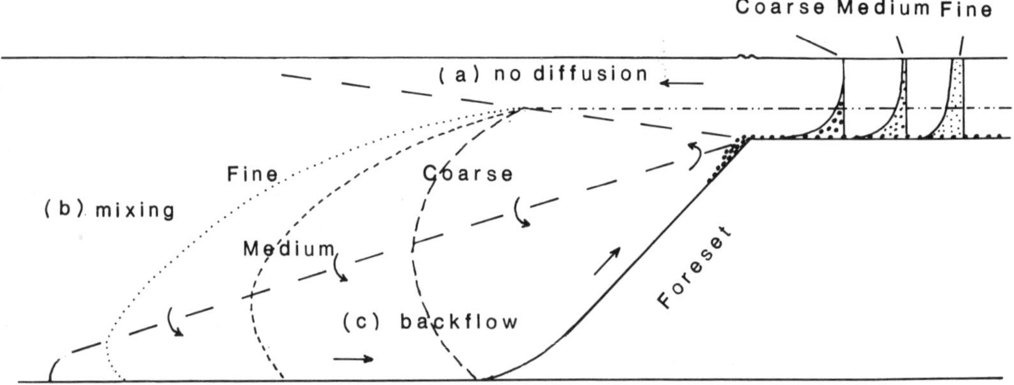

FIGURE 6.3 *The two-dimensional flow pattern over a delta profile. The zone of no diffusion can be considered as residual streamflow. The zone of mixing is characterized by strong eddies. The sorting of transported particles as a function of settling velocity is illustrated. (After Jopling, 1965a).*

flowing stream, and the depth ratio (depth of stream/depth of basin). The idealized vertical distribution of suspended sediment in a stream with a logarithmic vertical velocity profile is also logarithmic. At the delta lip, the flow separates and can be divided into three regions (Figure 6.3): (a) a zone of no diffusion, which may be thought of as residual stream flow and is not affected by turbulent eddies; (b) a zone of mixing, which is characterized by large-scale eddies; and (c) a zone of reverse circulation divided from the zone of mixing by a locus of zero velocity. Jopling described sedimentation over the delta front by plotting trajectories of particles as they transversed these changing flow-velocity fields. The trajectories were calculated by using the resultant of the forward motion of the grain imparted by the fluid flow and the settling velocity. In spite of a number of objections to this approach, the observed data fit the predicted grain-size distributions tolerably well (at least for sand grains). The vertical distribution of grain concentrations in the inflowing stream as well as the settling velocity of the particles is important in determining where any given grain will settle out beyond the foreset lip. The bulk of the bedload material will collectively settle out at the top of the foreset slope, while suspended material will be carried forward to be deposited lower on the foreset slope or as bottomset beds.

The geometry of the delta front is dependent upon the velocity of the stream relative to the sediment grain size and the depth ratio of the delta. If the grain size is large relative to the stream competence, then most of the sediment is transported as bedload and will be deposited at the top of the foresets. With continued loading, the slope will become unstable, and the foreset sediments will slump to form an angular contact with the bottomset or lake sediments. At higher water flow, or, more strictly, bed shear stress (grain mix constant), some grains will go into suspension and will be deposited further down the foreset slope; a tangential contact between foreset and bottomset beds will develop as toeset beds are formed. With continued increase in water velocity, the angle of the foreset beds decreases until all evidence for crossbedding is lost and horizontally laid beds result.

This model has only been laboratory tested for inorganic sediment grains. The question remains as to how applicable it is to the problems of plant-debris sorting in fluviolacustrine environments. The following two examples demonstrate that the concept also applies to the deposition of organic material in both low-and high-energy environments.

Cores taken through a small fluviolacustrine delta in an artificial lake at the Imperial College Field Station near Ascot, Berkshire, England, clearly demonstrate leaf deposition in a low-energy environment (Spicer, 1975). The sediment was composed of flocculent ferric hydroxide($Fe(OH)_3 . nH_2O$)together with some silt-size quartz grains. The flocculent nature of the sediment precluded any meaningful measurement of settling velocity, but the bulk of the inorganic sediment had a settling velocity much less than that of most of the plant material. A series of cores transecting the delta parallel to the stream flow revealed two distinct leaf concentrations (Figure 6.4). The lower leaf bed (A) is interpreted as being composed mainly of leaves deposited directly into the lake from the immediate local vegetation. Inorganic sedimentation in the open lake is minimal compared with the annual contribution of leaves, but a substantial thickness of leaves is prevented from forming owning to selective degradation by aquatic invertebrates and microorganisms. Nevertheless, the leaf layer deposited directly in front of the ad

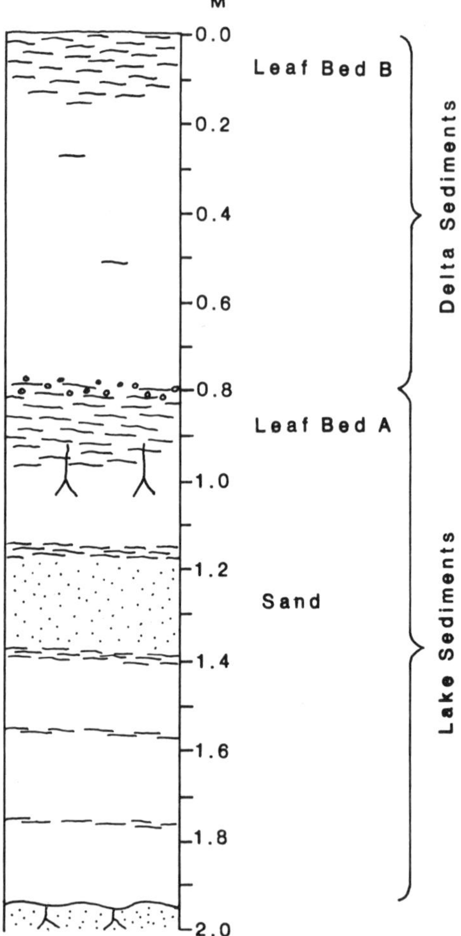

FIGURE 6.4 *An idealized vertical section through deltaic and lacustrine sediments deposited in a low-energy environment. (Based on data from 22 cores taken from Silwood Lake, Berkshire, England).*

vancing delta is rapidly buried owing to the increased sedimentation rate associated with the advancing delta and is thus preserved from immediate destruction. In the region of the forest slope, sedimentation is rapid, and the rain of locally derived plant debris is diluted. This dilution results in a body of sediment with a low concentation of organic material. A second leaf bed (leaf bed B) is formed at the top of the foreset slope as a result of collective settling of the stream bedload material and consequently is enriched with leaves derived from plants growing upstream, possibly some distance away from the depositional environment.

Figure 6.5 schematically illustrates the process of formation of the two leaf beds. At point A on the lake bed, the rate of leaf deposition is high relative to the sediment deposition rate; and leaf bed A results. At point B on the foreset slope, the inorganic sedimentation rate is high compared to the rate of leaf deposition. At point C leaf deposition again exceeds sediment deposition because the leaves transported as stream bedload collectively settle out and accumulate as the upper leaf bed (Figure 6.4 B).

The species composition of the two leaf beds therefore reflects the spatial distribution of the source taxa. Because the situation is a dynamic one, the relationship is not constant. As infilling of the basin proceeds, riparian plants will colonize the subaerial deltaic sediments; and the species composition of this community is likely to be similar to the lake and streamside vegetation. This vegetation not only contributes leaves to the stream flowing through it but it also filters out those leaves transported from the distantly growing trees. Consequently as infilling progresses and the size of this delta surface community increases, the species composition of both upper and lower leaf beds will become more alike. Both will be enriched with leaves derived from the local vegetation. This process is illustrated in Figure 6.6 The development of the hydrosere community may be detectable in the fossil record by a lateral change of species composition of the upper leaf bed in the direction of infilling.

Clearly the discrete nature of the two beds is also dependent on the quantity and hydraulic characteristics of the inorganic sediments versus the plant debris relative to the velocity of the stream. In higher-energy environments a large proportion of the plant material, as well as the finer in-

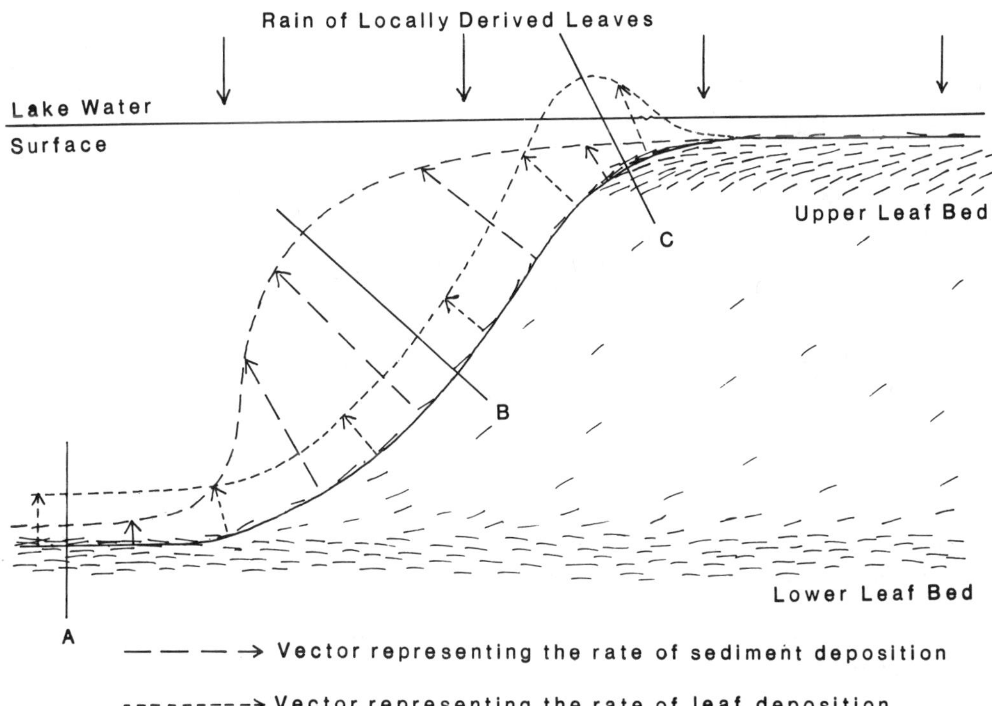

FIGURE 6.5 *A diagrammatic representation of leaf deposition in a fluviolacustrine environment. The rates of leaf and sediment deposition normal to the sediment surface are shown by the dashed (leaf) and dotted (sediment) vectors. At position A on the lake bed the rate of leaf deposition exceeds that of the sediment; at B the sediment deposition rate is greater than that of the leaves; while at C the leaf deposition rate is again greater.*

organic sediment grains, will be suspended. They will therefore be carried over the delta lip and deposited down the foreset slope and perhaps in the bottomset beds. The separation of the leaf beds will consequently be destroyed, and sorting of plant material according to hydrodynamic characteristics, such as settling velocity, will occur. An example of plant debris sorting in a high-energy environment is seen in flood deposits within Clair Engle Lake, northern California.

In December 1964 massive floods occurred in nothern California. Both the cause and the effect are well documented (California Department of Water Resources, 1965; Stewart and LaMarche, 1967). At Clair Engle Lake it appears that the bulk of the sediments, exposed in the lake basin as a result of the 1976-1977 drought, were deposited during that time. These sediments range from silt to coarse gravel. Systematic quantitative sampling of deposits derived from six separate drainages revealed significant sorting of plant remains, although two distinct leaf beds were not in evidence and plant debris was deposited over the foreset slope, toeset, and bottomset beds (Figure 6.7). The sorting under these circumstances, as suggested by Jopling's studies, was primarily the result of relative differences in the hydraulic characteristics of the various plant organs. Figure 6.8 shows the distribution of seeds and moss fragments across the delta front profile from one drainage.

Although the moss remains, because they have generally a low settling velocity in water, have been carried into the toeset and bottomset beds, the separation from the seeds is not distinct. This indistinctness arises, first, because the population of fruit and seed debris, as well as moss fragments, exhibits a wide range of settling velocities: we are dealing with a biological population to which physical parameters can only be assigned within statistical limits. Second, the vertical distribution of both seeds and moss fragments within the stream during transport will not be mutually exclusive, and consequently, using Jopling's model, their final deposition will also be mixed.

Another important difference between this flood assemblage and low-energy deposits is the number of different plant organs that are represented. The low-energy environment contained mostly leaves, whereas the flood deposits yielded large numbers of seeds, fruits, fern and moss fragments, twigs, cones, leaves, roots, and even logs. The only way that many plants, for example, herbaceous species, can be represented to any extent in the fossil record is in the form of seeds. Consequently, the flood deposit is likely to yield a greater number of taxa. For reasons already discussed, long-distance wind transport or low-energy water transport contribute a biased sample of the source vegetation to the depositional basin. This bias is likely to be less when forest floor litter is washed into the deposits by large-scale erosion during overbank discharge. When such erosion is coupled with a rapid burial rate at the site of deposition, a statistically more valid vegetation sample is likely to be preserved. Forest floor erosion during the 1964 flood is shown in Figure 6.9. The violent weather that frequently causes flooding

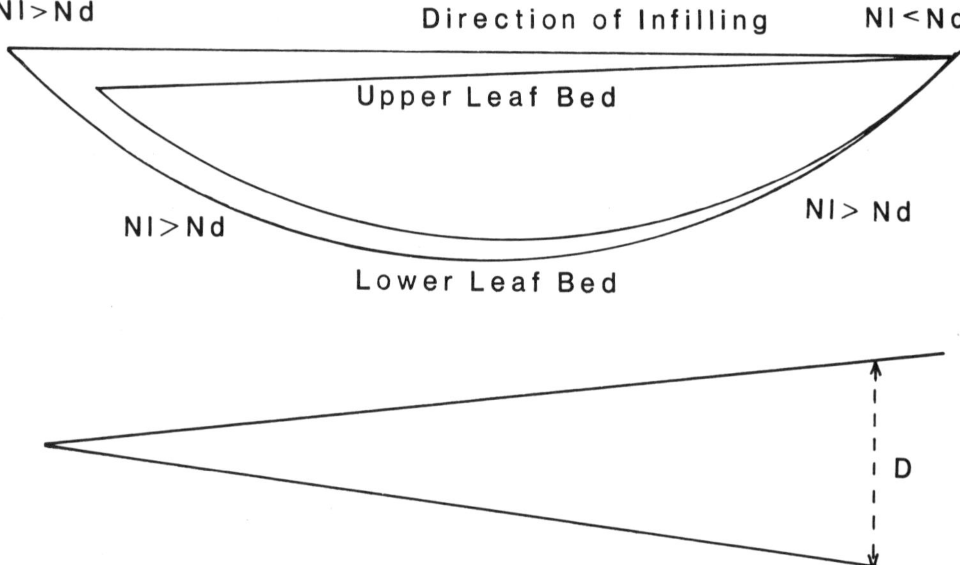

FIGURE 6.6 *Top: Longitudinal vertical section through a completely infilled lake basin. The species composition of the upper leaf bed changes from one dominated by a component derived from distantly growing source trees Nd in the early stages of infill, to an assemblage dominated by locally derived species Nl as vegetation develops on the subaerial deltaic sediments. Bottom: If a distance measure D is used to describe the species dissimilarity between the two leaf beds the situation described above may be represented by two converging lines.*

FIGURE 6.7 *Crossbedding within flood deposits of the Mule Creek drainage, Clair Engle Lake, Calif. Plant debris is scattered over topset, foreset, toeset, and bottomset beds.*

commonly damages the vegetation. Whole branches bearing leaves and reproductive organs may be torn from a tree and, if they become preserved, may provide the paleobotanist with excellent material for morphologic and taxonomic studies (Figure 6.10). Leaf fall is also accelerated under such conditions, with the result that much fresh and relatively undamaged material enters the fossil record. Although violent, the transport to the final site of burial is also rapid, and the remains become entombed before any biodegradation can take place.

A further example of the importance of rapid deposition can be seen in the low-energy environment. The core diagram (Figure 6.4) shows a sand body underlain by poorly preserved leaves and overlain by well-preserved leaves. This sand, which is interpreted as resulting from a short period of rapid stream discharge, buried partially decayed lake-deposited leaves. Moreover, the rapid water flow washed into the lake a population of relatively fresh leaves from the stream bed and woodland litter. These leaves settled out over the newly deposited sand body, but before they began to decay, they were buried by the settling fines. Not only were the leaves well preserved, but the leaf bed also had a species composition that was unique to the cores in that it contained *Alnus glutinosa* leaves normally lost by rapid decay.

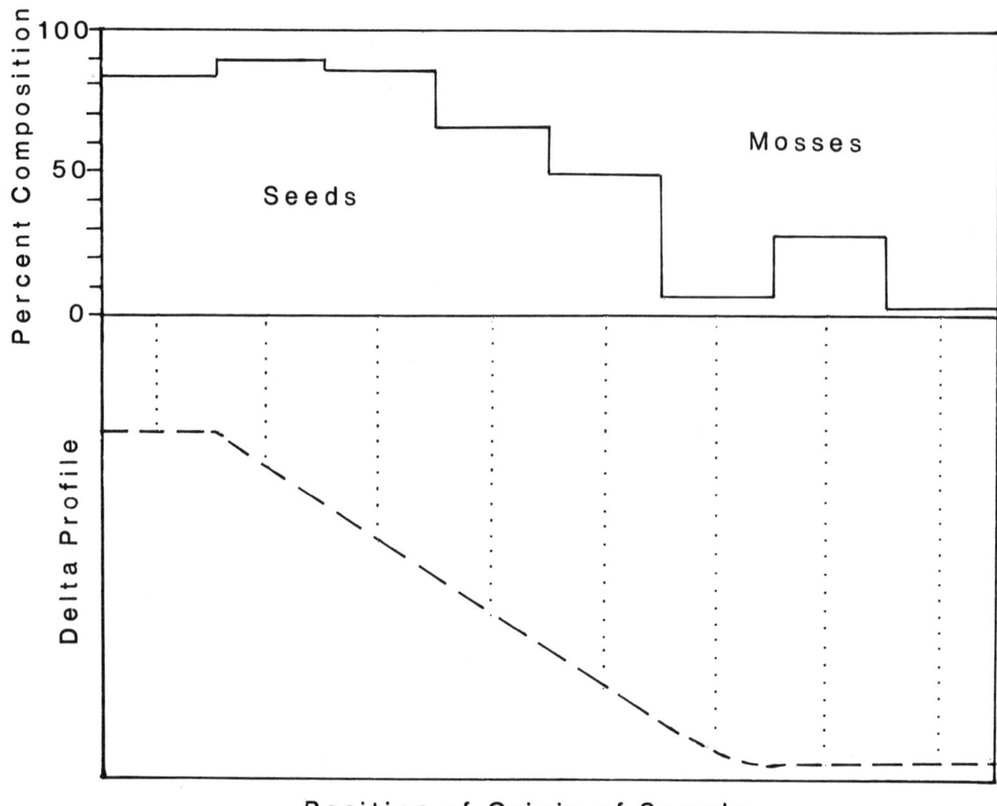

FIGURE 6.8 *The distribution of seeds and moss fragments over the delta profile of Mule Creek. Plant debris other than seeds and mosses has been omitted. The preponderance of seeds on the foresets and mosses in the toeset beds, bottomset beds, and lake clays attests to hydraulic sorting.*

CONCLUSIONS

The importance of depositional sorting to the biostratigraphy of plant megafossils is fundamentally twofold. Sorting can lead to highly erroneous conclusions; it can also be used to improve our paleoecological interpretations.

Sorting, because it is a potential source of gross error, can lead to the preservation of highly biased samples of the source vegetation, from which paleoclimates and relative ages have to be determined. Nevertheless, cross-correlation and age determinations of stratigraphic units can be accurately and relatively easily obtained using plant megafossils, provided that the effects of sorting are adequately understood. It is clear from the preceding examples that the suites of fossils that are to be used for such stratigraphic purposes should, under ideal circumstances, be collected from the same kind of depositional environments. In determining whether two environments are similar, such factors as basin size, basin geometry, and the energy of the system must be considered. In this respect, sedimentary structures, grain-size parameters, and mineralogy can supply important information. The finer the division of time-rock units under investigation, the greater the necessity to establish parity of depositional environments.

The second important aspect of sorting is apparent when detailed biostratigraphy is undertaken for paleoecological purposes. Here, sorting can be used to great advantage when, and if, sampling is designed to detect patterns of deposition.In low-energy environments, both the lateral pattern of plant debris across single bedding planes and the vertical changes in species composition of discrete leaf beds can be used to map spatial relationships among source communities. Vertical differences in species composition need not necessarily reflect temporal changes in vegetation. Similarly, lateral variations in species composition may reflect short-term temporal, not spatial, heterogeneity in the source communities. When changes in the species composition of fossil assemblages are associated with changes in lithology, and in particular grain size, this association may provide information as to which portions of the source vegetation are normally lost from the fossil record owing to rapid biodegradation.

The patterns of deposition, particularly those found laterally across bedding planes, may be extremely complex and not detectable by traditional methods of collecting and analysis. Quantitative systematic sampling, followed by multivariate statistical analyses, similar to those used in phytosociology, have been successfully used to identify these patterns (Spicer, 1975; Spicer and Hill, 1979). In addition, strict sampling regimes, although time consuming, do force closer inspection of the fossil material and therefore may yield a larger number of species that ultimately may be useful in a diagnostic context.

FIGURE 6.9 *Erosion of the forest floor during flooding. Photograph taken by Roger Wiborg, December 1964, 10 hours after the crest of the flood on Coffee Creek, Trinity County, California.*

In high-energy environments, paleoecologically useful patterns may not arise; but, initially at least, the deposits will contain a more diverse and statistically more reliable sample of the source vegetation as a whole. Moreover, the hydraulic sorting of plant debris over a delta front could be utilized to minimize sampling effort without sacrificing information important to the reliable interpretation of the paleovegetation. However, erroneous results may arise if sampling is restricted, for example, to only foreset or bottomset beds.

Taphonomic studies of sorting, long overdue, fortunately confirm many of the assumptions that historically have been used in paleoecological interpretation. And they go much further, in that they suggest new approaches that enable more reliable stratigraphic data to be recovered and facilitate more detailed interpretation of those data, while simultaneously indicating ways of reducing sampling effort to a minimum.

ACKNOWLEDGEMENTS

I am grateful for the assistance of the following persons: Dr. M. D. Muir and Dr. K. L. Alvin of Imperial College London and Dr. J. A. Wolfe of the U.S. Geological Survey. Part of this work was supported by a Natural Environment Research Council Studentship and part by a Lindemann Research Fellowship.

FIGURE 6.10 *Source vegetation of flood deposits. Violent weather and undercutting of stream banks during floods facilitate the transport of fresh plant material and often whole trees to the site of deposition. (Photograph taken during December 1964 flood on Coffee Creek by Roger Wiborg.)*

REFERENCES

Birks, H. H. 1973. Modern macrofossil assemblages in lake sediments in Minnesota. In "Quaternary Plant Ecology," *14th British Ecol. Soc. Symp. (Univ. Cambridge 1972), Proc.*, H. J. B. Birks and R. G. West, eds. New York: Halsted Press.

California Department of Water Resources, 1965. Flood! *California Dept. Water Resourc. Bull.* 161 48p.

Campbell, R. C. 1967. *Statistics for Biologists*. Cambridge, Mass.: Cambridge Univ. Press, 242p.

Chaloner, W. G. 1958. The Carboniferous upland flora. *Geol. Mag.* 95: 261.

Chaloner, W. G., and M. Muir. 1968. Spores and floras, in *Coal and Coal-Bearing Strata* D. Murchison and T. S. Westoll, eds. (Edinburgh, Inter. Univ. Geol. Congr., 1965, Newcastle-upon-Tyne) New York: Am. Elsevier, 1969. pp. 172-146.

Chaney, R. W. 1924. Quantitative studies of the Bridge Creek flora. *Am. Jour. Sci.* 8: 127-144.

Ferguson, D. K. 1971. *The Miocene Flora of Kreuzau, Western Germany.* 1, *The Leaf Remains.* Amsterdam and London: North Holland Publ. 298p.

Jopling, A. V. 1960. An experimental study of the mechanics of bedding. Ph. D. thesis, Harvard Univ., Cambridge Mass.; 358p.

Jopling, A. V. 1963, Hydraulic studies on the origin of bedding, *Sedimentology*, 2:115-121.

Jopling, A. V. 1965a. Hydraulic factors controlling the shape of laminae in laboratory deltas. *Jour. Sed. Petrology* 35(4):777-791.

Jopling, A. V. 1965b. Laboratory study of the distribution of grain sizes in crossbedded deposits. In "Primary sedimentary structures and their hydrodynamic interpretation. *Soc. Econ. Paleontologists Mineralogists Spec. Pub. No. 12*, pp. 53-65.

Kaushik, N.K. 1969. Autumn shed leaves in relation to stream ecology. Ph. D. thesis, Univ. Waterloo, Ontario, 219p.

Kaushik, N.K., and H. B. N. Hynes, 1971. The fate of dead leaves that fall into streams: *Arch. Hydrobiol.* 68:465-515.

McQueen, D. R. 1969. Macroscopic plant remains in Recent lake sediments. *Tuatara* 17:13-19.

Nevers, R. 1958. Upper Carboniferous plant-spore assemblages from the *Gastrioceras subcrenatum* horizon, North Staffordshire. *Geol. Mag.* 95:1-19.

Petersen, R. C., and K. W. Cummins. 1974. Leaf processing in a woodland stream. *Freshwater Biol.* 4:343-368.

Spicer, R. A. 1975. The sorting of plant remains in a Recent depositional environment. Ph. D. thesis, London Univ., London, 309p.

Spicer, R.A., and C.R. Hill, 1979. Principal Components and Correspondence Analyses of quantitative data from a Jurassic plant bed. *Rev. Palaeobot. Palynol.*, 28: 273-299.

Stewart, J. M., and V. C. LaMarche, Jr. 1967. Erosion and deposition produced by the flood of December 1964 on Coffee Creek, Trinity County, California. *U.S. Geol. Surv. Prof. Paper 422-K*, 22p.

7

VEGETATION CHANGE IN THE MIOCENE SUCCOR CREEK FLORA OF OREGON AND IDAHO: A CASE STUDY IN PALEOSUCCESSION

Ralph E. Taggart
Aureal T. Cross

SUMMARY

Palynological and macro-paleobotanical analysis of a series of samples from four stratigraphic sections in the Succor Creek area of Oregon and Idaho indicate a pattern of repetitive disruption of mid-Miocene forest communities. The probable cause of these periodic disruptions was direct ash falls and gas venting that resulted from the volcanic activity that produced the volcanically derived sediments in which the flora is preserved. Reestablishment of forest vegetation was apparently not a rapid process. Recovery of the forest vegetation was probably impeded by marginal rainfall and the requirement for mesic vegetation to move back into the area following disturbance. The result of this delay was the development of intervals dominated by pine and xeric woody and herbaceous plant species. At least two of these xeric intervals have been documented to date in the Succor Creek record.

The bottomland and low-elevation slope forests prior to disturbance are fairly uniform. The vegetation on these sites was dominated by a diverse array of forest tree and shrub species with oak and elm as codominants. It is possible to document hydrologic succession in these lowland communities, initiated by the impoundment of streams to form ponds and lakes, the gradual infilling of the bodies of water to form swamps, and the gradual transition of the swamp sites back to the typical bottomland forest cover. In three of the sections in the area of the classical Succor Creek macrofossil localities and in the upper part of the Type section for the Sucker Creek Foramtion, the pollen record indicates that upland sites were dominated by a montane conifer complex of spruce, fir, and hemlock. There is no evidence for extreme topographic relief. This, coupled with the fact that the montane conifers constitute over 50% of the pollen record from many samples, suggests that the climate was probably cool but highly equible. In the lower part of the Type section a record of deciduous forest pollen is preserved but the montane conifers are virtually absent, indicating that in early Type-section time the climate may have been slightly warmer, excluding the coniferous communities from the low hilltops. Additional evidence for climatic variation is found in pronounced oscillations between montane conifer and deciduous forest pollen percentages where both are represented.

The xeric vegetation zones occur between the upper and lower forest zones, stratigraphically, in the Type section, and in the upper zones of two sections to the south. The pollen and macrofossil records for these intervals indicate a landscape dominated by pine and a variety of herbaceous species. Deciduous woody species are poorly represented in these xeric intervals and suggest their source was from a limited riparian community dominated by oak, willow, poplar, and sycamore. Although the vegetation of the xeric intervals was characterized by some exotics, the overall aspect of the regional vegetation is quite similar to that presently found in the Succor Creek region. The existence of extensive successional areas dominated by plants adapted to edaphic aridity is of criticial importance in modeling the gradual modernization of vegetation in the area. Such community types would have been the sites for the evolution of many plant species, particularly grasses and forbs, that would become important in the climax vegetation of the region in the Pliocene .The repetitive nature of disturbance of the vegetation associated with the volcanic activity in the area suggests that modernization of the vegetation in the reigon may well have involved the gradual extension of the time required to reestablish forest vegetation, with the nature of the resulting forest determined by the climatic conditions prevalent during the successional sequence. Under this model, the xeric vegetation elements may well have been evolving on such sites throughout much of the Middle Tertiary, with a gradual increase in the aeral extent of such edaphically dry sites as rainfall gradually decreased in conjunction with the uplift of the Cascades. Rather then requiring the rather rapid evolution or migration of xeric vegetation types during the Pliocene, such a model involves the gradual expansion of xeric vegetation that had been evolving for a considerable period under pressure of repeated severe disturbances. Such xeric species could expand quite rapidly as rainfall became more limited, since they were already common on disturbed sites and xeric exposures throughout the region.

INTRODUCTION

The Succor Creek flora is preserved in rocks outcropping along the general course of Succor Creek in southeastern Oregon (Malheur County) and southwestern Idaho (Owyee County). These rocks were mapped as the Sucker Creek Formation by Kittleman et al. (1965). The first known collection of fossil plants from the area was made by Lindgren and the material was described by Knowlton (1898) in his treatment of the Payette flora. Subsequent analyses of Succor Creek material appear in Brooks (1935), Arnold (1936a,b, 1938), Smith (1938, 1939), Chaney and Axelrod (1959), and Graham (1965). Graham's study involved a revision of the macroflora and a preliminary study of palynomorphs preserved in the matrix of leaf-bearing shales from some localities. Analysis of the Succor Creek plant material and mammalian remains from the type section for the Succor Creek Formation (Downs, 1956) indicate a late middle to early late Miocene age (Barstovian) for the flora.

Radiogenic dating has proved inconclusive, in part because of the error spread inherent in the nature of the dated materials and, in some cases, because of uncertainty as to the precise locality data for dated samples. Everden et al. (1964) report a K-Ar date of 16.7 m.y. for a basalt sample collected about 9 miles north of Sheaville, Oregon, somewhere near the point where U.S. Highway 95 crosses Succor Creek. Unfortunately, the original collection data are imprecise and the sample cannot be placed in any reasonable stratigraphic context, in part because the area of collection is characterized by considerable faulting. The most reliable dates up to the present time were obtained by Kittleman on two co-existing samples from Unit 20 of his type section(Kittleman et al., 1965). One of these samples yielded a date of 18.5 ±1.7 m.y. (UO-128KArI), based on glass shards; the second sample based on sanidine, yielded a value of 15.4 ± 0.9 m.y. (UO-128 KArII) (Kittleman, pers. comm, 8 May 1974 and 9 March, 1978). The older of the two dates is probably the more reliable due to the greater possiblity of gas migration in the sanidine. Although considerably more dating of material from measured sections will be required to assess the utility of radiogenic dating in providing a local time framework, it is reasonable to assume that most of the fossil material studied falls within the range of 16 to 19 m.y.

Although the floristic date from the fossil plants collected are quite sound, data from previous studies cannot be translated directly into meaingful observations on the vegetation dynamics during Succor Creek time. This is due to a built-in limitation in the data base, which the Succor creek shares with the vast majority of the Tertiary floras of the Pacific Northwest. The Succor Creek flora is not, in fact, a meaningful floristic entity. It is a composite data assemblage derived from over a dozen florules collected from as many localities within the Succor Creek region (Figure 7.1). The localities differ from one another in the diversity of plant materials recovered from the rocks and in the nature of the rocks themselves. The presence of common elements at the various localities and the fact that the localities occur within a limited geographic area (somewhat less than 100 square miles) justified the historical summarization of the florule data into the Succor Creek "flora." The data from these florules contain two sources of variation. The first, implicit in the geographic distribution of the various localities, is variation in the source vegetation through space. The second involves variation in the source vegetation with time. None of the previous studies in the area has involved the stratigraphy of the florule sites. This lack of stratigraphic control is one of the major elements requiring attention if the Succor Creek data are to be useful in ecological modeling.

The authors have been involved in a detailed paleoecological study in the Succor Creek area using available macrofossil data and Graham's (1965) pioneering study of the palynomorphs as a starting point. Our study involved the detailed measurement and sampling of stratigraphic sections in the region, the analysis of palynomorph distribution in the measured sections, correlation of macrofossil finds within and associated with the sections, and an attempt to correlate the various "classic" localities within the developing picture of regional stratigraphy. Although the study is still under way, a number of interesting facets have begun to emerge. One of these concerns the role of succession in the communities of Succor Creek time, whereas the second deals with the role of successional plant communities on the Late Tertiary modernization of vegetation within the region. The focus here will be on these two aspects of our ongoing study.

METHODS

Stratigraphy

Four study sections will be considered: the Rockville, Shortcut, Valley, and Type sections (Figure 7.1). The Rockville, Shortcut, and Valley sections were measured by the senior author during the 1970 field season. The Type section, which is the type section for the Sucker Creek Formation, originally was measured by Kittleman (Kittleman et al., 1965) and included several covered intervals where strata were obscured by

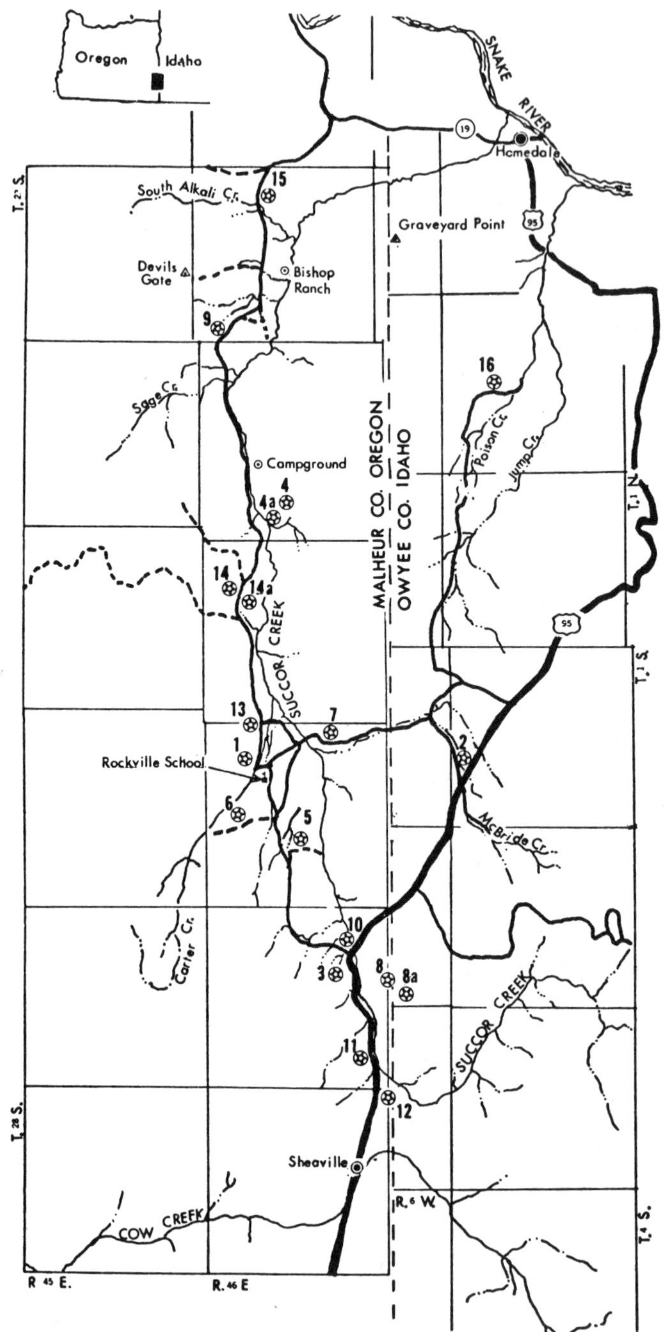

FIGURE 7.1 *A map of the Succor Creek region showing the location of measured sections and florule localities presently under study. The sites discussed here include 1, the Rockville section; 2, the Shortcut section; 3, the Valley section; 4, the Type section; 5, the Quarry locality; 6, the Maple Ridge locality; and 7, the Pine locality.*

weathered talus. This section was remeasured by the authors during the 1973 season and we were successful in documenting a complete stratigraphic sequence at the type site. All sections were measured by means of hand level, tape, and Brunton compass.

Paleobiology

The paleobiological phase of the study involved the collection of as much fossil data from the measured sections as possible. As each section was measured, samples were collected for palynological analysis. Each distinctive rock layer was sampled at least once and often at several levels where the thickness of the unit made multiple samples appropriate. Each of the units was examined for the presence of plant macrofossils and collections were made from productive zones. Fossil wood fragments were abundant in all of the study sections but much of this material was derived from stumps and logs at a stratigraphically higher level. Collections were rarely made of isolated fragments but were obtained from intact stumps or logs or from "nests" of wood fragments representing the weathering products of in situ stumps. Insects and fish and mammal bones were collected where noted in the course of the field study.

Laboratory treatment of palynological samples followed standard practice with one exception. The samples themselves contained large amounts of vitric ash and altered ash products, resulting in a strong exergonic reaction when hydrofluoric acid (HF) was added to macerate the material. Our observations indicated that considerable quantities of pollen were destroyed in processing, presumably due to oxidation at the air-acid interface as the sample boiled during the addition of the acid. Pollen yields improved dramatically when the sample was mixed with crushed ice prior to introduction of the acid, presumably because "boiling" was eliminated. The use of strong oxidants such as Schultze's Reagent was also avoided to minimize degradation of pollen and spores. The complete processing schedule is described by Taggart (1971).

Laboratory treatment of compression, impression, and fossil wood material followed standard practice. Macrofossil specimens, palynological samples and residues, and prepared slides are retained in the Paleobotanical Herbarium at Michigan State University.

RESULTS

Stratigraphy

The detailed stratigraphic sequence of the Rockville, Shortcut, and Valley sections is described by Taggart (1971). An emended description based on our remeasurement of the Type section is presently being prepared for publication. Kittleman's original description (Kittleman et al., 1965) is adequate for the present discussion however.

The rocks of the study sections consist almost entirely of volcanic sandstones, vitric and altered tuffs, ash beds, lignitic shales and siltstones, and lignites. For the sake of brevity, terms such as sandstone, shale, siltstone, etc. will be used throughout the remainder of this chapter. It should be assumed that in each case the clastic component is derived from volcanic ash or the weathering products of such materials. In some cases the materials fell directly into the basin of deposition with size sorting controlled by wind direction and the distance of the basin from the volcanic source. In most cases, however, the clastic component appears to have been derived from water-transported materials washed from ash deposited on the surrounding landscape. The study sections appear to represent both lacustrine and fluviatile environments, depending in part on the course of local stream channels and on the degree of empoundment based on the status of natural containment features such as natural levees and dams formed by ash falls, lava flows, or other volcanic extrusives.

Low-energy depositional environments such as ponds and lakes are characterized by the admixture of varying quantities of organic detritus with the clastic particles Typically this results in banding in the sands and shales and varying organic content in the siltstones and mudstones. Low-grade lignites are developed in all sections, usually in association with highly organic mud, silt, and claystones.

Stratigraphic Palynology

Thirty-four taxa at either the generic or family level are represented in the pollen and spore tabulations from the four study sections. In many cases each taxon contains more than one pollen or spore type so that at least 64 different biological entitites are represented. Additional taxa were encountered in scanning the prepared slides but these were not sufficiently common to be represented in the quantitative tabulations. The pollen and spore profiles from each of the sections are therefore quite complex, even though only a few types actually dominate the assemblage at any given level. In order to simplify the data presentation, the various palynomorph taxa are grouped according to a generalized ecological model. Six different categories are employed, each containing taxa that can be assumed to have had some degree of association based on the distribution patterns in modern plant communities in which the same pollen and spore groups occur. The six different categories or ***paleoassociations*** include a *Montane Conifer* complex (spruce, fir, and hemlock), a *Bottomland/Slope Forest* complex (deciduous trees and shrubs comparable to extant deciduous forests in eastern North America and eastern Asia), a *Swamp* complex, a *Pine* complex, a *Xeric* complex (herbaceous plants and some trees and shrubs commonly associated with xeric or disturbed sites), and an *Other* category containing undertemined and unknown pollen types as well as those that cannot be unambiguously assigned to one of the other categories. The latter category is completely artificial; the Pine category is at least partially so. The wide ecological range of modern pines makes it reasonable to segregate the group because placement of the pollen in groups within the genus is tenuous at best.

Table 7.1 indicates the assignment of the various taxa to these paleoassociations and the representation of each of the taxa in the various study sections. Although the paleoassociations are quite generalized, they do provide useful differentiation without the high degree of error that would characterize a more detailed system. The initial analysis will concentrate on the patterns of abundance of these paleoassociations in each of the study sections, followed by the internal variations in each paleoassociation as represented in the various study sections.

The Valley Section

The pollen data for the Valley section are presented in Figure 7.2. The major feature of the Valley sequence is an oscillation in the relative importance of the Montane Conifer and Bottomland/Slope Forest paleoassociation during Valley time. There is also a siginificant peak in the Swamp paleoassociation in the upper half of the section. The broad peak in swamp pollen is due almost entirely to pollen of the Taxodiaceae and occurs in a thick shale sequence between 15 and 40 meters. This unit, informally known as the Valley Plant Beds, yields fine collecting for *Glyptostrobus* fossils. A second narrower peak occurs near the top of the section and at this level fossil stumps assignable to the Taxodiaceae occur rooted in the lignites and organic shales. Pine pollen is variably represented but is generally subordinate to the Montane

TABLE 7.1 *Major community assignments for taxa included in the quantitative tabulations for the four study sections. The number of pollen types in each taxon and the presence or absence of taxa in each of the study sections are also indicated.*

Paleoassociation	*Taxon*	*Valley*	*Rockville*	*Shortcut*	*Type*
Montane conifer	*Abies* (2)	+	+	+	+
	Picea (2)	+	+	+	+
	Tsuga (1)	+	+	+	+
Bottomland/slope forest	Spores (10)	+	+	+	+
	Acer (2)	+	+	+	+
	Alnus (1)	+	+	+	+
	Betula (1)	+	+	+	+
	Carpinus (1)	+	+	-	+
	Carya (1)	+	+	+	+
	Castanea (1)	+	+	+	+
	Ericaceae (1)	-	-	+	+
	Fagus (1)	+	+	+	+
	Liquidambar (1)	+	+	-	+
	Juglans (1)	+	+	+	+
	Mahonia (1)	+	+	+	+
	Podocarpus (1)	+	+	-	-
	Pterocarya (1)	+	+	+	+
	Populus (1)	+	+	+	-
	Quercus (2)	+	+	+	+
	Salix (1)	+	+	+	+
	Tilia (1)	+	+	-	-
	Ulmus (1)	+	+	+	+
Pine	*Pinus* (5)	+	+	+	
Swamp	Taxodiaceae (1)	+	+	+	+
	Nymphaeaceae (1)	+	-	-	-
	Potamogeton (1)	+	+	+	+
	Typha (1)	+	+	-	+
Xeric	Chenopodiaceae/ Amaranthaceae (2)	+	+	+	+
	Sarcobatus (1)	-	+	+	+
	Compositae (9)	+	+	+	+
	Gramineae (3)	+	+	+	+
	Leguminosae (1)	-	-	-	+
	Malvaceae (2)	+	+	+	+
	Onagraceae (2)	+	+	-	+
Other	*Undifferentiated bisaccate gymnosperms, unknowns, undetermined pollen types.*				

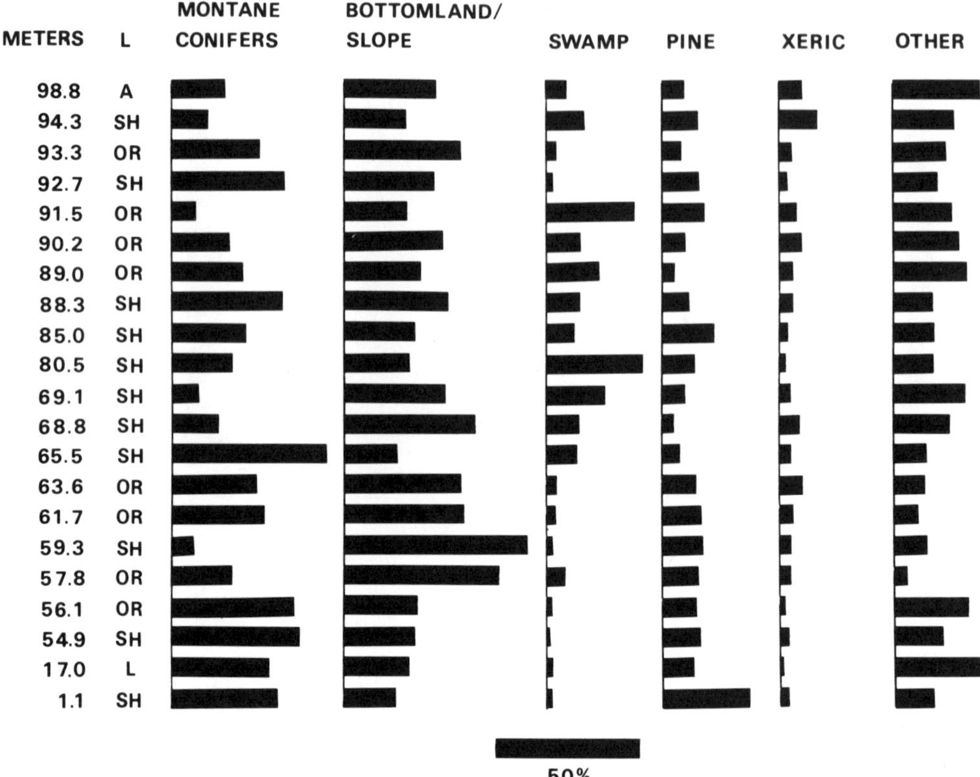

FIGURE 7.2 *Relative frequency of the various major groups of pollen and spore taxa (see Table 7.1) recovered from samples from the Valley section. The Meters column indicates the position of each sample relative to the base of the measured section. The L column indicates the general type of rock comprising each sample. This column is keyed as follows: A, ash; SS, sandstone; SH, shale; L. limestone; and OR, organic sediment. With the single exception of the limestone, the clastic component of these rocks is derived from volcanic ash.*

Conifer and Bottomland/Slope Forest association. Xeric pollen is a minor component except for a slight increase at the top of the section. The Other category is variably represented and includes undifferentiated bisaccate gymnosperms, a relatively small suite of unknown pollen types, and a variable number of pollen grains of undeterminable identity due to the quality of preservation in specific samples.

The Rockville Section

Pollen and spore data for the Rockville section are shown in Figure 7.3. The lower 5 meters of this section are remarkably like that of the Valley section in terms of the representation of the various associations. This zone is characterized by oscillations in dominance between the Montane Conifer and Bottomland/Slope Forest pollen with Pine and Swamp pollen variable and of secondary importance. Xeric pollen is insignificant in lower Rockville time and the Other category is made up largely of undifferentiated bisaccate pollen and some of the same unknowns found in the Valley sequence. The situation in middle and late Rockville time is considerably different, however, for the pollen of the Montane

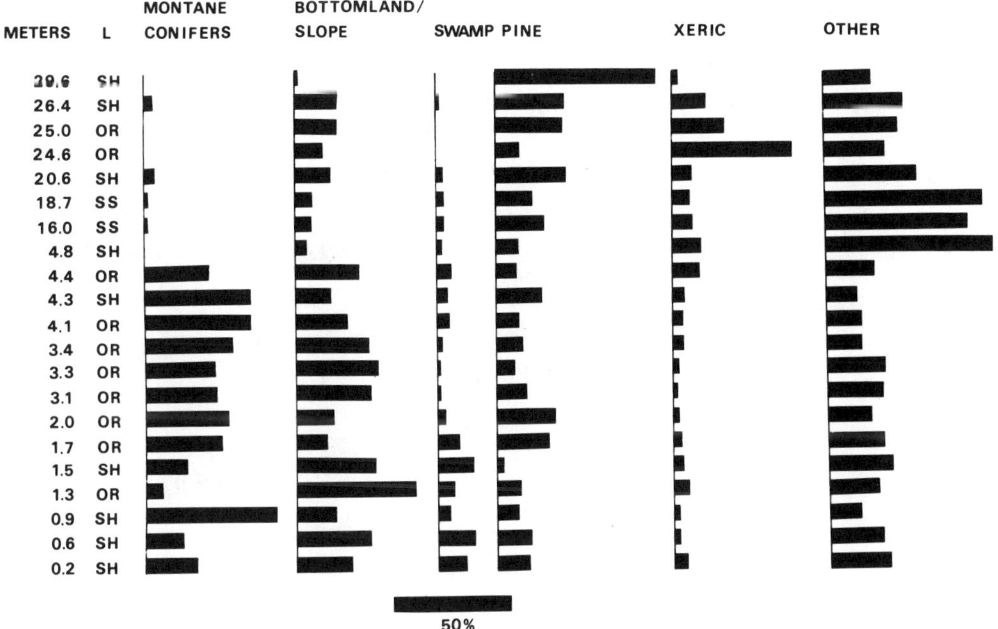

FIGURE 7.3 *Relative frequency of major groups of pollen and spores recovered from samples from the Rockville section. The lithology key is identical to that of Figure 7.2.*

Conifer paleoassociation declines to insignificance coupled with a decline in the relative importance of the Bottomland/Slope pollen. The latter declines in both quantity and diversity above the base of the section. Coincident with the reduction of these two paleoassiciations, pollen of the Pine and Xeric paleoassociations assume dominance, with Pine particularly important near the top of the Rockville section. It is interesting to note that the Pine locality (7 of Figure 7.1) which is near Rockville and which correlates with the upper part of the Rockville section, yields large numbers of pine needles. Pine needles are quite common at this site in contrast to their rarity elsewhere. The Other category also undergoes a change with the dominance shift above 5 meters. Above the transition the paleoassociation is composed of a number of unknown pollen types (Other) that may represent the pollen of herbaceous plants (Xeric) of uncertain affinity. These taxa are not found in the Valley section or the lower 5 meters of the Rockville section and they do not represent any genera now known from the Tertiary of the Pacific Northwest. The dominance transition in the Rockville pollen sequence is abrupt and occurs across a short sedimentary interval that gives no indication of a depositional hiatus.

The Shortcut Section

Although there are fewer productive samples in the Shortcut section (Figure 7.4) compared to the Rockville, the two are essentially identical in their pollen profiles. The section begins with the Montane Conifer and Bottomland/Slope Forest elements as codominants. The dominant pollen types abruptly shift to Pine, then to Xeric and Other. The Other category is represented by the same suite of unknown pollen types found in the upper Rockville and, like the Rockville section, the transition zone cannot be correlated with a

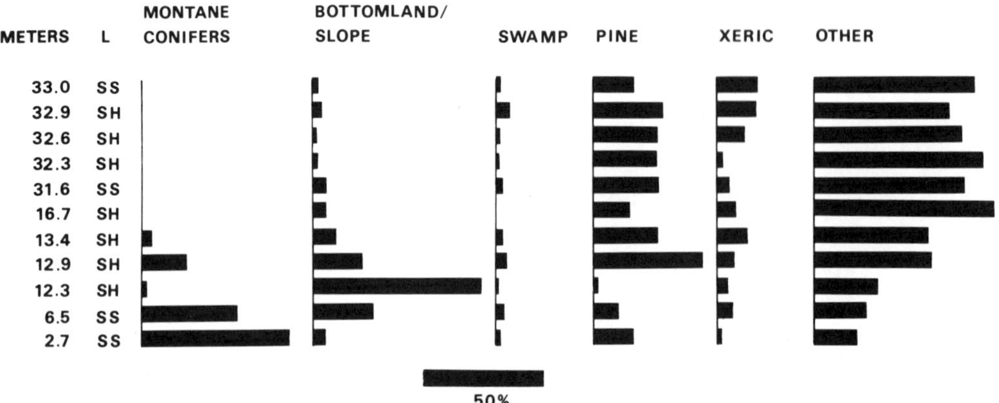

FIGURE 7.4 *Relative frequency of the major groups of pollen and spores recovered from samples from the Shortcut section. The lithology key is identical to that of Figure 7.2.*

significant hiatus in the depositional record.

The Type Section

The record for the Type section begins with a period of dominance of the profile by the Bottomland/Slope Forest element with the Montane Conifer association virtually absent (Figure 7.5)-a different situation from that observed in the Valley and the lower Rockville and Shortcut sections. This is followed by an abrupt change to dominance by Pine and Xeric pollen tyes, with a marked reduction in the number and diversity of Bottomland/Slope Forest pollen. The Pine-Xeric zone is followed by an increase in the Bottonland/Slope Forest pollen, this time in conjunction with the Montane Conifer paleoassociation. The upper Type section is quite similar to the Valley and the lower Rockville and Shortcut sections. There is some indication of another transition to Pine-Xeric dominance at the very top of the Type section. A single sample at this point yielded too few grains for quantitative analysis but it is perhaps significant that the only grains encountered were Pine and Xeric association pollen types.

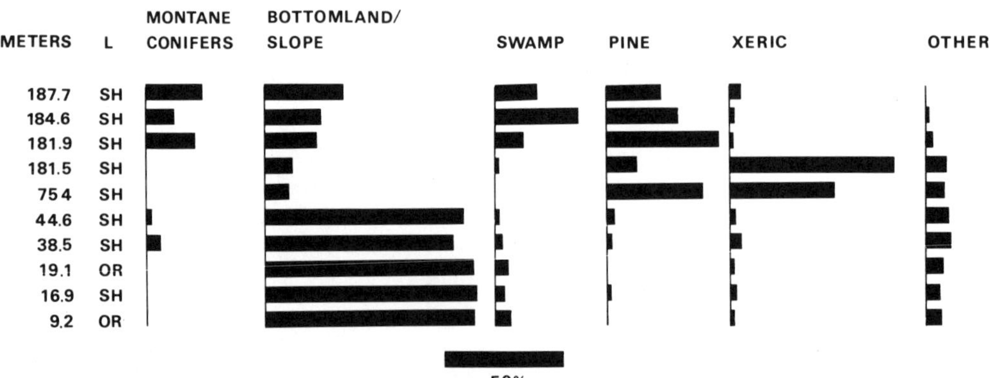

FIGURE 7.5 *Relative frequency of major groups of pollen and spores recovered from samples from the Type section. The lithology key is identical to that of Figure 7.2*

Variation Within the Pollen Associations

Montane Conifer Paleoassociation

Table 7.2 is a summary of the relative importance of the three taxa making up the Montane Conifer paleoassociation as present in the Valley, lower Rockville and Shortcut, and upper Type sections. Of the total pollen tabulated for each section, the Montane Conifer paleoassociation averaged 24.8% with lows of 7% in the Valley, 6.5% at Rockville, 0.5% at Shortcut, and 12% in the Type section and highs of 53%, 56.5%, 61% and 24% in the Valley, Rockville, Shortcut, and Type sections respectively. Looking at the relative composition of the paleoassociation at all sites, *Picea* ranks first with 82.0%, followed by *Abies* at 12.8%, and *Tsuga* at 5.2%. Although there is some variation from section to section, this relative ranking characterizes the Montane Conifer paleoassociation in all sections and florule samples in which it occurs.

Bottomland/Slope Forest Paleoassociation

Data on the importance and composition of the Bottomland/Slope Forest paleoassociation in the Valley, lower Rockville and Shortcut, and upper and lower Type sections are tabulated in Table 7.3. It comprises 39.9% of the total pollen for all sections, ranging from 5 to 69%, where dominance is shared with the Montane Conifer paleoassociation, and from 77 to 88% in the lower Type section, where the Montane Conifer paleoassociation is virtually absent. The most significant tree and shrub pollen types are *Ulmus, Alnus, Quercus, Carya, Pterocarya, Betula, Acer, Fagus,* and *Juglans.* Oak and elm are the arboreal dominants, with elm the most significant component in all but the upper Type section where it is overshadowed by oak. Although there is some variation in the ranking of he various pollen types from section to section, the composition of the paleoassociation appears to be relatively constant throughout.

Deciduous Vegetation in the Pine-Xeric Intervals As noted previously, the relative diversity and importance of the Bottomland/Slope Forest paleoassociation declines during the intervals of Pine-Xeric dominance in the upper Rockville and Shortcut and middle Type sections. Table 7.4 summarizes the importance and composition of the paleoassociation during these intervals. Mean representation for all sections is only 7.8%, ranging from lows of 1% at Rockville and Shortcut and 10% in the middle Type section to highs of 5% at Shortcut, 11% in the middle Type section,

TABLE 7.2 *Representation and internal composition of the Montane Conifer Paleoassociation in the Valley, lower Rockville and Shortcut, and lower and upper Type sections.*

	Valley	*Rockville*	*Shortcut*	*Type*	*Mean for All Sections*
Range [a]	7-53	6.5-56.5	0.5-61	12-24	
Mean [a]	25.8	30.3	24.2	19.0	24.8
Internal Composition [b]					
Picea	74.5	85.2	82.2	86.0	82.0
Abies	18.9	9.5	10.7	12.3	12.8
Tsuga	6.6	5.3	7.1	1.7	5.2

a Percent of Total pollen
b Percent of total.

TABLE 7.3 *Representation and internal composition of the Bottomland/Slope Forest paleoassociation in the Valley, lower Rockville and Shortcut, and lower and upper Type sections. Ranking of taxa is based on mean internal composition for all sections.*

	Valley	*Rockville*	*Shortcut*	*Upper Type*	*Lower Type*	*Mean for all Sections*
Range a	17-63	13-51	5-69	21-33	77-88	
Mean a	32.1	27.0	25.4	25.7	84.4	39.9
Composition b						
Ulmus	21.3	43.0	26.0	18.2	80.8	27.9
Alnus	31.8	16.8	28.3	-	13.5	18.1
Quercus	13.1	9.0	9.4	50.6	6.4	17.7
Carya	3.4	3.6	3.5	5.2	11.4	5.4
Pterocarya	5.3	8.2	5.1	5.2	3.1	5.4
Betula	6.1	3.6	6.3	1.3	4.3	4.3
Spores	3.1	3.3	5.1	3.9	0.9	3.3
Acer	4.4	1.2	1.2	7.8	1.4	3.2
Fagus	3.7	0.8	5.1	1.3	4.3	3.0
Juglans	1.4	1.2	2.8	5.2	2.6	2.5
Salix	2.5	4.2	5.5	-	0.2	2.5
Corpinus	0.4	0.7	-	1.3	7.3	1.9
Mahonia	-	-	-	-	7.8	1.9
Liquidambar	1.6	0.3	-	-	4.0	1.2
Populus	1.1	3.4	0.4	-	-	1.0
Castanea	0.5	0.7	0.8	-	1.4	0.7
Ericaceae	-	-	0.4	-	0.2	0.1
Tilia	0.2	0.1	-	-	-	0.1
Podocarpus	-	0.1	-	-	-	0.1

a Percent of total pollen
b Percent of internal composition.

TABLE 7.4 *Representation and internal composition of the Bottomland/Slope Forest paleoassociation during the intervals of Pine-Xeric dominance (Figures 7.3-7.5) in the upper Rockville and Shortcut and middle Type sections. Taxa are ranked on the basis of mean internal composition for all sections.*

	Rockville	Shortcut	Middle Type	Mean for All Sections
Range a	1-18	1-5	10-11	
Mean a	10.3	2.6	10.5	7.8
Internal Composition b				
Quercus	44.4	14.0	71.8	43.4
Populus	16.0	30.7	-	15.6
Salix	21.4	19.3	5.0	15.2
Spores	7.7	8.9	18.6	11.7
Acer	3.7	16.7	-	6.8
Alnus	3.9	10.4	-	4.8
Pterocarya	-	-	4.5	1.5
Castanea	2.1	-	-	0.7
Tilia	0.9	-	-	0.3

a Percent of total pollen
b Percent of total.

and 18% at Rockville. Oak is the dominant of the paleoassociation during these intervals (43.4%), followed by poplar (15.6%), willow (15.2%), maple (6.8%), and alder (4.8%). The few additional pollen types are represented at insignificant levels.

The Swamp Paleoassociation

The Swamp paleoassociation represents plants of widely different habitats. Pollen of the Taxodiaceae (standing-water swamps or wet bottomland sites) comprises approximately 98% of the pollen for the paleoassociation, followed in importance by *Potamogeton* (pondweed) at about 2%. The latter is characteristically found in ponds and lakes. The source plants for the waterlily pollen (Nymphaeaceae) were probably found in similar habitats, although the waterlily pollen is quite rare. Marsh habitats are represented by the pollen of *Typha*, a rare component of the paleoassociation.

The Pine Paleoassociation

Five pine pollen types were recognized in assembling the quantitative data. One of these is common in the zones dominated by the Montane Conifer and Bottomland/Slope Forest paleoassociations. Another type is rare in those intervals and more common in the intervals of Pine-Zeric dominance. The remaining three types are common in and confined to the Pine-Xeric intervals. The Pine paleoassociation thus shows a variable qualitative as well as quantitative representation in the study sections.

The Xeric Paleoassociation

The pollen of the Compositae is the most significant component of the Xeric paleoassociation, followed in importance by the Gramineae, Chenopodiaceae - Amaranthaceae, and Malvaceae. The Leguminosae and Onagraceae are rare elements. The paleoassociation is diverse during the intervals of Pine-Xeric dominance and low in both diversity and representation during Montane Conifer-Bottomland/Slope Forest intervals. Most of the Xeric element taxa are restricted to the Pine-Xeric intervals.

Composition of the Flora

As useful as the palynological data are in interpreting the flora, the pollen data alone provide only a small measure of the diversity of the flora. The quantitative pollen counts are dominated by pollen of anemophilous plants while some pollen types, although present, are too rare to register in the quantitative tabulations. Although the present paper is not directly concerned with the floristics of the Succor Creek complex, Table 7.5 is provided to give some idea of the diversity of the flora at the generic level, based on palynological and macrofossil data. The disclosure of new florules in the study of additional sections may be expected to further expand our knowledge of the composition of the flora.

DISCUSSION

Stratigraphy

Extensive faulting within the Succor Creek region, coupled with the isolated nature of some of the depositional basins makes correlation of measured sections extremely difficult. Two sections however, the Rockville and Shortcut, can be correlated with some confidence. These two sections are located across the valley of Succor Creek from one another and each is clearly visible from the top of the other section despite the 8 miles that separate them. Throughout most of their thickness they show the same sequence of rock units, even to the presence of a thin shale zone containing insect body parts and some leaves and seeds. This lithological similarity, coupled with the strikingly similar pollen profiles, makes it fairly certain that both were deposited in the same basin system during essentially the same

TABLE 7.5 *Families and genera presently recognized in the Succor Creek complex; S, seeds; F, fruits; L, leaves; St, stem impressions; W, wood; C, cells or cysts; Sp, spores; P, pollen.*

	Family	Genus	Macrofossil a	Microfossil a
ALGAE	Botryococcaceae	*Botryococcus*	—	C
	Hydrodictyaceae	*Pediastrum*	—	C
FUNGI		numerous	—	C
LYCOPSIDA	Lycopodiaceae	*Lycopodium*	—	Sp
SPHENOPSIDA	Equisetaceae	*Equisetum*	L,St	Sp
PTEROPSIDA	Blechnaceae	*Woodwardia*	L	Sp
	Polypodiaceae	*sensu latu*		*Sp*
		Polypodium	L	
	Osmundaceae	*Davallia*	L	Sp
		Osmunda	L	Sp
GYMNOSPERMS	Ginkgoaceae	*Ginkgo*	L	—
	Cupressaceae	*Thuja*	L	—
	Pinaceae	*Abies*	S	P
		Cedrus	—	P
		Keteleeria	L	—
		Picea	S	P
		Pinus	L,S	P
		Tsuga	—	P
	Podocarpaceae	*Podocarpus*	—	P
	Taxaceae	*Cephalotaxus*	L	—
	Taxodiaceae	*sensu latu*	W	P
		Taxodium	L	
		Glyptostrobus	L	—
	Ephedraceae	*Ephedra*	—	P
MONOCOTYLEDONS	Gramineae	*sensu latu*		P
		Cypericites	L	
	Potamogetonaceae	*Potamogeton*	—	P
	Typhaceae	*Typha*	L	P
DICOTYLEDONS	Aceraceae	*Acer*	*L,S*	*P*
	Aquifoliaceae	*Ilex*	L	P
	Araliaceae	*Oreopanax*	L	—
	Berberidaceae	*Mahonia*	L	P
	Betulaceae	*Alnus*	*L,F*	*P*
		Betula	L	P
		Carpinus	L	P
		Ostrya	L	P
	Buxaceae	*Pachysandra*	—	P
	Caprifoliaceae	*sensu latu*		P
	Chenopodiaceae/ Amaranthaceae	*sensu latu*	—	P
	Compositae	*sensu latu*	—	P
		Ambrosia	—	P
		Artemesia	—	P
	Cornaceae	*Cornus*	L	P
	Ebenaceae	*Diospyros*	L	—
	Elaegnaceae	*sensu latu*	—	P
		Shepherdia	—	P
	Ericaceae	*sensu latu*		*P*
		Arbutus	L	
		Vaccinium	L	
	Fagaceae	*Castanea*	L	P

Family	Genus	Macrofossil [a]	Microfossil [a]
	Fagus	L	P
	Quercus	L,S	P
Hamamelidaceae	*Liquidambar*	L	P
Juglandaceae	*Carya*	L	P
	Juglans	L	P
	Pterocarya	L	P
Lauraceae	*Persea*	L	—
	Sassafras	L	—
Leguminosae	*sensu latu*		P
	Gymnocladus	L	
Magnoliaceae	*Hiraea*	L	—
	Magnolia	L	—
Malvaceae	*sensu latu*		P
	Anoda	L	
	Sphaeralcea	—	P
Meliaceae	*Cedrela*	L,S	
Nymphaeaceae	*sensu latu*		P
	Nymphaea	L,St	
Nyssaceae	*Nyssa*	L,F	P
Oleaceae	*Fraxinus*	L,S	P
Onagraceae		—	P
Platanaceae	*Platanus*	L	—
Rosaceae	*Amelanchier*	L	—
	Crataegus	L	—
	Pyrus	L	—
Rutaceae	*Ptelea*	F	—
Salicaceae	*Populus*	L	P
	Salix	L	P
Saxifragaceae	*Hydrangea*	L	—
Simarubaceae	*Ailanthus*	L	—
Tiliaceae	*Tilia*	L	P
Ulmaceae	*Celtis*	L	?
	Ulmus	L	P
	Zelkova	L	?
Umbelliferae	*sensu latu*	—	P

[a] Data derived from Chaney and Axelrod (1959), Graham (1965), Taggart (1971, 1973), and umpublished data from our laboratory.

time interval. Although several principal lithologic units of the two sections are readily correlated, some of the total exposure at each site is not duplicated at the other. The basal sand of the Shortcut section is not exposed at Rockville, making that unit somewhat older that the oldest strata exposed at Rockville. Similarly, the indurated shale sequence at the top of the Rockville section is somewhat younger than the youngest sediments preserved below the modern erosional surface at the Shortcut outcrop.

The relative position of the Valley section is more difficult to deal with because most of the units in the Valley sequence have no direct analogs in the Rockville-Shortcut sequence. The 10-meter white-ash bed at the top of the Valley section is remarkably like the prominant white-ash bed in both the Rockville and Shortcut sections. Although the sequence of organic sediments and shales below the ash is considerably thicker than a similar sequence below the Rockville-Shortcut ash bed, this may simply reflect greater sediment input in the Valley area. The pollen profile of the Valley section is consistent with that of the lower Rockville and Shortcut sections, so that at this time the tentative assumption is that the Valley sequence represents a thicker analog of the lower Rockville and Shortcut, which perhaps extends somewhat further back in time. This tentative correlation is supported indirectly by the fact that the thick basal sand of the Valley, which is not greatly different from that of Shortcut, was evidently deposited quite quickly as indicated by the lack of any pronounced change in pollen spectra below, within, and immediately above the sand unit.

The rocks of the Type section show very little similarity to those exposed in the study sections to the south. They may represent a completely different unit of time, an appealing suggestion considering the extensive faulting between the Type section and the other study sections, or they may be essentially contemporaneous rocks accumulated in another basin system. The latter view finds some support in the similarity in the pollen spectra of the upper Type section and those from the Valley and lower Rockville and Shortcut sections. Confirmation of a return to Pine-Xeric dominance at the very top of the Type section would firm up the hypothesis and would indicate that much of the Type sequence represents an earlier sedimentary interval relative to the sections to the south. Further studies of the Type section and other sections in the same area (Figure 7.1) are currently underway in an attempt to determine the relative time-stratigraphic position of the Type sequence vis a vis sections to the south. Despite the present uncertainty, the Type section data do contribute significantly to our understanding of vegetation dynamics during Succor Creek time.

Model for Succor Creek Vegetation Dynamics

The pollen spectra from the Valley section are typical of many Miocene pollen suites and could be expected as a logical extension of what was known based on Graham's study of florule matrix samples (1965). All Graham's samples showed a mixture of what we have classified as Montane Conifer and Bottomland/Slope Forest pollen types and the sequence of pollen spectra from the Valley section is simply a dynamic expression of the interaction of these paleoassociations. A similar situation exists with regard to the lower Rockville and Shortcut sections. The surprising find in the Succor Creek study was the change from mesic forest pollen types to Pine-Xeric dominance in the upper Rockville and Shortcut sections.

These changes seem too abrupt to support a hypothesis of cliseral change in vegetation structure. A simpler explanation may be found in the events that have preserved the flora. This is the volcanic activity that characterized the region during Succor Creek time and continued on into the Pliocene, with the sedimentation of the Deer Butte and Grassy Mountain formations overlying the Succor Creek (Kittleman et al., 1965). Chaney and Axelrod (1959) comment on the importance of this widespread volcanic activity in preserving a record of mid-and late-Tertiary vegetation in the Pacific Northwest, yet comparatively little attention has been paid to the possible effects of such volcanic activity on the local vegetation. A major exception has been the study of the sequence of fossil forests in the Amethyst Mountain section of Yellowstone National Park in Wyoming (Dorf 1960, 1964; Fisk 1976). Here the forests were clearly destroyed repeatedly by successive volcanic eruptions. If activity were intense and widespread, gas venting could also be expected to have a destructive effect on local vegetation. In the case of the Yellowstone sequence, the landscape was apparently re-populated by the same array of species that characterized the original forest (Fisk 1976). In the Succor Creek area the landscape was occupied by successional vegetation of more xeric aspect than the preceding forest-a vegetation array producing the Pine-X-eic pollen signature seen in the upper Rockville and Shortcut pollen spectra. The data from Rockville and Shortcut left open the question as to whether the xeric vegetation continued to occupy the sites indefinitely or whether they were part of a successional sequence that would eventually lead back to the establishment of more mesic vegetation types (Taggart and Cross 1974). Now that data are available for the Type section, the weight of the evidence clearly favors interpretation of the Pine-Xeric intervals as a successional sequence.

The model that we have developed to account for the data involves the disruption of climax forest vegetation by volcanic ash falls and gas venting, the initiation of a successional sequence in which xeric forbs and shrubs and pine play a pivotal role, followed by reestablishment of climax forest communities of Succor Creek time (*sensu latu*) involving mesic dediduous forest communities in the valley bottoms and on low slopes, with the development of montane conifer forests on higher slopes. The extent of the development of the conifer forest communities would depend on local topography, slope exposure, and prevailing climatic regime. The region no longer supports such vegetation arrays, due primarily to the rain shadow of the Cascades, which were uplifted in the late Tertiary. It is quite possible that the mesic forest arrays of Succor Creek time were already under some moisture stress but the integrity of the communities could be maintained (barring disturbance) by a degree of microclimatic amelioration. At intervals, however, the forests were subjected to disturbance in the form of widespread volcanic eruptions and following such disturbance the landscape was dominated for a period by herbaceous plants and pine. Eventually, the more mesic forest vegetation would become reestablished, possibly with the migration of plants up-the stream or river valleys as weathering gradually ameliorated the edaphic dryness of the ash-covered landscape. Using this model as a guide it is possible to interpret the distribution and interaction of a number of different community types during Succor Creek time.

Climax Vegetation Types

Lowland and Slope Paleocommunities

Bottomland Forests The macrofossil record from diverse Succor Creek florules essentially preserves a record of the lowland forest vegetation in the immediate vicinity of the basins of deposition. These forests were dominated by a variety of oaks, followed in importance by maple, chestnut, walnut, black gum, basswood, holly, and beech. Willow and sycamore were locally important on floodplains and wetter sites. Understory vegetation included ferns, dogwood, and blueberry, but relatively few herbaceous plants.

Slope Forests The vegetation of the adjacent slopes was somewhat drier in aspect than the bottomland forests and is poorly recorded in the macroflora. The pollen record indicates that the low slope forests were dominated by elm in conjunction with oak, hickory, and wingnut *(Pterocarya).* Oregon grape *(Mahonia)* was probably an understory component in such paleocommunities. These taxa are poorly preserved in the macroflora, with the exception of a new florule from a shale bed above a thin lignitic layer at the top of the Type section. As will be discussed later, the paleocommunities at this site may have been developed on a somewhat more open and hence drier exposure. In any case, the florule provides excellent collecting for elm and the only leaves of hickory that have been collected to date. Various herbaceous plants may have grown in the ground-layer vegetation in the low slope forests, particularly on southern exposures; but this supposition is based on pollen data alone and has yet to be corroborated with macrofossils. The florule at the top of the Type section, however, does contain a number of interesting seeds of undetermined affinity that may have a bearing on the question.

Hydrologic Succession The pollen record, particularly in the Valley section, documents episodes of swamp development and closure that undoubtedly represent a record of hydrologic succession at the lowland sites. Diatoms and other algae such as *Botryococcus*, pondweed, and waterlily provide direct evidence for the existence of ponds or lakes in the area. These bodies of water may have been formed in a number of ways. If the Succor Creek drainage systems were relatively mature with small gradients, lakes and ponds could have formed by meander cutoffs or behind natural leaves. Such lakes would have been flooded at intervals as flood waters temporarily breached the natural levees. Some Succor Creek shales are varvelike and lignites commonly show thin banding due to the presence of sand, suggesting that they may have formed in such flood-plain lakes. A second mode of lake or pond formation involves the impoundment of distributary channels by ash falls or lava flows. Such impoundments would be vulnerable to sudden drainage if the containment features were breached, an event that appears to have been fairly common, based on the number of small Swamp pollen peaks in the study sections.

Marshy areas on lake and pond borders supported stands of cattail and horsetail, much as do virtually all such habitats in North America today. The ponds or lakes were evidently shallow and the process of progressive infilling would begin with the accumulation of clastic sediments and organic debris as vegetation encroached on the margins. Eventually such sites would assume the aspect of standing-water swamps or bogs. Such sites were dominated by *Glyptostrobus* (Chinese Water Pine) in late Type-section and Valley, Rockville, and Shortcut time. *Taxodium* (bald cypress), although common in the Weiser flora to the northeast (Shah, 1968; Smiley et al., 1975), apparently did not occur in the Succor Creek area at this time, although recently discovered macrofossil material from sediments below the Type section indicates that it was present in early Type-section time. It was in these swamps that organic debris accumulated that would later develop into the lignitic zones and organic shales that occur in all of the measured sections. Tree stumps are found in situ in such zones in the Valley, Rockville, and Shortcut sections and all have been shown to represent wood of the Taxodiaceae. The upper part of the Valley section records a major phase of swamp development and it is in these beds that abundant *Glyptostrobus* leaf and cone material has been obtained. Most Swamp pollen peaks are considerably smaller, reflecting either limited geographic extent or termination of swamp development by premature drainage. Such drainage would result in a rapid shift from swamp to bottomland forest vegetation. The erratic distribution of alder pollen may indicate a role as a swamp-margin plant for this genus. Leaves similar to those of the modern water tupelo *(Nyssa aquatica)* in the macrofossil record may indicate that the water tupelo developed in shallow standing-water swamps, much as it does in the margins of *Taxodium* swamps in southern Illinois today.

Montane Paleocommunities

The existance of upland communities dominated by spruce, fir, and hemlock, as documented by the pollen record, was first reported by Graham (1965) and represents a major vegetation type that is essentially unrecorded in the macroflora, with the exception of a few seeds attributed to spruce and fir. Other conifers may have been important as well but so far are unrecorded; juniper and Douglar fir are two taxa that might be expected on the basis of present upland vegetation in the west. Some species of pine may also have had a montane distribution but this is a differentiation that cannot be made on the basis of the pollen record. The precise contribution of spruce, fir, and hemlock to the community is difficult to determine, because of the three taxa, spruce pollen possesses the best aerodynamic qualities and thus is more likely to enter the basins in larger numbers. This transport bias probably results in an under-representation of fir and most particularly of hemlock. The conifer forest line must have been quite close to the basin at times as shown by the fact that the total montane conifer pollen input exceeds 50% in several samples from the Valley, Rockville, and Shortcut sections, reaching a high of 61% in one Shortcut spectrum. Although most of the poplar pollen in the record is attributed to floodplain plants, some of it was undoubtedly derived from aspen in theMontane Conifer-Bottomland/Slope Forest ecotone. This is based on the presence of leaves assigned to *Populus pliotremuloides* (a new record for the flora), recovered from a florule at the top of the Type section. This flora also yielded leaves of *Betula thor* (cf. B. *papyrifera)*, indicating that this birch was probably a member of the transition forest as well.

Paleoclimate

Based on the macrofossil record the Succor Creek climax paleocommunities have been interpreted as indicative of warm temperature climatic conditions in which subfreezing temperatures would have been rare and of short duration (Graham, 1965; Shah, 1968). The importance of the Montane Conifer component of the pollen record however, makes it necessary to reevaluate this assessment. There is no evidence for extreme topgraphic relief in the area during Succor Creek time and it is probable that the landscape consisted of low hills dissected by stream-valley systems. Any reconstruction of climate must take into account the fact that where Montane Conifer pollen types are present they average 26% of the pollen record and often exceed 50% in specific samples. Such close juxtaposition of disparate community types is best explained by postulating a cool but highly equable climatic regime for the region. Such a hypothesis is consistent with the representation of montane conifer forests on the low hills, whereas frost-sensitive plants such as *Cedrela*, *Oreopanax*, *Persea*, and *Magnolia* existed in the bottomland forests.

The pronounced oscillation in the relative percentage of Montane Conifer and Bottomland/Slope Forest pollen that characterizes the Valley, Rockville, Shortcut, and upper Type sections is probably the result of relatively small temperature fluctuations controlling the position of the conifer forest line relative to the basin of deposition. If slope gradients were low, a slight altitudinal shift induced by temperature changes would have the effect of altering the position of the conifer forest border by a considerable degree, resulting in a noticable change in the conifer pollen input. The Montane Conifer-Bottomland/Slope ecotone would move upslope with small temperature increases and downslope with a temperature decrease. The lateral extent of such movements would be effectively amplified by the low slope gradients, resulting in pronounced shifts in the position of community margins with relatively small fluctuations in temperature.

The forest vegetation of lower Type-Section time is distinctly different, primarily due to the absence of montane conifers. The relatively smaller input of montane conifer pollen in the upper Type-section record relative to the Valley, Rockville, and Shortcut sections may indicate that topographically the Type section site may have been at a somewhat lower elevation than the basins to the south. If this were so, a slightly higher average temperature during lower Type-section time may well have been sufficient to exclude the montane forests from the low hilltops. Slightly warmer temperatures, coupled with somewhat lower elevation, may explain the presence of *Taxodium* in the swamp forests of lower Type-section time in contrast with its apparent absence in the florules from the upper Type section and the localities to the south. A slight cooling trend from early to late Succor Creek time is consistent with the widely accepted model for climatic deterioration during the Miocene and would match the cooling trend documented for the Weiser floral sequence to the northeast (Shah 1968, Smiley et al., 1975).

If the climate were indeed cool and equable throughout much of Succor Creek time, rainfall of 80 to 100 cm. would be sufficient to support the postulated climax vegetation types. Although it has been assumed that rainfall was evenly distributed throughout the year (Smiley et al., 1975), the possible existence of pine stands on south-facing slopes and successional communities of xeric aspect may indicate a slight trend toward a summer-dry distribution pattern.

Successional Vegetation in the Region

The source vegetation indicated by the pollen data from the pine-xeric zones of the Rockville, Shortcut, and Type sections is quite different from that of the climax communities. The overall aspect is one of a landscape dominated by pine, grasses, and a variety of herbaceous dicots of which the composites were the most important. The depauperate deciduous tree component was probably restricted to narrow zones along the stream or river channels. These riparian communities were dominated by oak, willow, poplar, and sycamore with occasional individuals of more mesic forest genera such as basswood and chestnut. This distribution pattern is not unlike that postulated by MacGinitie for the Miocene Kilgore flora of Nebraska (1962).

The profound nature of the transition to pine-xeric cominance raises the question of why this phenomenon was not recognized earlier, given the intensive study the flora has received. Apparently the macrofossil data do in fact provide a useful record if viewed from the proper perspective. Analysis of Graham's leaf-count data (1965) and our own field experience indicate that there are two major types of leaf localities represented in the complex of florules known from the Succor Creek area. One type, the diverse type, contains representatives of most of the known macrofossil taxa. In virtually all cases where correlation was possible with known sections, these "diverse" localities are equivalent to Valley or lower Rockville and Shortcut time and are thus representative of the diverse mesic forest vegetation that surrounded the basins of deposition. Other localities, such as the Maple Ridge and Quarry sites (Graham 1965), while productive in terms of the number of leaf fossils, actually are quite low in diversity The Quarry site is extremely productive in terms of the number of specimens collected but the vast majority of the material represents oaks, willows, and *Cedrela.* A similar suite can be recovered from the indurated shale sequence at the top of the Rockville section. The Maple Ridge indicator fossils are not maples at all but ae sycamore, found in association with oaks and willows. Most of the oaks from these low-diversity localities *(Quercus dayana* and *Q. hannibali)* have smaller coriaceous leaves that are quite like those of scrub oaks in appearance. It thus appears probable that the low-diversity localities record the limited riparian forest development characteristic of the pine-xeric intervals. Without stratigraphically controlled pollen data, this low diversity had previously been accounted for on the basis of preservation factors.

The change in pollen dominance with the onset of the pine-xeric intervals indicats a profound disruption of existing forest vegetation. Direct ash falls and gas venting are the most likely causative agents. The ash-covered landscape, large-y denuded of forest vegetation, was then rapidly covered by herbaceous plants, largely composites and grasses. If the large number of unknown pollen types characteristic of these intervals is any indication, other herbaceous plants of uncertain affinity may also have been involved in this pioneer succession. Over a period of time several species of pine colonized these areas. Along the stream courses the riparian communities dominated by oak, willow, and sycamore were developing and gradually increasing in extent. Fluctuation in the Pine association curves indicate that the encroachment of pine was often delayed and even set back, possibly by additional disturbance; but if uninterrupted, pines eventually covered much of the landscape, as indicated by the Rockville pollen data. The existence of such an extensive pine state is further supported by macrofossils from the Pine locality (Figure 7.1); which has been correlated with the uppermost Rockville beds. Here, large number of pine needles may be collected, including both two and three needle types, in sharp contrast to the rarity of pine needles at other localities. The associates of pine at this locality are oaks and willows.

Events following the pine stage are difficult to document since only the upper Type section records the eventual reestablishment of mesic forest vegetation and the number of transition samples is too small for good resolution. It would appear however that oak, hickory, and sweetgum were involved in the transition communities. It is perhaps significant that the latter two taxa often are found in almost pure stands in montane Mexico where they comprise successional communities established after disturbance. Sweetgum also occurs in successional communities in southern Illinois. These communities are transitional back to moist bottomland forests associated with bald cypress *(Taxodium)* swamps, and almost pure stands of sweetgum may be found.

Excellent analogs of the Succor Creek successional communities may be found in the present vegetation of the Owyhee Mountains east of the Succor Creek area. In the Silver City area between 1800 and 2500 meters elevation, there is an interesting complex of community types that is quite similar to that postulated for the successional intervals of Succor Creek time. Here, streamside thickets of poplar, willow, and alder give way to slopes dominated by sagebrush and other xeric shrubs, grasses, and a variety of forbs. Juniper and stands of Douglas fir are found on some exposures (Figure 7.6). Spruce and fir are found in protected coves at the 2500-meter level.

Paleotopography During Succor Creek Time

Axelrod (1968), in his evaluation and interpretation of the Tertiary floras and topographic history of the Snake River Basin, attempted to integrate structural data with altitudinal data derived from fossil floras to reconstruct the Cenozoic topographic history of the region. According to Axelrod's interpretation, the region was originally part of a major Eocene volcanic plateau, extending from northern Idaho into Nevada. The axis of

FIGURE 7.6 *The general aspect of modern vegetation at an elevation of approximately 1850 meters in the Owyhee mountains east of the fossil localities. The streamside flora is dominated by willow, poplar, alder, and currant. South-facing slopes at this elevation are covered by sagebrush and a variety of xeric shrubs and forbs. North-facing slopes support juniper and stands of Douglas Fir with grasses; shrubs, including oregon grape; and forbs in the understory. Below this elevation, the conifer element drops out and the riparian complex becomes less diverse. Above 2300 meters, spruce and fir enter the conifer assemblage. The overall aspect of this mosaic of community types is remarkably like that postulated for the successional intervals during Succor Creek time.*

the plateau, with an altitude of more than 1200 meters, was covered by subalpine conifer forest in Eocene time. The subsidence that was to form the Snake River Basin began in the Oligocene and was well advanced in the Mocene, with the trough forming in the west and advancing eastward. The macrofossil data from the Succor Creek, Payette, and Weiser floras led Axelrod to characterize the regional Miocene vegetation as deciduous-slope hardwood forest. In plotting Miocene vegetation distribution he created a major salient in the hardwood forest distribution extending from the Oregon-Idaho boundary well into central Idaho. Areas north and south of the developing basin were considerably higher in elevation and were considered to support subalpine conifer forests of the type indicated by the Trapper Creek flora (Axelrod, 1964). The lowland nature of the Succor Creek flora was further reinforced by Smiley et al. (1975), who used the lack of montane conifer macrofossils in the lower Weiser flora (Sucker Creek Formation, *sensu latu)* as a justification for concluding that montane conifer forests were absent from the Weiser area during Succor Creek time.

The importance of montane conifers in the Succor Creek pollen record requires a reconsideration of the topographic setting for these floras. Axelrod (1968) suggested that the basin margins would have been formed from the dissected edges of the old plateau, but he evidently considered such bordering uplands to have been located somewhat south of the Succor Creek area. It seems, in fact, that the flora was developed on just such a dissected terraine, which would place the border of the basin somewhat closer to its axis. As noted earlier, montane conifer representation in the upper Type section is somewhat lower than that of the Rockville, Shortcut, and Valley sections. It is quite reasonable to attribute this reduced representation to a somewhat lower elevation for the Type-section area relative to the localities to the south. Since the distribution of these localities is north-south, a line roughly normal to the axis of the Snake River Basin, such a distribution pattern for montane conifers is quite reasonable. Palynological studies in the Weiser area could well contribute to a precise determination of the geographic position of the edge of the basin during Succor Creek time by careful analysis of the importance of the montane conifer record.

It appears that the Succor Creek uplands were remnants of the Eocene volcanic plateau and that they supported montane conifer forests whose origins may be traced to an upland Eocene vegetation complex. The development of the complex array of Succor Creek communities on a sloping dissected terraine dropping off into the Snake River Basin to the north results in their being essentially transitional between the cypress swamp-bottomland forests of the John Day basin to the east and the upland floras of the Trapper Creek type in south-central Idaho.

The Role of Secondary Succession in Tertiary Floras in the Region

The first question to be addressed is to what extent we might be able to recognize secondary succession in Tertiary floras. Our experience with the Succor Creek macroflora would indicate that such recognition is unlikely, except in exceptional cases, in studies supported only by macrofossil data. Diversity studies of macrofossils can provide useful input with adequate stratigraphic control and representation, but too few floras have been studied in sufficient detail. One major exception is the Amethyst Mountain section in Yellowstone National Park (Wyoming), which has been studied by Dorf (1960, 1964) and more recently, in palynological detail by Fisk (1976). Neither the macrofossil nor palynological data provide a clear successional picture. Revegetation of disturbed areas involved essentially the same species that characterized the climax, predisturbance forest. What representation of successional communities did exist was probably a function of differential recolonization rates for different forest species. It would seem that distinct plant species characterized by pioneering strategies simply had not evolved on upland sites by Eocene time. Such disturbed sites were increasingly common in the Pacific Northwest during the Oligocene and represented unexploited habitats that served as the selective force for the appearance of pioneering strategies, particularly in herbaceous plants whose short life-cycles would be highly adaptive in such exploitive situations. If the Succor Creek flora is any indication, a number of pioneering grasses and forbs were present by the middle Miocene. Disturbed sites were still active evolutionary stages, however, if the large number of unknown pollen taxa is considered. It appears most probable that these unknown pollen types represent herbaceous plants that had simply not evolved to the point where they can be placed in modern families or genera based on pollen features. The genesis of distinctive plant communities associated with secondary succession on upland sites may well date to the middle Tertiary, with increasing diversity and partitioning of niche space in such communities into the late Tertiary.

The existence of well-developed secondary successional communities of xeric aspect may well provide some insight into the problems of modeling the pattern of late Tertiary modernization of vegetation in the region. In addition to a general cooling trend through the Tertiary, the major climatic factor affecting the intermountain region was the gradual development of an intensifying rain shadow as the Cascades were uplifted to the west. The conventional model for modernization involves the gradual exclusion of mesic vegetation elements and the evolution and migration into the area of species with xeric adaptations in response to this regional drying trend. Although evolution and adaptation in response to increasing aridity certainly did occur, the Succor Creek successional communities indicate that the process of modernization was probably far more complex. The evolution of species adapted to pioneer strategies on disturbed and edaphically dry sites probably began in the middle Tertiary and these groups showed considerable diversity by Succor Creek time. Secondary successional sites were in all probability the stage for the evolution of many taxa that would later assume a prominent position in the climax vegetation of the region during the Pliocene and Quaternary. These successional taxa, evolving on disturbed sites in a period when regional vegetation still consisted of mesic forest vegetation, were in essence preadapted to the regional dryness that would later characterize the region. Although additional study will be required to test this hypothesis, it is quite possible that Miocene successional communities played a major role in the modernization of vegetation in the region. As annual rainfall gradually decreased, the successional sequence would require longer periods for the transition back to forest vegetation following disturbance, with an increased possibility for preclimax stages to become permanently established on some sites. The mesic climax forests would gradually become less diverse with successive reforestation under conditions of increasing aridity. In some cases edaphically dry sites might never fully recover, being characterized by more xeric vegetation types. The areal extent of such xeric communities would gradually increase as water stress gradually excluded many taxa on a regional basis, leaving "successional" communities as "climax" communities by default.

The incorporation of a secondary succession component into the model for modernization is far more satisfactory than relying entirely on climatically induced selection to account for the evolution of the majority of the modern vegetation elements in the Pliocene or Pleistocene.

A great many floras require additional analysis with palynological and stratigraphic control in order to adequately assess the impact of secondary succession on Tertiary vegetation history. The data to date indicate that origin of distinctive successional communities on upland sites may be traced to the middle Tertiary. What the situation may be in regard to lowland and coastal sites remains to be determined. In any case, an increased awareness of the existence of distinctive successional communities can only serve to increase our understanding of paleovegetation dynamics and may well provide additional data that will bear on critical problems in the history of plants.

Considering the diversity of Tertiary vegetation and the distrubition of community types in response to moisture, elevation, and other environmental gradients, and the rate, character, and extent of recovery of such communities from catastrophic events, there is little reason to consider the dynamics of Tertiary vegetation to be any less complex than that to be observed today.

REFERENCES

Arnold, C. A. 1936a. The occurrence of Cedrela in the Miocene of Oregon. *Am. Midland Naturalist* 17:1018-1021.

Arnold, C. A. 1936b. Some fossil species of Mahonia from the Tertiary of eastern and southeastern Oregon *Contr. Mus. Paleontol. Univ. Michigan* 5:57-66.

Arnold, C. A. 1938. Observations on the fossil flora of eastern Oregon. *Contr. Mus. Paleontol. Univ. Michigan* 5:79-102.

Axelrod, D. I. 1964. The Miocene Trapper Creek flora of southern Idaho. *Univ. California Pub. Geol. Sci. 51*:1-180.

Axelrod, D. I. 1968. Tertiary floras and topographic history of the Snake River basin, Idaho. *Geol. Soc. Am. Bull.* 79:713-734.

Brooks, B. W. 1935. Fossil plants from Sucker Creek, Idaho. *Ann. Carnegie Mus.* 24:275-336.

Chaney, R. W., and D. I. Axelrod. 1959. Miocene floras of the Columbia Plateau. *Carnegie Inst. Wash. Pub. 617*, 237p.

Dorf, E. 1960. Tertiary fossil forests of Yellowstone National Park, Wyoming. *Billings Geol. Soc. 11th Ann. Field Conf. Guidebook;* pp. 253-260.

Dorf, E. 1964. The petrified forests of Yellowstone National Park. *Sci. American* 210(4):107-112.

Downs, T. 1956. The Mascall fauna from the Miocene of Oregon. *Univ. California Pub. Geol. Sci.* 31:199-354.

Everden, C. F.; D. E. Savage; G. H. Curtis; and G. T. James. 1964. Potassium-argon dates and the Tertiary floras of North America. *Am. Jour. Sci.* 262:945-972.

Fisk, L. H. 1976. Palynology of the Amethyst Mountain "fossil forest," Yellowstone National Park, Wyoming. Ph.D. Dissert., Loma Linda Univ., Loma Linda, Calif., 340p.

Graham, A. K. 1965. The Sucker Creek and Trout Creek floras of southeastern Oregon. *Kent State Univ. Bull 53;* 147p.

Kittleman, L. R. ; A. R. Green; A. R. Hagood; A. M. Johnson; J. M. McMurray; R. G. Russell; and D. A. Weeden. 1965. Cenozoic stratigraphy of the Owyhee region, southeastern

Oregon. *Bull. Mus. Nat. Hist. Univ. Oregon* 1:1-45.

Knowlton, F. H. 1898. The fossil plants of the Payette Formation. *U.S. Geol. Surv. 18th Ann. Rpt.*, pt. 3 pp721-744.

MacGinitie, H. D. 1962. The Kilgore flora—a late Miocene flora from northern Nebraska. *Univ. California Publ. Geol. Sci.* 35:67-158.

Shah, S. M. I. 1968. Stratigraphic paleobotany of the Weiser area, Idaho. Ph.D. dissert., Univ. Idaho, 166p.

Smiley, C. J.; S. M. I. Shah; and R. W. Jones. 1975. Guidebook of the later Tertiary stratigraphy and paleobotany of the Weiser area, Idaho. *Ann. Meet. Geol. Soc. Am. Guidebook*, in Cooperation with Idaho Bur. Mines and Geol., 13p.

Smith, H. V. 1938. Some new and interesting late Tertiary plants from Sucker Creek, Oregon-Idaho boundary. *Bull. Torrey Bot. Club* 65:557-564.

Smith, H. V. 1939. Additions to the fossil flora of Sucker Creek, Oregon. *Papers Michigan Acad. Sci. Arts Letters* 24:107-120.

Taggart, R. E. 1971. Palynology and paleoecology of the Miocene Sucker Creek flora from the Oregon-Idaho boundary. Ph.D. dissert., Mich. State Univ., E. Lansing 196p.

Taggart, R. E. 1973. Additions to the Miocene Sucker Creek flora of Oregon and Idaho. *Am. Jour. Bot.* 60:923-928.

Taggart, R. E., and A. T. Cross. 1974. History of vegetation and paleocology of upper Miocene Sucker Creek beds of eastern Oregon. *Symp. Strat. Palynol., Birbal Shani Inst. Paleobot., Spec. Pub.* 3:125-132.

8

BIOSTRATIGRAPHIC ANALYSIS OF EOCENE CLAY DEPOSITS IN HENRY COUNTY, TENNESSEE

Frank W. Potter, Jr.
David L. Dilcher

SUMMARY

Excellently preserved plant fossil assemblages are found in middle Eocene, Claiborne Formation, sediments within the upper portion of the Mississippi Embayment in southeastern North America. Over the past century these deposits have been studied by various investigators and provide a great deal of information about early Tertiary angiosperms. As part of a current reinvestigation of the fossil plants from these sediments, material has been collected from over 25 localities in western Kentucky and Tennessee. These localities are clay lenses isolated from one another and are presently of uncertain relative age within the middle Eocene. This is the major problem addressed in this paper, which is an attempt to begin to outline reliable relative dates for these numerous small disjunct clay and lignite lenses.

An understanding of the relative ages of these clay deposits can be approached best through a study of their depositional history as interpreted by the nature of the sediments surrounding the clay lenses and details of the biostratigraphy within the lenses themselves. These deposits are sediments that filled abandoned river channels on an ancient flood plain.

Palynomorph content and relative abundances have been used to date some of the localities. This information has been important in establishing their stratigraphic affinities with the Claiborne Formation. However, attempts based on palynomorphs have not yet produced results sufficiently reliable for ranking the isolated lenses sequentially within the Claiborne Formation. These attempts, based on limited samples from several localities in the area, have resulted in conflicting results. The pollen diversity changes vertically at one locality through the thickness of a single clay lens, which suggests that depositional history, flood-plain vegetation distribution patterns, and ecological factors are valid explanations for palynomorph variation previously attributed to fluctuations of stratigraphic significance.

The distribution of megafossils from the localities studied indicates that specific angiosperm taxa are restricted to certain localities. Also, cuticular analysis demonstrates that some angiosperm leaves common to several localities can be distinguished by special trichome types or epidermal cell forms unique to specific clay pits. This is regarded as evolutionary diversity expressed during the middle Eocene. As presently understood, the variation of megafossils is insufficient to establish floristic (time) zones necessary to interpret the history of the deposition of these clay lenses. However, the megafossils combined with other floristic features and sedimentary features provide data that can be applied to the history of deposition of the clay lens under consideration. We suggest a model for this deposition here.

The depositional model presented is based primarily upon the vertical relation of the clay lenses to one another. It also provides an explanation for the vertical variation observed in the pollen profiles of these clay lenses. There are some changes in floristic composition, leaf morphology, and cuticular anatomy observed between deposits. Also the depositional history of each clay lens can be traced through phases, which resulted in similar sedimentary sequences or parts of sequences for most clay lenses studied. Support for this model of the ancient flood-plain sediments in western Tennessee is also provided by evolutionary trends of select leaf types.

INTRODUCTION

Clay lenses are commonly distributed along the sedimentary beds of the Mississippi Embayment that are exposed in a narrow zone running northeast by southwest through western Kentucky and Tennessee. The age of these sediments ranges from Upper Cretaceous to middle Eocene. Following the Embayment axis (northeast-southwest) the Cretaceous sediments are exposed along the extreme eastern margin of this zone and the youngest sediments lie to the west. This report is concerned with those clay lenses in western Kentucky and Tennessee which are considered to be part of the Claiborne Formation deposited during the middle Eocene (Figure 8.1).

We have examined the sediments from numerous clay pits in the area shown in Figure 8.2. Nearly 25 clay pits have yielded fossil leaves, fruits, seeds, wood, flowers, and pollen. Each clay body represents an isolated depositional event in space; and the question to be addressed is, do these clay deposits also represent isolated events in time? And, if these depositional events do represent limited and different segments of middle Eocene time, can some orderly arrangement of their sequence of deposition be discovered? This paper is an attempt to arrange these clay pits in a stratigraphic sequence. The evidence available is the preserved palynomorphs, megafossil material, and the geomorphology of the clay bodies. The work presented here is an attempt to establish a preliminary sequence as a working model to be refined and affirmed or corrected by further and more detailed study. The importance of establishing such a sequence is that it allows the use of the Mississippi Embayment sediments as a biostratigraphic tool to finely divide and sort out in chronological order vegetational changes of plant ecosystems and evolutionary modifications of plant form in the structurally preserved fruits, seeds, leaves, flowers, and wood common in these sediments.

Many leaves, fruits, seeds, and flowers are restricted to particular clay pits; a few fossil forms are common to nearly all pits. Palynomorphs, as well, may be restricted to particular clay pits or even to particular zones within a clay pit (Potter, 1976). A careful comparative study of the pollen types of various clay pits has not been done, but it would yield a great deal of useful data. Isolated pollen samples have been analyzed and numerous megafossils have been collected from these clay pits. The data available from these studies are used in this report.

Recent pollen studies of the Mississippi Embayment region have been summarized by Elsik (1974); Elsik and Dilcher, (1974), Fairchild and Elsik (1969), Frederickson, (1969), Potter (1976) and Tschudy (1973a, b, 1975).

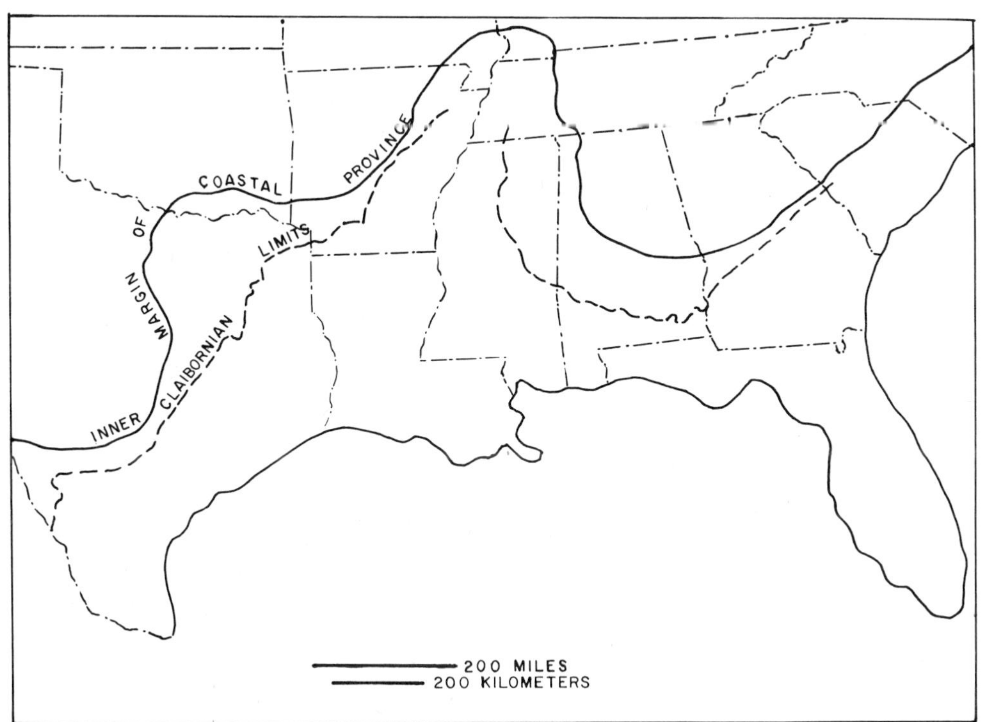

FIGURE 8.1 *Generalized map of approximate limits of Claibornian and Cretaceous transgressions in northern Gulf coastal province (redrawn from Murray, 1961).*

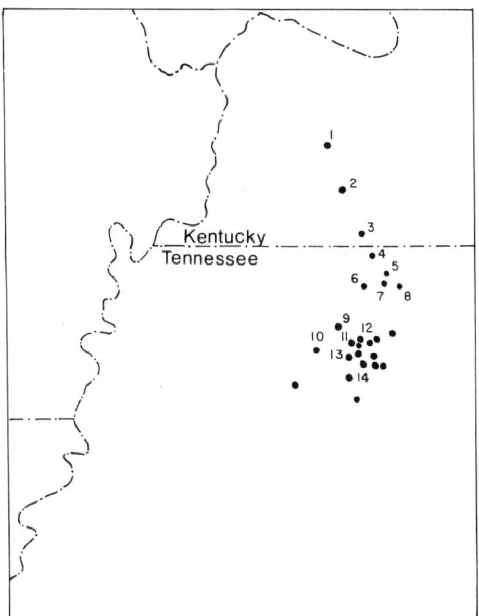

FIGURE 8.2 *Pit locations in western Kentucky and Tennessee; 1, Lamkin pit; 2, South-40 pit; 3, Bell City pit; 4, New Puryear pit; 5, Puryear pit; 6, Cottage Grove area; 7, Foundry Hill pits; 8, Buchanan pit; 9, Warman pit; 10, Gleason area; 11, Lawrence pit; 12, Miller pit; 13, Atkins pit; 14, Liberty Church area.*

These investigators recognize the usefulness and importance of the pollen profiles of these isolated clay deposits and some establish particular zones based on the presence or absence and the abundance of various pollen types. Basic pollen types have been recognized for the Eocene sediments of the embayment area and Fairchild and Elsik (1969); Tschudy (1973b); and Elsik (1974), present index palynomorphs for the Gulf Coast (based on presence or absence and abundance of pollen types) that allow an investigator to discriminate among lower, middle, and upper Eocene sediments. Tschudy charts the occurrences of the palynomorphs in the Mississippi Embayment indicating limited stratigraphic ranges of several within the middle Eocene; Elsik also plots relative abundances of palynomorphs for the Claiborne sediments of the Texas Gulf Coast. These data suggest that palynomorphs can be used to place these isolated clay lenses in the Wilcox, Claiborne, or Jackson (lower, middle, upper Eocene) formations and possibly also provide precise information as to subdivision of member age equivalent for middle Eocene time.

The ability to subdivide Claiborne time using palynomorphs (Tschudy, 1973b) gave some hope that each of the numerous isolated clay lenses in western Tennessee and Kentucky could be placed stratigraphically if pollen samples were analyzed. Before such a general project of sampling was undertaken, however, we felt that it was critical to examine the nature of a single clay lens in some detail. Therefore, the palynomorphs of the Miller clay pit were examined from precisely located samples (vertically at 10-cm intervals) from the base up to the top of a 10-m-thick seam of clay-lignite-clay exposed in open pit mining. As a result of this work, Potter (1976) found that the relative abundance of pollen types present varied as the lithology of the sediments varied, indicating a strong environmental control influencing this aspect of the pollen profile. Potter questioned the stratigraphic usefulness of the abundance of palynomorphs and indicated that some pollen types may be more important as environmental indicators than as indicators of the subdivision of Claiborne time. Detailed pollen profiles of several clay pits open in this area should be completed. Such studies will provide important data to test the model proposed here of the depositional history of these lenses.

Early paleobotanical reporting was done by Berry (1916, 1930) on the floras of over 200 Paleogene plant-bearing localities (many of which were from clay deposits) in southeastern North America. He suggested a paleoenvironment for deposition of such plant-bearing sediments generalized from a locality he examined in Texas. He concluded that these clay deposits represent near-marine paleoenvironments with strand vegetation while some others represent drowned river valleys Berry (1916, 1930).

Dilcher (1971) considered the origin of the clay lenses only in western Kentucky and Tennessee. The information applied to this question was derived from outline maps of known clay deposits, the form of their cross-section, and the nature of sediments underlying the clays. Potter (1976), concentrating upon a single clay lens, carried this analysis further. The results of these studies demonstrate that the small clay lenses typical of Henry County, Tennessee, and nearby areas are narrow, linear-elongated, most often curved, river-channel shaped in cross-section, surrounded on the sides and undersides by cross-bedded sands (occasionally fine gravels), and consist of uniform fine clays with an occasional lignitic seam or cap. Primarily upon this information, we proposed that these small,curved, elongated clay lenses represent clay plugs in oxbow lakes on an ancient flood plain. Plant megafossil remains are common in several clay lenses, absent in some, and pollen is present except where oxidized. Plant megafossils are not often uniformly distributed through the clay lenses but are more abundant in some layers.

GEOLOGY OF THE CLAIBORNE FORMATION

Regional

In the Mississippi Embayment the maximum marine transgression extended into southern Illinois during the Late Cretaceous (Figure 8.1). Repeated marine transgressions occurred throughout the Paleogene in the embayment region. Because the embayment contained a variety of depositional environments over exceedingly broad areas in the common coasts of the embayment, several distinctive marine sediments can be traced using fossils over broad areas. Non-marine sediments also are found covering broad areas, several of which have been related to particular formations chiefly by the application of palynology to the spore-and pollen-bearing sediments (Fairchild and Elsik, 1969; Frederiksen, 1969, 1973; Tschudy 1973b; Elsik 1974; Elsik and Dilcher 1974).

The Claiborne Group of middle Eocene sediments is best characterized as a time rock unit that encompasses a large variety of lithologic types that can be unified only by establishing a common age based upon fossils (Murray, 1961). This had been done successfully for the predominantly marine and near-marine sediments exposed in the southern reaches of the Mississippi Embayment. The non-marine sediments bordering the margins of the marine, consisting of a diverse array of lithologies, and have always presented a problem for identification as Claiborne.

The extensive marine transgressions between the Wilcox, the Claiborne, and the Jackson help to provide regional recognition for these basic groups of sediments (Figure 8.3). Within the Claiborne the numerous transgressive-regressive phases are easily recognized in the southern areas of the embayment. Both Tschudy (1973b) and Elsik (1974) use pollen data from non-marine intervals in Texas, Arkansas, Louisiana, Mississippi, and Alabama to establish pollen profiles that they feel confident represent regional time unit differences. These localities are bounded by marine sequences where the pollen types are first established. Then they have extrapolated to areas where non-marine sediment dominate (such as Kentucky, Tennessee, and some areas of Arkansas) and identified Wilcox, Claiborne, and Jackson sediments, as well as upper and middle Claiborne sediments.

The basis for assigning non-marine sediments (mostly plant-bearing clays) to Wilcox, Claiborne, or Jackson was on the basis of the presence or absence of specific pollen types. As Tschudy (1973b) demonstrates, several pollen types (e.g., *Thomsonipollis* spp.) end abruptly at the Wilcox-Claiborne boundary while others are first found in the Claiborne. The Claiborne-Jackson contact is similarly identified, although there fewer pollen types end so abruptly and fewer begin as new types.

Tschudy (1973b) presents 16 pollen types which begin their occurrence in Claiborne time. Both Tschudy and Elsik (1974) differentiate the formations recognized in non-marine sediments by the relative abundance of several pollen types and presence or absence of a few pollen types. Each suggests that this can be applied upon a regional basis throughout the embayment area. They have both examined clays from Henry County, Tennessee, and indicated (Elsik, 1974; Elsik and Dilcher, 1974; Tschudy,

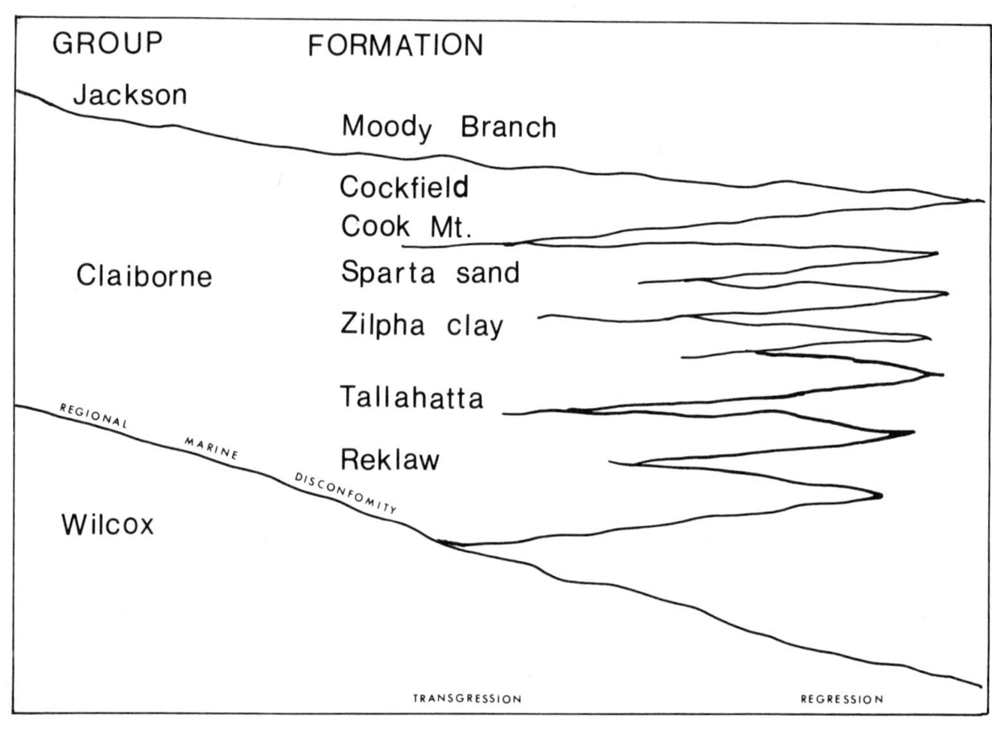

FIGURE 8.3 ***Transgression-regression schematic (not to scale).***

personal communication) an upper Claiborne (Cockfield Fm., and perhaps Cook Mt. Fm. or lower Cockfield Fm.) age for them.

Local

The marine transgressions of the Eocene (Figure 8.3) can not be identified in the Henry County area of western Tennessee (Parks, 1971). The more extensive transgressions such as those separating the Wilcox, Claiborne, and Jackson groups can be identified only in the southwestern corner of Tennessee (Stearns, 1957; Murray, 1961). The Claiborne Formation in Henry County, Tennessee, consists mainly of non-marine sands and isolated disjunct lenses of clay. As already discussed, there are numerous plant fossil-bearing lenses of clay in this area. Data from a regional consideration of the pollen flora, where it can be tied into marine sequences, indicates that these lenses are best considered to be late-middle Eocene in age (Cockfield Formation of the Claiborne Group) (Elsik, 1974; Elsik and Dilcher, 1974; Tschudy, personal communication). Differences in the floras of particular clay pits in Henry County, Tennessee, and near-by areas are obvious by a comparison of the composition of the floras from each pit and by the specific nature of the epidermal cells, trichomes, and aspects of the overall leaf form of several taxa. Two explanations for these differences are: ecological variability or evolutionary variability. It is important to attempt to reconstruct the paleoenvironments and relative ages of these clay pits in order to understand the cause of the variations observed in the fossil record between these sites.

The ecology of most deposits seems to have repeated a relatively similar history for each clay lens. As discussed earlier in this paper and by Dilcher (1971) and Potter (1976), the nature of these sediments indicates a typical oxbow-lake depositional system. Each clay lens repeats a similar environmental history with some variation in the extent of associated lignite and relative abundance of plant fossils. Thus environmental factors most probably do not account for the differences observed in the plant fossils from clay pit to clay pit.

The most logical explanation for the observed differences of the types, relative abundance, form, and anatomy of the plants at the various localities is that each locality represents a slightly different segment of time. The time interval of evolution preserved in each clay pit is very short (500 to 1500 years) because oxbow lakes on a flood plain are recognized as short-term events when viewed against geological time (Crisman and Whitehead, 1975). The tools available at the present, such as palynology and knowledge of the megafossils are probably not sufficient to rank order these individual clay pits within Cockfield Formation time. The data provided below is presented in an attempt to establish a rank order for these clay pits within late-middle Eocene time.

DEPOSITION OF FOSSIL-BEARING SEDIMENTS

The low-lying features of the Mississippi Embayment allowed for easy migration of river systems across broad flood plains; as sediments brought in by the rivers accumulated. Deposition of the flood-plain sediments, discussed in this paper, resulted from the meandering of the "Appalachian" River system entering the embayment in Kentucky and building a poorly defined delta in North Central Mississippi (Grim, 1936).

Based upon field examination; the clay pits can be differentiated into two types. One type has relatively massive deposits of clay and some associated lignites. Often these lignite seams overlie the clays, although in some localities the lignite is interspersed with the clays. No good megafossils have been found in these clays. The lignites are sufficiently decomposed to obscure plant structure and in some localities numerous root channels extend from the lignite into the underlying clay. These features suggest a low-lying area of shallow water receiving fine-grain sediments interspersed with some swamp environments with accumulation of organic material.

The second type of clay pit consists of small isolated clay lenses, often accompanied by overlying lignite seams. Generally, these lenses have 5 to 10 m of light tan clay lacking micro-and megafossils, overlain by dark brown to gray clay rich in plant fossils. The overlying lignites consist of compacted leaves, wood, fruits, and seeds; and occasionally above the lignites are scattered layers of light tan, gray, to pink clays or sands. The clay deposits are underlain by cross-bedded channel sands (as seen at the Puryear site) and unsorted pea-size stream gravels (the Warman site); and, in outline and cross-section, they are characteristic of abandoned channels apparently representing meander cutoff sloughs, oxbow lakes, and stream reaches (Figure 8.4). Field examination indicates a common depositional history for most of the lenses of clay. The fossil-bearing clays are a product of fine sediments, which filled the oxbow lakes on the ancient flood plain. The lignites accumulated as the lake was filled with clay, forming a wet swampy environment. These events probably took about 500 to 1500 years. Several clay lenses contain fossil assemblages representing both of the above paleoenvironments, some lack any evidence of lignite accumulation, but all represent short individual segments of time.

Sufficient sites are present in western Tennessee to permit comparison of vertical position as determined from topographic maps and overburden exposures observed in the field (Figure 8.4). Although not all sites align precisely, an initial eastward meandering-channel shift is suggested. Once reaching the eastern margin of the embayment, continued deposition shifted the channel back toward the basin axis. This resulted in a sequence of channel deposits ascending in age, oldest near the base of the series along the west edge of the study area, the youngest also at the western edge but at a higher elevation (Figure 8.5). The eastern pits, e.g., Puryear, are of intermediate age. Thus the sequencing of the relative ages of the pits is based upon their relative geomorphic position in the flood plain. The whole series of clay lenses represents a geologically short time span, the last sweep of the river having occurred during late-middle Eocene (Claiborne Group) deposition.

The lithology of the two pit types provides supporting evidence, the basal clay and lignite sites are believed to represent an almost flat lowland-swamp flood-plain environment sloping approximately three feet per mile southward (Figure 8.6a). These lowland, swampy areas were dissected and buried by the sediments of the shifting river system that produced the ascending series of clay plugs (Figure 8.6b). The amount of silt and clay accumulation and organic content of these plugs depended on the degrees of isolation from the main channel as the river returned to its original course (Figure 8.6c).

This model provides a straightforward explanation for the data presently available regarding the nature of the fossil plant record and sedimentology of these clay lenses. The fossil assemblages can be examined for evolutionary changes and environmental influences independently of their use to establish the time controls.

Selected megafossil distribution among the clay deposits is presented in Figure 8.7. This figure is presented to

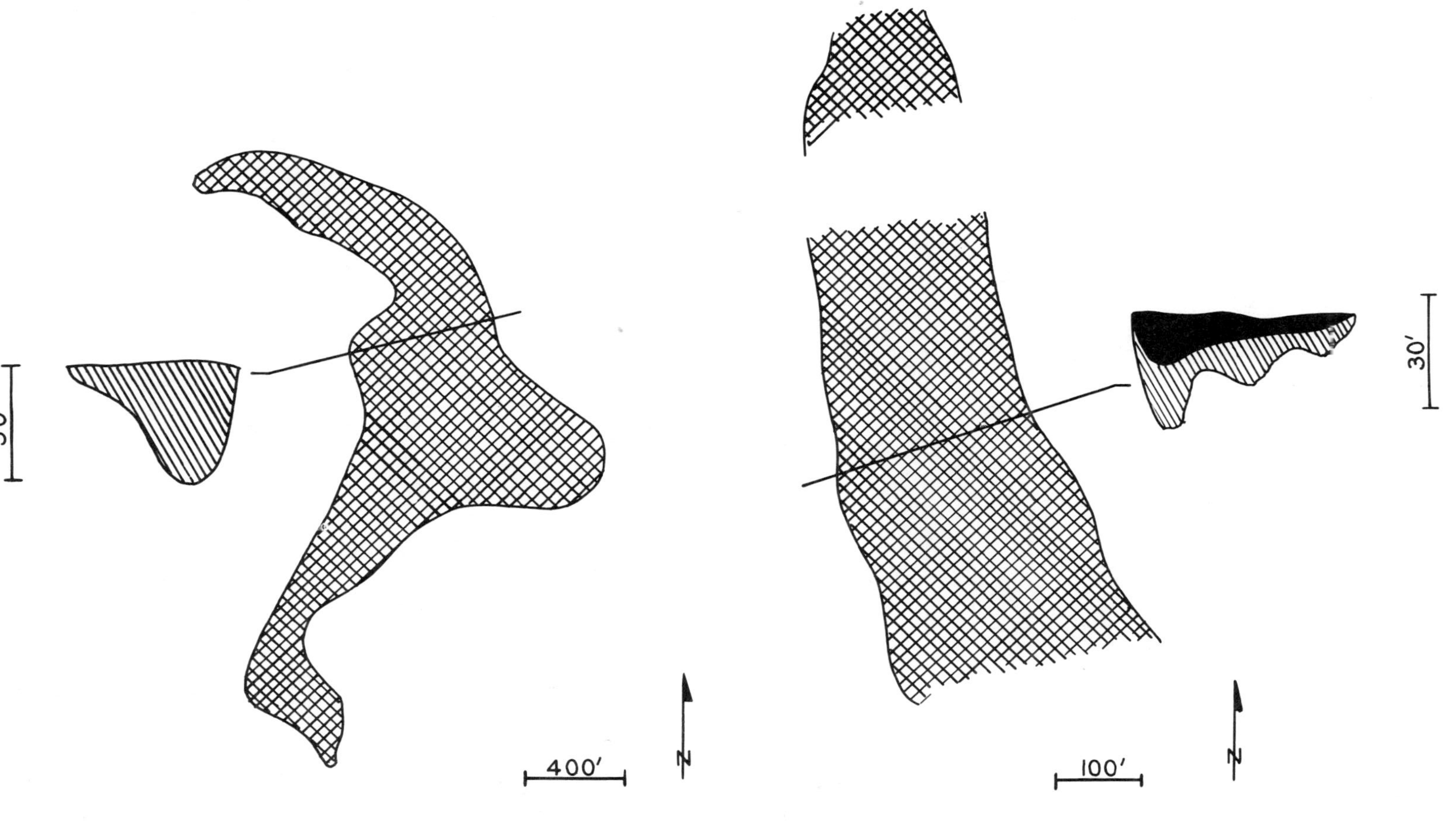

FIGURE 8.4 *Outline and cross-section for two clay pits in western Tennessee. Diagonal lines indicate clay sediments; the solid area indicates lignite.*

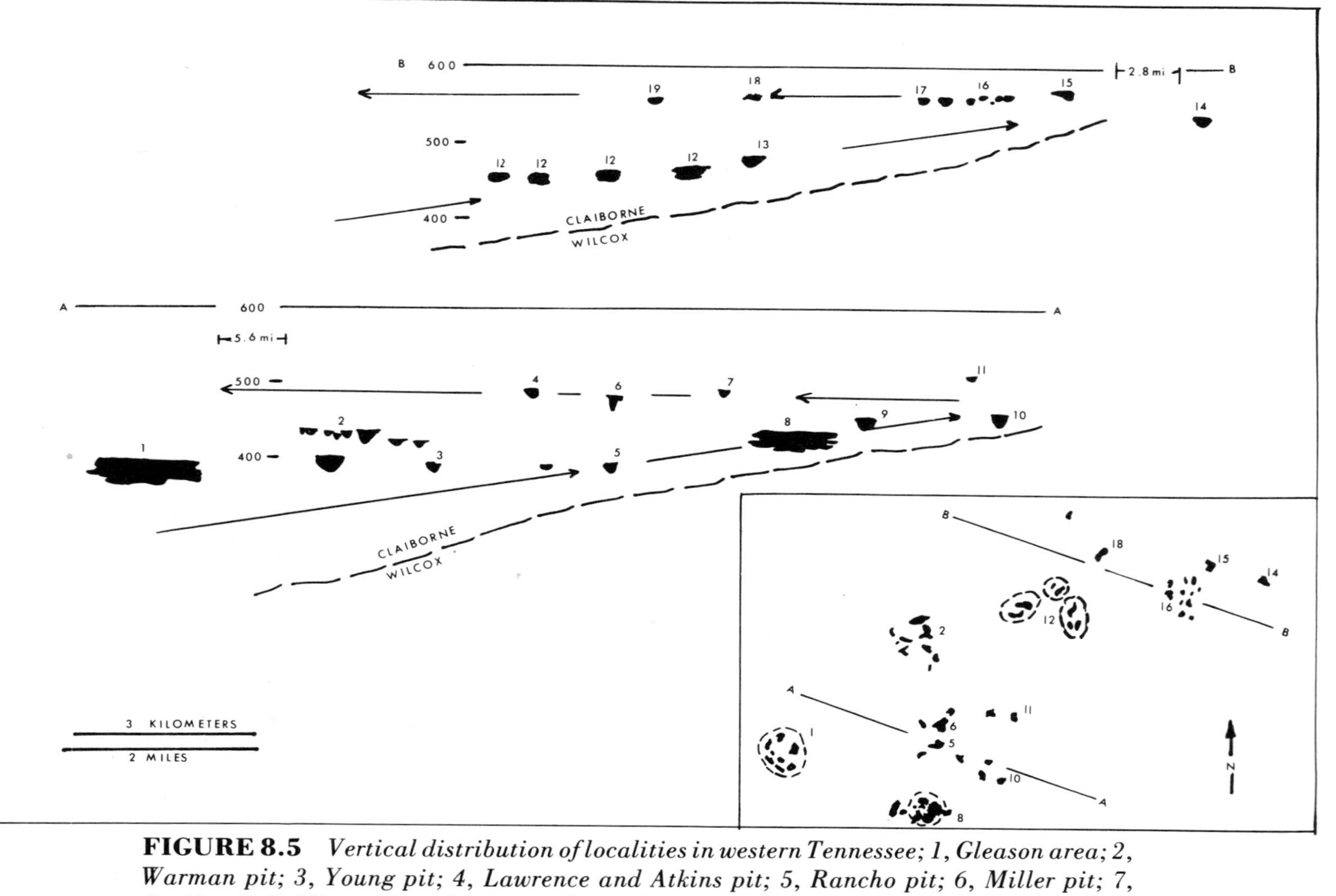

FIGURE 8.5 *Vertical distribution of localities in western Tennessee; 1, Gleason area; 2, Warman pit; 3, Young pit; 4, Lawrence and Atkins pit; 5, Rancho pit; 6, Miller pit; 7, Grable pit; 8, Liberty Church area; 9, Carauthers pit; 10, Haynes pit; 11, Rushing pit; 12, Cottage Grove area; 13, Martin pit; 14, Buchanan pit; 15, Puryear pit; 16, Foundry Hill pits; 17, Spinks pit; 18, New Puryear pit; 19, Breedhive pit.*

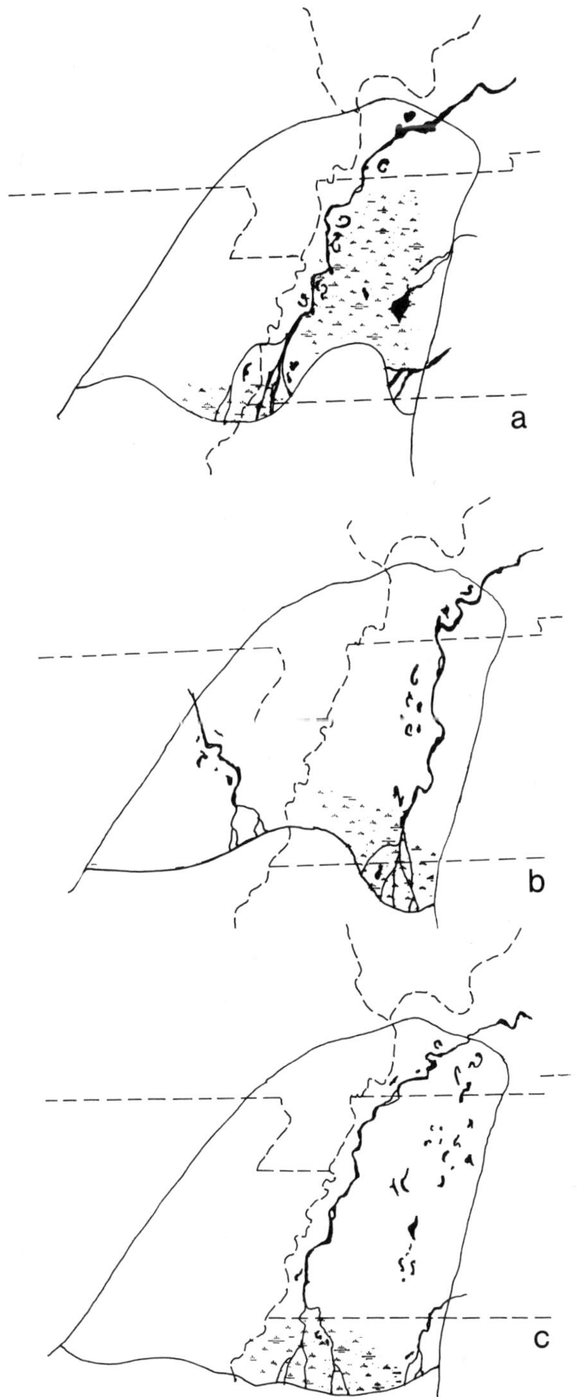

FIGURE 8.6 *Meander sweep of the Appalachian River system during Late-Middle Claiborne time. a, initial position; b, eastern extent of migration; c, end of migration cycle.*

MEGA-FOSSILS / PITS	Philodendron	Palms: Costapalma	Palms: Palustrapalma	Palms: Palmacites	Rubiaceophyllum small	Rubiaceophyllum large	Apocynophyllum Reticulatus	Apocynophyllum Laminatus	Palaeodendron	Knightiophyllum form 1	Knightiophyllum form 2	Engelhardia	Dendropanax	Annonaceous	Ocotea	Paleooreomunnea	Paraoreomunnea	Paraengelhardia	Podocarpus
NEW PURYEAR																			●
PURYEAR				●		●			●	●					●	●		●	
FOUNDRY HILL AREA																			
(BUCHANAN)																			●
MILLER						●		●						●					
LAWRENCE					●	●					●	●	●	●			●		
ATKINS					●		●	●						●					
WARMAN		●				●				●		●	●				●		●
(LAMKIN)		●				●	●					●		●	●	●	●		●[1]
(SOUTH-40)																			
RANCHO	●			●	●														
YOUNG	●																		

FIGURE 8.7 *Distribution of selected megafossils among the various pits in western Tennessee. Pits in brackets are not in the study area. Fossils at 1 were recorded as dispersed cuticle.*

TABLE 8.1 *Suggested sequence of deposition: (See Figure 8.5).*

B________B′	YOUNGEST	A________A′
Breedhive pit New Puryear pit Spinks pit Foundry Hill pits Puryear pit	═ All very close in age ═	Rushing pit Lawrence pit & Atkins pit Miller pit Grable pit
Buchanan pit Martin pit Cottage Grove pits		Haynes pit Carauthers pit Liberty Church pits Rancho pit Young pit Warman pit Gleason area pits
	OLDEST	

show that more data than is presently available is required in order to use megafossils as reliable indicators for arranging these clay deposits in an ordered sequence. Much more work is needed to sort out ecological versus stratigraphical control over the occurence of megafossils.

To relate the clays to the regional stratigraphy of the embayment, the most reliable and useful fossil material in these terrestrial clay deposits is pollen. The megafossil material reflects local paleoenvironments. The variability observed in angiosperm leaf cuticles provides recognizable differences between elements of the floras that otherwise are difficult to distinguish. Pollen from paleoenvironmentally equivalent samples of the clay deposits discussed here have not been studied. Perhaps if this were done, pollen would be as useful for local as well as regional stratigraphy.

SEQUENCE OF DEPOSITION

The suggested sequence of deposition presented in Table 8.1 is primarily derived from the information provided in Figure 8.5. The vertical distributions of these clay deposits in western Tennessee all appear to be restricted to the upper-middle Eocene (Cook Mt. and Cockfield formations, Figure 8.3). Their preservation probably results from their burial by subsequent sediments laid down in the rather extensive transgression marking the boundary between the Claiborne and the Jackson groups (Figure 8.3). These clays deposited in a flood-plain environment avoided destruction until their final burial. Many of these clay deposits are similar in age, while those deposited deeper into the flood-plain sediments are most probably older. This depositional sequence is the first approximation of such events in the Mississippi Embayment and is proposed as a model for subsequent consideration.

POST DEPOSITIONAL HISTORY

Individual clay lenses often consist of clays that range in color from dark gray, brown, tan, cream, white, pink to red. Previously, some field geologists held the idea that the color of this clay could be used as a field guide to differentiate sediments of the Wilcox Group from those of the Claiborne Group. The palynology and the megafossils in the clays demonstrate that each deposit is a single unit of a single age. The differences in color are of no value to mark one group or formation from another in these deposits. The variations in color represent differences of postdepositional history that each clay lens and each part of a clay lens has undergrone.

The gray and brown colors are the result of unoxidized organic material; the lighter colors indicate either a total lack of organic materials or some degree of oxidation. Subsequent coloring by iron is common, producing pink or red clays. Once deposited, the organic matter in a clay lens will not oxidize unless the water table is lowered. If only the upper portion of the clay lens extends above the water table, only the upper portion will oxidize. If the entire lens is above the water table it will oxidize from the top, bottom, and sides, slowly discoloring the organic-rich clays. Oxidation of such a lens of clay will proceed unevenly and probably take a great deal of time to be completed. Often a wide (4-10 feet) band of gray or brown clay may extend through the central portion of a lens surrounded by tan and cream colored clays as a result of depressed water tables.

Changes in the water table surrounding the clay lenses in Tennessee are possibly an ancient phenomenon and may have been a seasonal event. Some oxidation of these clay lenses may be the result of Recent water-table fluxuations in Henry County, Tennessee.

CONCLUSIONS

The leaf floras, fruits and seeds, dispersed cuticle, and palynology of the clay deposits in Henry County, Tennessee indicate a similar (but not identical) age and paleoecology for these sediments. These same fossils also demonstrate some unique differences among the individual clay deposits. Many of these differences can be related to the slightly different ages of the individual clay lenses. But very few of these differences are related to the paleoecology of individual lenses when similar sediments of two deposits are compared. Thus, while these differences can be recognized for each deposit, it is extremely difficult to use this data to order rank the clay deposits from oldest to youngest. In order to circumvent this problem and establish a working model of the relative ages of these clay deposits their ages have been correlated with their relative elevation of deposition upon a common ancient flood plain.

The following conclusions have been discussed above.

1. Many of the clay pits in Henry County, Tennessee, represent clay plugs in oxbow lakes on an ancient flood plain.
2. Probably the clay pits represent a last major sweep of the "Appalachian" River system through its flood plain near the end of the middle Eocene.
3. The basin continued to subside throughout middle and post middle Eocene time, burying the river-channel features by later transgressions of the Embayment and thus effectively preventing their loss by further flood-plain activity.
4. The extent of the entire depositional events discussed here was confined to a moderately short span of geologic time, of sufficient length however, to provide an excellent model for detailed evolutionary studies between individual localities.
5. Each individual clay pit represents a very-short-term event lasting approximately 500 to 1500 years.
6. The fossil plants preserved within the clay lenses in Henry County, Tennessee, contain a record of predominantly flood-plain vegetation.
7. Many clay pits represent more than one paleoenvironment (clay sediments represent open oxbow lakes and lignites closed swampy environments).
8. The postdepositional history of these clay lenses reflects the influence of changes in the water table allowing oxidation of those clays that contain organic material. The oxidized clays are light in color; those rich in organic material are darker.
9. Some floristic variations observed may be ecological while the variations observed in the leaf form and anatomy of the cuticle of similar plant fossils found in several clay pits probably represent evolutionary changes, illustrating differences in age of individual pits.

REFERENCES

Berry, E. W. 1916. The Lower Eocene floras of southeastern North America. *U.S. Geol. Surv. Prof. Paper 91*, 481p.

Berry, E. W. 1930. Revision of the Lower Eocene Wilcox flora of the southeastern United States. *U.S. Geol. Surv. Prof. Paper 156*, 196p.

Crisman, T.L., and D. R. Whitehead. 1975. Environmental history of Hovey Lake, Southwestern Indiana. Am. Mid. Nat. 93:198-205.

Dilcher, D. L. 1971. A revision of the Eocene flora of southeastern North America. *Palaeobotanist* 20:7-18.

Elsik, W. C. 1974. Characteristic Eocene palynomorphs in the Gulf Coast, U.S.A. *Palaeotographica, B* 149:90-111.

Elsik, W. C., and D. L. Dilcher, 1974. Palynology and age of clays exposed in Lawrence clay pit, Henry County, Tennessee. *Palaeontographica, B* 146:65-87.

Fairchild, W. W., and W. C. Elsik. 1969. Characteristic palynomorphs of the Lower Tertiary in the Gulf Coast. *Palaeontographica, B* 128:81-89.

Frederiksen, N.O. 1969. Stratigraphy and palynology of the Jackson Stage (upper Eocene) and adjacent strata of Mississippi and western Alabama. Ph.D. dissert. Univ. Wisconsin, 356p.

Frederiksen, N.O. 1973. New Mid-Tertiary spores and pollen grains from Mississippi and Alabama. *Tulane Stud. Geol. Paleontol. 10:65-86*

Grim, R. E. 1936. The Eocene sediments of Mississippi. *Mississippi Geol. Surv. Bull.* 30:1-238.

Murray, G. E. 1961. Geology of the Atlantic and Gulf Coastal province of North America. New York: Harper, 692p.

Parks, W. S. 1971. Tertiary and Quaternay stratigraphy in Henry and northern Carroll Counties, Tennessee. *Jour. Tennessee Acad. Sci.* 46:57-62.

Potter, F. W. 1976. Investigations of angiosperms from the Eocene of southeastern North America: Pollen assemblages from Miller pit, Henry County, Tenn. *Palaeontographica, B* 157:44-96.

Stearns, R. G. 1957. Cretaceous, Paleocene, and lower Eocene geologic history of the northern Mississippi Embayment. *Geol. Soc. Am. Bull.* 68:1077-1100.

Tschudy, R. H. 1973a. *Complexiopollis* pollen lineage in Mississippi Embayment rocks. *U.S. Geol. Surv. Prof. Paper 743C,* 15p.

Tschudy, R. H. 1973b. Stratigraphic distribution of significant Eocene palynomorphs of the Mississippi Embayment. *U.S. Geol. Surv. Prof. Paper 743B,* 24p.

Tschudy, R. H. 1975. Normapolles pollen from the Mississippi Embayment. *U.S. Geol. Surv. Prof. Paper 865,* 42p.

9

THE CHEMISTRY OF FOSSILS: BIOCHEMICAL STRATIGRAPHY OF FOSSIL PLANTS

Jim Brooks
Karl J. Niklas

SUMMARY

Bio-organic geochemistry has important applications in stratigraphic studies of fossil plants, and it is now possible to examine the paths and occurrences of organic carbon compounds in natural *(Biopolymers)* and geological *(Geopolymers)* samples.

In recent years, the scientific developments of separation and identification procedures have provided us with means to examine complex mixtures of organic compounds encountered in rocks. In addition to the products and morphologic remains of a single species of plant or animal *(geological fossils)*, it is now possible to study the biochemical and geochemical cycles of individuals or groups of chemical components *(chemical fossils)*, and the origin, occurrence, history, and distribution of chemical fossils in recent and ancient sediments.

All the numerous organic compounds produced by living organisms are not equally preserved in sediments, and this paper discusses the possible origin, diagenesis, and occurrence of these chemical fossils in sediments, and evaluates the scope and variation of these components and their potential and relative usefulness in biostratigraphy. Using geochemical statistical data, it is now possible to interrelate chemical fossils, to postulate direct relationships with modern or extinct organisms, and also examine the diagenetic paths of chemical, geochemical and microbiological alterations during sedimentation.

INTRODUCTION

The study of stratigraphy and sediments leads to the reconstruction of past geological events and related environments, while the study of fossil remains allows for the reconstruction of related biological history. Recently, with advances in chemical, biochemical, and geochemical information and techniques, together with the use of computers, it has become possible to combine these two approaches in an interdisciplinary study of biostratigraphy. Each area provides information that is complementary, and in particular circumstances one can often provide much more information than the other (Swain, 1969; Swain, Bratt and Kirkwood, 1967; Han et al., 1968; Wehmiller, Hare, and Kujala, 1976; Casagrande and Siefert, 1977). This article will attempt to outline some of the basic principles and mechanisms operative in fossilization, as well as provide some examples where paleobiochemical and stratigraphic data mutually support a consistent interpretation (see Welte, 1965).

Many groups of organisms are found as fossils with their more resistant components preserved (Schopf, 1975); the various organic constituents and their numerous permutations of mixture collectively result in complex organic chemical profiles. Three broad categories of remains (referable in some instances to the degree of chemical homogeneity) can be defined on the basis of size and technique of isolation from associated inorganic matrices:

1. Macroscopic fossil remains, such as leaves, seeds, and fruits, that are large enough to be collected and studied with purely physical methods.
2. Microscopic fossils such as pollen grains, pores, and various algae, that require special methods for extraction (e.g., floatation, ultracentrifugation)
3. Chemical fossils that require specific methods of chemical extraction, separation, and identification (e.g., gas chromatography-mass spectroscopy).

Clearly, the first two categories of size do not preclude the presence of the third, since organic constituents have been isolated from all types of fossil remains. Each of these groups has, however, advantages and limitations when used in biostratigraphy.

The chemistry of fossils deals with the characterization of organic matter at the molecular level as it occurs in sediments and other environments. The current studies on chemical fossils mainly deal with lipids originating from geological sources (*geolipids)* and with geopolymers, whose structures can be interrelated and compared with biologically formed materials (Gelpi et al., 1970). Organic matter in sediments is a potentially valuable source of information about biological systems and terrestrial environments and as such holds much information that is useful in biostratigraphy.

Variations in the chemical structure and properties of a chemical fossil can reflect variations in the nature of its biological origin and environment of formation (Brooks 1977 a,b,c). Various criteria are used to identify the presence of former living systems in sediments:

1. The presence of microorganisms (palynology and paleobotany).
2. The presence of extractable soluble organic compounds with chemical structures and occurring in ratios characteristic of biologically produced chemicals.
3. The presence of unstructured or partly structured insoluble organic matter that can be related to materials of known biological origin. This material is usually the most important quantitative organic constituent of sediments and often accounts for up to more than 98 percent of the total amount.

The relative stability of the insoluble organic matter and many of the soluble organic compounds, together with their relative abundance and widespread occurrences, makes them important parameters in biostratigraphic studies.

Organic carbonaceous materials are present in sedimentary rocks as old as 3.7 x 10^9 years (see Barghoorn, Meinschein, and Schopf, 1965; Brooks and Shaw, 1973; Brooks and Muir, 1977; Nagy and Nagy, 1968); thus "natural products" organic chemistry, coupled with organic geochemistry and paleochemistry, embraces the study of the fate and distrubition of carbon compounds in nature over millions of years. Scientists are often surprised to learn of the large amounts of chemicals trapped in sediments. Of the estimated total of 6.4 x 10^{12} metric tons of organic carbon in or on the earth's crust, only 3 x 10^{12} metric tons exist as contemporary organisms in the biosphere. The organic carbon in sediments and related environments far outweighs that in the biosphere, even if some allowance is made for the graphitic carbon in the earth's surface (Welte, 1974).

It has been known for centuries that organic matter was present in geological specimens, but it was not until 1934, when the German chemist Triebs (1935, 1936) isolated and identified biologically important organic compounds from crude oils, shales, and coals that the chemistry of fossils was first studied on a qualitative and quantitative basis. These compounds, the *porphyrins*, which accounted for only trace amounts of the total organic matter identified in the samples, were the first organic geochemicals to be directly related to known biochemical materials. Porphyrins are large organic molecules with characteristic physical properties that are easily measured, and their geochemical significance is that they are identifiable as degradative products of chlorophylls—the naturally occurring green pigments in all photosynthesizing plants (see Boylan, Alturk; and Eglinton, 1968), as well as various animal products, e.g., cytochromes, phycobilins. Triebs (1935) recognized their significance and suggested that (a) crude oils and organic matter in shales were of biological origin, and (b) the conditions of formation of the oil and thermal history of the shales could not have involved high temperature (often quoted as 200 degrees C). The implication of these suggestions proved far reaching and the basic principles, methods, and conclusions put forward have provided a foundation for organic geochemistry. These findings precipitated the systematic study of organic compounds in sediments; crude oils; coals; lignites; soils; fossil resins; earth waxes; and, more recently, individual and groups of fossils (Hodgson and Peake, 1961; Casagrande and Hodgson, 1974).

The chemistry of fossils concerns comparative biochemistry, studied not only for its relevance to the present day but also to past eras. Paleochemistry involves the examination of fossil specimens in the hope that some information is still present in the form of the structure of the biological polymers (e.g., proteins, polysaccharides) and secondary metabolites. A further approach, evolutionary biochemistry, has had considerable success in comparing the amino-acid sequences of certain key proteins isolated from different living species (Boulter et al., 1970; Eck and Dayhoff, 1969), and recent work (Niklas, 1976a,b; Niklas and Gensel, 1977) has shown chemotaxonomic relationships between some nonvascular and vascular plants.

The term "chemical fossil" or "biological marker" is generally applied to organic substances that show pronounced resistance to chemical change and whose molecular structure gives a strong indication that it could have been created in significant amounts only by biological processes. One can usefully compare them with such hard parts of organisms that ordinarily persist after the soft parts have decayed. For example, hydrocarbons, the compounds consisting only of hydrogen and carbon, are comparatively resistant to chemical and biological attack. Unfortunately, many other biologically important molecules such as proteins, nucleic acids, and polysaccharides contain many chemical bonds that hydrolyze, or cleave, readily; these molecules rapidly decompose after the organism dies and becomes incorporated into a sediment (Figure 9.1). Organic compounds originally synthesized by living organisms and more or less modified have been found in many different rocks. Figure 9.2 shows several of the biopolymers, hydrocarbon and oxygenated compounds, and nitrogenous compounds that have been identified in rocks of different geological age. In studies on fossils it is usually accepted that the more detailed the morphology and structure of the material, the better it is for identification and study. In studies of the chemistry of fossils, simple chemical structures such as alkanes are often used.

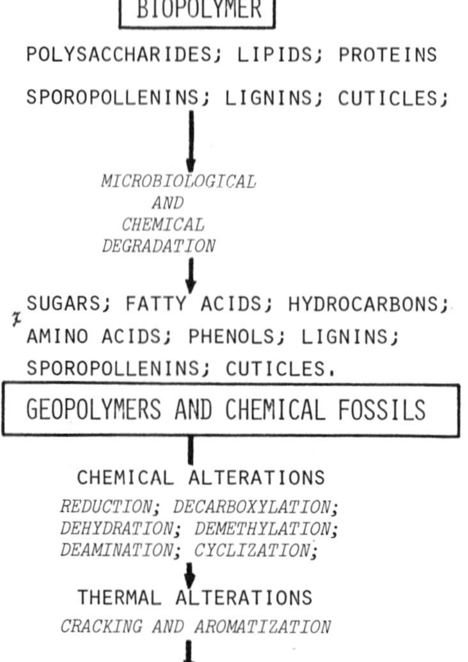

FIGURE 9.1 *Chemical changes in fossils during sedimentation.*

There are several good reasons. Alkanes are generally prominent components of the soluble lipid fraction of sediments. They survive geological time and conditions because the carbon-hydrogen and carbon-carbon bonds are strong and resistant to hydrolysis and most biological agents.

The use of constituents as biological markers must, however, be approached with caution, since numerous compounds may be abiogenically synthesized. Of the commonly used constituents, the normal alkanes, the methyl alkanes, and the isoprenoid alkanes, such as pristane and phytane, have been shown to be synthesized by the Fisher-Tropsch process. The steroids, triterpenoids, and tetraterpenoids, as well as other organic constituents, show a high level of reliability in paleochemotaxonomic studies due to their great specificity in structure. Various features of organic molecular structure favor the use of a particular class of compounds, either because of their resistance to thermal degradation or because their optical activity indicates a biological origin.

CHEMISTRY OF LIVING AND FOSSIL MATERIALS

Organic chemistry deals with plant and animal organisms and was first introduced into chemical terminology as a convenient classification of substances derived from plant or animal sources. Living organisms rely on the element carbon for the make-up of their building materials, as well as on oxygen, and nitrogen, sulphur, phosphorous, and other elements. Some knowledge of organic chemistry is, therefore, required for an appreciation of the aims, methods, results, and applications of chemical studies of fossils and their use in biostratigraphy.

Recently, chemists and biochemists have provided data on the chemical composition and structure of natural, high molecular weight compounds. Studies of their biosynthesis from monomeric precursors, and investigations of the products of biological decompositions, have assisted in understanding the manner of their transformation in natural processes. Geochemists and paleobiochemists, having developed a basic interest in fossil organic substances, are now using these data on the chemical structure of the parent material that took part in the formation of organic substances in sediments.

Proteins play crucial roles in virtually all biological processes and are high molecular weight polymers composed of monomeric units called *amino acids.* Proteins consist of giant molecules of molecular weights ranging from about 12,000 to several million. Fibrous proteins include fibroin (silk), collagen (connective tissue), and keratin (skin, hair, wool, horn, feathers, nails, etc.); globular pro-

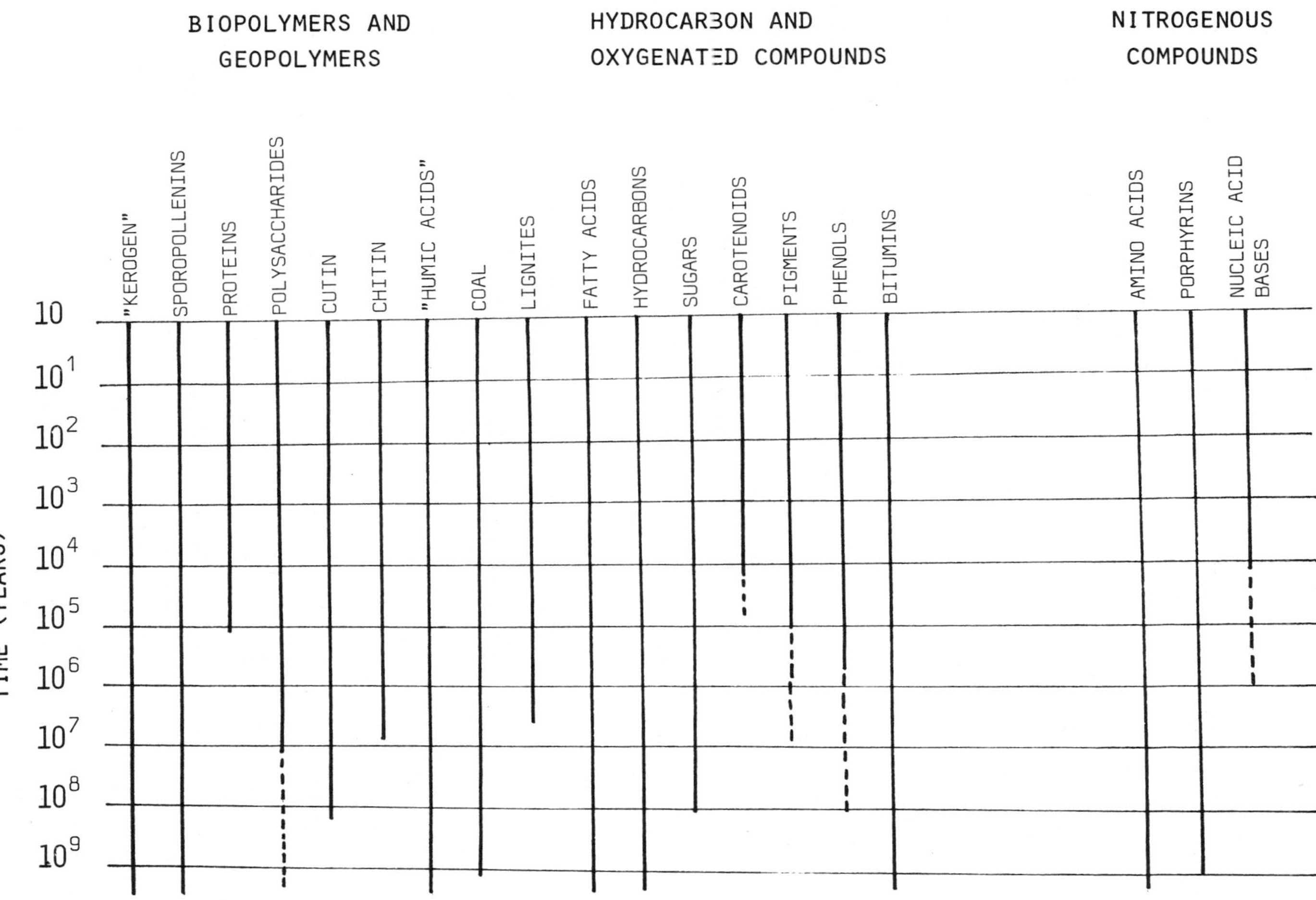

FIGURE 9.2 Presence of chemical fossils in sediments.

teins are more numerous than fibrous materials and consist of such important biochemicals as enzymes, glucoproteins, lipoproteins, and tissue components; plasma proteins occur in blood, serum, and cellular contents. It was thought for a long time that proteins were always completely decomposed by microbiological (enzymes in microorganisms) and chemical hydrolysis in aqueous media. Recently, however, investigations have shown that transformation of proteins in natural processes is very complex and that the role played by these compounds in the formation of fossil organic substances is quite large (see for example, Wehmiller, Hare, and Kujala, 1976).

Carbohydrates, including sugars, are among the most abundant consistuents of plants and animals, in which they serve many useful functions. They are a source of energy; they form supporting tissues of plants and some animals in the same way that proteins are used by the majority of animals. Polysaccharides belong to two general groups: those that are insoluble and form the skeletal parts of plants and of some animals; and those that constitute reserve sources of simple sugars in cells. Both types are high molecular weight polymers, often built up from a single pentose or hexose unit. In this respect they differ from proteins, which are high molecular weight substances containing several different units.

Cellulose is the most widely distributed skeletal polysaccharide. It constitutes approximately half of the cell wall of wood and other plant products. *Hemicellulloses* occur in association with cellulose and are found in large amounts in cereal straws and brans. *Chitin* is a polysaccharide that forms the hard shell of crustaceans and insects, and is composed of glucosamine units. *Starch* is the reserve carbohydrate in the majority of plants, and *glycogen* is the reserve carbohydrate of animals. These high molecular weight carbohydrates and their decomposition products all participate in the complex reactions accompanying the transformation of organic substances in geological processes.

Lignin is a high polymer arising from an enzyme-initiated dehydrogenation of trans-coniferyl 1, trans-sinapyl 2, and trans-p-coumary 1, 3 alcohols. It is highly variable in structure, depending upon the nature of the plant and its ontological development. Lignin occurs in the lowest vascular plants having weak woody tissue (*Equisetum*, ferns), and in wood of coniferous plants, but appears to be absent in lower plants. Lignin plays an important role in the formation of humic substances, peat lignites, and coal, whose chemical composition often depends on the original plant source. Investigations show that lignin undergoes appreciable chemical alterations in the formation of coalified wood and lignites. Aromatic compounds of other types, such as lignans, flavones, and anthocyanins, are widely distributed in nature and participate in the formation of fossilized organic substances.

Cuticle is a term given to the lipophilic layer of material external to and superimposed on the cell walls of epidermal cells. This term takes on a different meaning when applied to fossil material and must be broadly defined in the paleobotanical context, since compression results in the incorporation of many nonepidermal constituents. Even in living material, the term "cuticle" has been found to be imprecise (Martin and Juniper, 1970). Since the inner portions of the outer epidermal cell walls are a cellulose/pectin material, with a reversed cutin/cellulose content gradient, it is perhaps more desirable to use the term *cuticular complex*. In the cuticular complex, cuticular waxes represent the lipophilic fraction most easily extracted by organic solvents. The biopolymer, *cutin*, is the major component of the non-cellular membrane of the aerial parts of land plants. Cutin in most taxa is composed of interesterified and polymerized hydroxy-fatty acids.

Sporopollenins are probably the most resistant organic materials of direct biological origin found in nature and geological samples. Sporopollenins are the chemical components that make up the outer wall of pollen grains, spores, and various microorganisms. Their resistance to microbiological, chemical and physical attack permits study of the morphology and microstructure of pollen, spores, and-microorganisms and is the basis of the science of *palynology*. It has been shown that a component of sporopollenins is formed by oxidative polymerization of carotenoids. These polymeric carotenoids may represent a significant fraction of the high molecular weight polymeric material in sedimentary rocks.

Lipids generally occur as the major components in natural products and are also very widely distributed as geolipids in sediments. The term *lipid fraction* is defined as those chemical components that are soluble in organic solvents; therefore, the term is not synomous with "lipids" since it covers a large number of compounds with very different chemical structures. The lipid fraction is usually subdivided into five major groups on the basis of structure and the initial building units of the molecules. These subdivisions are: acyclic or polyacetate compounds (including n-alkanes, fatty acids, alcohols, and hydroxy-fatty acids); acyclic or isoprenoid compounds (including isoprenoid hydrocarbons, phytane, pristane, phytol, carotenoids, squalene); cyclic compounds (such as terpenoids, compounds from fossil resins, triterpanes like gammacerane, lupane, hopane); aromatic compounds (including derivatives from cyclic terpenes, steroids, lignites and sporopollenins); and pigments.

Pigments that occur in natural products and that have been isolated from geological samples include *porphyrins; chlorins; carotenoids;* and, to a lesser extent, *quinones*. Naturally occurring plant pigments, such as *flavones* and *anthocyanins*, are not normally sought because of their tendency toward rapid alteration in the geological environment (see, however, Niklas and Giannasi, 1977 a,b). The chemical structure of porphyrins and chlorins confers great stability on the molecules and they are found in sediments of all ages. The precursor in plants of these molecules is *chlorophyll*, which itself is unstable in geological conditions. *Carotenoids* are produced in vast amounts in the oceans (1.2 x 10^{17} tons per annum) and also in about the same order of magnitude on land, but their relative instability is probably reflected in the lack of reports of their occurrence in sediments older than 20,000 years (see however, Watts and Maxwell, 1977).

Kerogen is the result, due to the effects of physical and chemical factors attending fossilization, of the gradual alteration of an organism's original biochemistry into usually an amorphous organic residue that is resistant to organic solvent extraction. The chemical composition of kerogen is of great interest since (1) it represents the major part of sedimentary organic matter; (2) there exist many similarities between kerogen, coal, and organic matter derived from sediments and soils (Tissot and Bienner, 1974); (3) the relationships between temperature and pressure may be extrapolated from physiochemical classifications of kerogens; and (4) an understanding of these residues allows for the progressive understanding of fossilization (Durand and Espitalie, 1976) The physiochemical properties of kerogens are roughly time-dependent and, therefore, of potential use in determining small-scale stratigraphic relationships.

It is estimated (McIver, 1967) that whereas all the intercellular carbon in the biosphere amounts to 3 x 10^{17} grams, kerogen is at least 10,000 times more

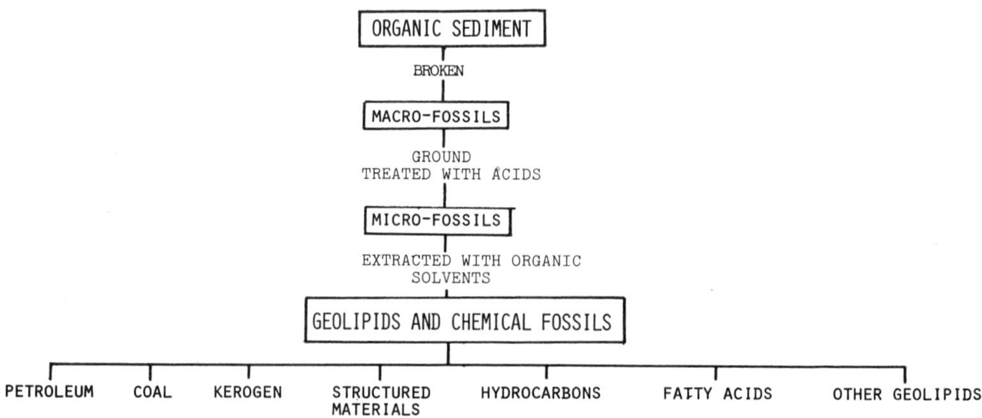

FIGURE 9.3 *General classification of fossils and chemical fossils.*

abundant. A general classification of organic matter in sediments is shown in Figure 9.3. Kerogen is present in all types of sediments and its properties may depend on biological sources; on chemical, physical, and microbiological environments of deposition; and on diagenetic processes (Niklas and Chaloner, 1976). Kerogens can originate from different source materials (e.g., blue-green algae, red algae, green algae, sporopollenins, lipid concentrates) and in a variety of environments (e.g., marine, inland sea, and lake and lacustrine) (Durand and Espitalié, 1976). With regard to type and origin of kerogens, two extreme situations exist, with all possible transactions in between:

i. A sediment deposited in a marine environment can receive solely autochthonous plankton.

ii. A nonmarine sediment can receive only detritus of higher plants.

Type (i) organic matter is predominantly composed of various proteins, carbohydrates, and lipids. Type (ii) organic matter is made up of lignins, cellulose, and scleroproteins. In the original state of types (i) and (ii) there is a difference of hydrogen to carbon ratios. The marine organic matter has a higher H:C ratio and is more paraffinic, whereas the nonmarine organic matter has a lower H:C ratio and is aromatic in nature (Figure 9.4). As shown by Brooks and Shaw (1977), there is also to be expected a change in composition of the organic matter as a function of geological age. Due to biological evolution, very old sediments contain only remnants of primitive organisms. For instance, in Cambrian sediments kerogens containing only sporopollenin-type material are found (Welte, 1974).

Composition and properties of some representative kerogens have been suggested by various worker (Brooks and Shaw, 1969; Burlingame et al., 1969; Djuricic et al., 1971; Schmidt-Collerus and Prien 1974; Cane, 1976; Yen, 1976; and Philip and Calvin, 1976). A useful model for keorgen studies based upon the chemical structure of sporopollenin has been used by Brooks (1970) and Libert (1974). The detailed structure(s) of kerogen is not yet known, but gradually research work is giving knowledge of an average structure that is proving helpful in interpretations of a number of geochemical, stratigraphical, and technological problems.

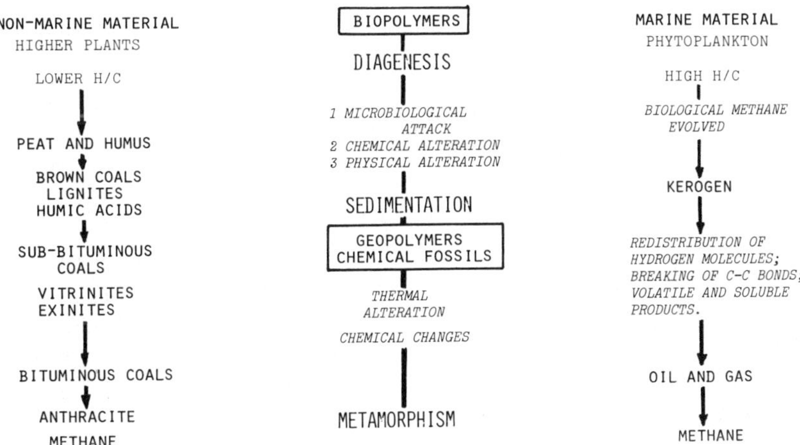

FIGURE 9.4 *Diagenesis of chemical fossils in sediments.*

DIAGENESIS OF ORGANIC MATTER IN SEDIMENTS

Studies show that physical, chemical, and microbiological agents are capable of altering or totally destroying organic matter in unlithified sediments at surface temperatures and pressures. These alterations are generally called *diagenetic processes*. The physical, chemical, and microbiological changes that operate during deposition and within the upper one to two meters of burial are defined as *early diagenesis*. Subsequent alterations of longer duration and often less intensity, that occur during and after lithification, but prior to metamorphism, are defined as *late diagenesis*. In the period of early diagenesis, the least resistant parts of the organic matter are degraded by microbiological and chemical reactions (Brooks, 1977b,c). During late stages of diagenesis, changes continue at a relatively slower rate as additional sediment accumulates and interstitial water is gradually squeezed out of the sediment. After expulsion of most of the interstitial water by compaction, diagenesis may cease.

Chemical and microbiological processes involved in the alteration of organic matter have been studied by many geochemists (see Tissot and Bienner, 1974; Brooks, 1977a) and various definitions and pathways have been given to the processes taking place. It is therefore important that a short general review is given of those processes that directly and indirectly affect the sedimenting organic matter.

The nature of the original biological material, the environment of deposition, microbiological alterations, chemical changes, temperature, pressure, and nature of ground-water solutions all play important roles in the sedimentation, diagenesis, and thermal alteration of organic matter into chemical fossils. The changes in the organic matter involve various important steps (Figure 9.4) and can be summarized:

a. microbiological degradation
b. chemical degradation (mainly hydrolysis)
c. chemical polymerization
d. organic diagenesis (reduction, decarboxylation, dehydration, demethylation, deamination and cyclization)
e. thermal alteration ("cracking" and disproportionation of the carbon-carbon bonds)

f. organic metamorphism (including further "coalification" processes)

The organic materials deposited in sediments consist mainly of naturally occurring biopolymers (polysaccharides, lipids, proteins, sporopollenins, lignins, cuticles, pigments, etc.; Figure 9.2.) When these biopolymers are incorporated into the sedimentary cycle, microorganisms partially or fully degrade them through the use of enzymes; breakdown the biopolymer into smaller units, often down to the monomers (e.g., polysaccharides to sugars; proteins to amino acids); and in turn add polymers indicative of their own biosynthetic capacities. These microbiological and chemical products can be identified in recent and ancient sediments and are often called *biological markers* or *chemical fossils*.

Alternatively, these degradation products can chemically join together (condensation or polymerization reactions) to form geopolymer complex molecules (kerogen, sapropelic matter, humic acids, etc., named depending upon the original source of the organic material). These microbiological and chemical changes appear to take place only in the upper meter or so of the sediment at near-surface temperatures and are considered early diagenetic stages.

Organic diagenesis of the sedimentary organic matter involves a large number of chemical reactions that take place in the upper meters of the sediment at temperatures from surface temperatures to about 50°C. The main chemical reactions taking place are:

a. breaking or cleavage of carbon-carbon bonds (e.g., chlorophyll degrades to give one porphyrin and one phytol molecule)
b. reduction of unsaturated carbon-carbon bonds (e.g. alkenes to saturated alkane molecules)
c. disproportionation reactions (redistribution of the hydrogen atom within the molecule)
d. aromatization reactions (aliphatic and acyclic hydrocarbons are changed into aromatic structures)
e. loss of ammonia, hydrogen sulphide, carbon dioxide, methane, oxygen, nitrogen, and sulphur from the organic molecules (ammonia and hydrogen sulphide are evolved from nitrogen and sulphur-containing molecules respectively)

After these initial diagenetic processes, there are a series of geothermal reactions leading to products of increasing thermal stability. Thermal alteration of organic matter (often called *coalification* or *carbonization)* involves the cracking of large molecules to form small organic compounds. These processes occur mainly in the temperature range 50° to 150°C. The major geochemical reactions taking place during thermal alteration of the organic matter may be considered to be:

a. disproportionation and redistribution of the hydrogen atoms and the break of some carbon-carbon bonds that leads to smaller molecules of increasing volatility and hydrogen content (with methane as the final product)
b. loss of hydrogen giving a carbonaceous residue of decreasing hydrogen content and increasing carbon content (with graphite as the end product)

The last step, organic metamorphism, occurs mainly at temperatures above 150°C; ultimately these processes change the organic matter to methane and carbon.

METHODS OF STUDY

Studies on the chemistry of fossils require specialists from various scientific disciplines who are able to focus their particular techniques on the analysis of geological samples in order to reach conclusions on the nature of the components present in the rock or sediment, the fate of these components since burial, and the original form (if any) of the material in a living system. The complexity of the chemicals extracted from rocks and sediments requires applications of the most modern sophisticated analytical methods to unravel the story and information laid down through the years.

Extraction and Isolation of Organic Matter

Insoluble Organic Matter (Kerogen)

Organic matter can be studied either *in situ* in the sediment, or else it can be freed from the sedimentary matrix. Figure 9.3 shows the procedure typically used to isolate organic matter from sediments.

It is often important to be able to determine textural relationships of the organic matter and its associated mineral matter; in this case, the organic matter must be examined *in situ.* If the organic matter present in the fossil is in a poor state of preservation, its structure may not survive the extraction processes usually employed, and therefore must be examined in the rock matrix. This is especially the case with metamorphosed or very ancient microfossils. Brooks, Muir, and Shaw (1973) examined lightly etched, cut surfaces of Precambrian chert and were unable to observe microfossils that did survive extraction processes. Separation of the organic matter is , however, a convenient concentration method and permits a more representative sample to be obtained.

Transmission electron microscope (TEM), scanning electron microscope (SEM), electron diffraction, and electron probe microanalysis have all been used to study the organic components present in sediments. In the electron microscope, and in particular the SEM, the morphologies of organic-walled microfossils and amorphous residues can be studied and related to the rocks in which they were found. The associated electron-beam analytical techniques of X-ray microanalysis and cathodoluminescence can provide elemental and chemical information from very small samples. The electron microscope also provides insight into the processes by which organic matter gradually increases in rank by following progressive changes in ordering (by electron diffraction in the TEM, and luminescence spectra in the SEM). The electron microscope is used extensively to provide information for stratigraphic dating of sediments and in determining paleoenvironmental parameters, the results of which have been used in exploration for economically important deposits, such as coal, oil shale, petroleum, and mineral deposits, as well as to provide a better understanding of the nature of kerogen (see Muir and Giles, 1977).

Almost all of the organic materials studied in the electron microscope are of plant origin, although some more resistant animal substances, e.g., chitin in trilobite cuticles (Darlingwater, 1973) in chitinozoans (Urban, 1972) and collagen fibrils in graptolites (Towe and Urbanek, 1972), are subjects of active morphological and paleobiological research using electron microscopy and assisting in the studies on the chemistry of fossils.

Separation of insoluble organic materials from inorganic minerals is necessary because the latter tend to interfere with the determination of most chemical and physical properties of the organic matter. The most common and, usually, the most satisfactory method for isolating chemical fossil material and removing inorganic materials is successive chemical treatment of the ground, extracted rock. For many organic-rich samples containing 10 percent organic matter, treatments with HCI (to remove carbonates) and with HF (to remove silica and silicates) are sufficient to give a more or less mineral-free organic residue. When the organic content is low and the pyrite (FeS_2) content is high, subsequent removal of FeS_2 is necessary.

The amount and properties of the organic matter isolated from a sample may vary considerably with the solvent used initially to remove the soluble matter, because all undissolved organic material remains part of the "insoluble" organic matter. For an accurate definition of the term "insoluble," the solvent and conditions of extraction need to be specified.

The organic residue remaining after removal of the inorganic matrix and subsequent extraction is generally a dark-brown amorphous (or partly structured) polymeric material, which can be studied by pyrolysis (Welte, 1965, 1972; Brooks and Shaw, 1969; Gransch and Eisma, 1968; Cummings and Robinson, 1972; Allan and Douglas, 1974), oxidation with chronic acid (Burlingame et al., 1969; Simoneit and Burlingame, 1973), permanganate (Djuricic et al., 1974) or by reduction in an atmosphere of hydrogen in the presence of various catalysts (Dungworth et al., 1971). The study of the insoluble organic components is important since it is very likely to represent indigenous material and not contamination (Brooks and Shaw, 1969). The techniques of extraction and analysis described above are generally applied to the study of chemical fossils, which include kerogens, humic acids, polyfunctional aromatic materials in soils, sporopollenins, cuticles, lignites, coal macerals, and individual fossil fragments.

Extractable Organic Matter

The extraction, separation and isolation procedures for the study of geolipids are those used in conventional analytical chemistry of naturally occurring products and are, in general the analytical methods used in chemistry, biochemistry, and geochemistry. The geological sample (rock, sediment, or fossil) is finely ground into a powder (100-200 mesh size) and extracted in a soxhlet apparatus or ultrasonic extractor with a mixture of benzene-methanol or other suitable organic solvents. Fractionation of the material from evaporation of the extract is carried out using thin-layer, column, or high-pressure liquid chromatography or a combination of these.

Analytical Methods

The use of classical chemical analytical methods is often limited because studies on geolipids and chemical fragments of fossils are frequently carried out on microgram quantities. In recent years analytical techniques and methods have been refined and improved, so that it is now possible to utilize the major analytical techniques to identify individual components in mixtures isolated from geological samples and from degradation products of fossil material. Elemental composition, ultraviolet, infrared and nuclear magnetic resonance spectroscopy, mass spectrometry, (MS), optical rotation, and gas chromatography (GC) are the most important methods routinely employed in the chemical studies on fossils. Mass spectral methods are the most widely used in the study of the chemistry of fossils, since they provide structural information on submicrogram quantities of individual lipids derived from sediments and from fossil degradation products (see Burlingame and Schnoes, 1969). Gas chromatography connected to low-resolution mass spectrometry has been used extensively to identify numerous chemical components derived from various geological samples. High-resolution mass spectrometry, often with computor-controlled data-handling equipment, is used to provide detailed structural information on components isolated from different sediments and strata. The most efficient and accurate method for the separation, identification, and characterization of microquantities of sedimentary organic matter is combined gas chromatograph (high efficiency capillary columns to give good separation of the components) and low-resolution mass spectrometry (to give detailed structural information about the separated components). Methods are being developed so that gas-chromatagraph-high-resolution mass spectrometry can be used in the studies.

Infrared analysis identifies functional groups present in the chemical fossil, but is less useful for hydrocarbons without functional groups. Esters, fatty acids and alcohols, unsaturated and hetero-organic compounds are easily distinguished and identified by infrared spectroscopy. Compounds containing chromophores absorb in the ultraviolet region of the spectrum. Many unsaturated aliphatic, aromatic, and heterocyclic systems readily absorb, showing high sentitivity and possibilities of identifications of trace amounts of chemicals. Nuclear magnetic resonance (^{1}H, ^{13}C especially) determines the atomic environment, enabling the identification of many chemical compounds (both low and high molecular weight materials) from geological environments. Optical rotary dispersion is a technique sometimes used in the studies of organic geochemistry to assist in the identification of certain compounds possessing an asymmetric centre. X-ray crystallography has been used to elucidate the structures and absolute stereochemistries of a number of geological triterpene hydrocarbons and also of geological polymers such as cuticle and sporopollenins.

Measurements of stable carbon isotopes (^{12}C and ^{13}C) can provide information about the cycle of carbon in nature and this method is now used extensively in geological studies to define and correlate environments in which the sedimentary organic matter was formed and deposited (see Degens, 1969).

Photosynthesis is very characteristic of many living systems, including algae, dinoflagellates, mosses, ferns, higher plants, and certain bacteria (e.g., the purple sulphur bacteria, Thiorhodaceae; the purple nonsulphur bacteria, Anthiorhodaceae; and the green sulphur bacteria, Chlorobacteriaceae). Photosynthetic processes not only provide many living systems with the biochemicals necessary for growth and development, but have recently become an important tool in the study of chemical fossils as a parameter to identify chemicals that have been produced by photosynthesis in different environments. Biosynthetic pathways in living systems tend to discriminate against ^{13}C, and selectively absorb $^{12}CO_2$ rather than $^{13}CO_2$. In effect, the plant behaves as an isotope separator concentrating the ^{12}C isotope in the plant. In this way, photosynthesis exerts a major control either directly or indirectly on the distribution of the stable carbon isotopes within plants. Marine, nonmarine, and terrestrial organisms show different stable carbon isotope values (expressed as ^{13}C in o/oo, based upon a standard carbonate, belemnite from the Peedee Formation, South Carolina, USA). Plants extract carbon from two main sources, atmospheric CO_2 and molecular carbonate and bicarbonate from aqueous environments. The differences in isotope composition of the organisms will be influenced largely by the ^{13}C value of the carbon source used for photosynthesis. The differences in ^{13}C in plants can generally be attributed to the fact that during photosynthesis marine organisms utilize carbonate and bicarbonate from oceans. The nonmarine organisms use carbonate and bicarbonate from fresh water, whereas land plants use the isotopically lighter carbon dioxide in the atmosphere; and this is mirrored in the different values for the chemical components present in modern and fossil plants from different environments.

PALEOBIOCHEMISTRY, PALEOCHEMOTAXONOMY, AND BIOSTRATIGRAPHIC APPLICATION

The conclusion that must be drawn from geochemical analyses of various organic debris is that diagenetic mechanisms are a multivariate function of fossil type (original biochemistry and morphology) and geologic environment (state or preservation, thermal history, etc.). Chemical technology is just now beginning to unravel various aspects that define fossilization. This section will attempt to outline the apparent relationships between the degree of diagenesis and various physical parameters (e.g., temperature, pressure) and specific examples where organic chemical data have been useful in paleontological, geological, or stratigraphic problems. Two extreme catagories of geochemical study may be defined: (1) the determination of chemical vertical profiles through sequences of fossil sediments and (2) the isolation of putatively taxonomic criteria referable to specific fossil taxa. These two approaches have as their respective objectives (1) the derivation of mechanisms leading to fossilization, the generation of fossil fuels, and/or geochemical chronologies, and (2) the chemotaxonomic definition of generic or suprageneric fossil groups.

As described in pervious sections, the techniques of optical examination (light, SEM, TEM), elemental analyses, infrared spectroscopy, and GC-MS have been used to characterize various fossil residues. Each method generates its own body of data, and has its own advantages and limitations. Ideally, all of the abovementioned techniques ought to be used wherever possible.

Color changes have been used to determine time-temperature histories of strata from which organic residues have been recovered; perhaps the most subjective is a thermal-alteration index based on the coloration of spores, pollen, and algal remains (Harper, 1971; Robbins, 1977). Color change apparently results from the progressive loss of nitrogen, sulphur, and oxygen, and concomitant enrichment of carbon leading to increased opaqueness. Data suggest, however, that biflagellate marine algae (dinoflagellates) respond to heat unevenly, whereas they also take longer to respond than pollen, Similarly, many aquatic plants (e.g., *Botryococcus)* commonly associated with sedimentary structures, show a color and opaqueness variability when living. Color alteration of visible organic debris as a function of burial can, however, provide a useful indication of the probable maximum burial temperatures and have a high correlation with the potential for hydrocarbon generation (Staplin, 1969; Burgess, 1974). The application of color change in conjunction with vitrinite reflectance is well documented (see, for example, Teichmüller and Teichmüller, 1968).

A related approach to thermal-color indices is the use of infrared spectra. Figure 9.5 shows the infrared spectra taken from a diverse collection of fossil plant "cuticles" and some extant taxa. A characteristic absorption spectrum composed of invariate peaks may be seen, while variate peaks can be used at times to characterize vascular and presumed vascular plants (Figure 9.5 A-N) from algal or non-vascular plants (Figure 9.5 O-V). Thus, to a limited extent, both previous thermal histories as well as broad taxonomic categories may be drawn from infrared spectra. Figure 9.6 shows the absorption spectra from levoglucosan, of the tar from cotton cellulose, the residue from cotton (after 16% loss by volatilization), oxidized cellulose, and its residue after 20% loss by volatilization. There is a close

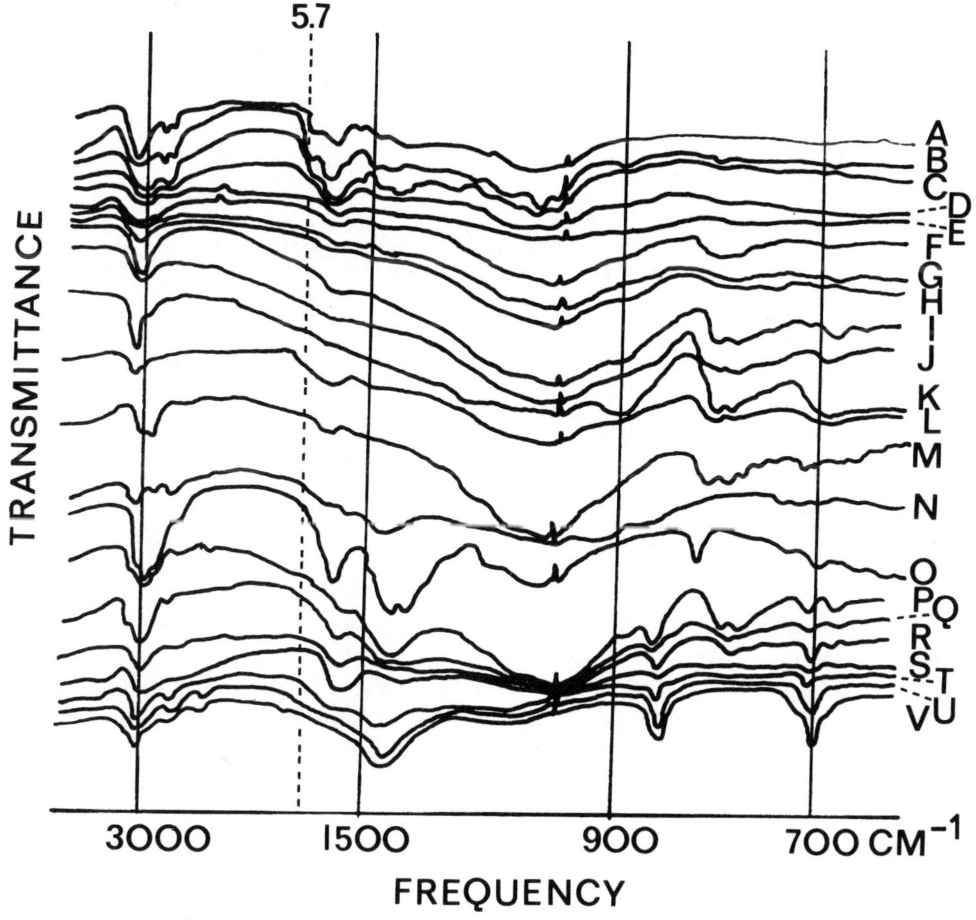

FIGURE 9.5 *Infrared spectra of selected vascular (A-N) and nonvascular plant fossils (O-V) (the transmittance of K_2Br disks are plotted against the frequency of light, CM-1; the lack of a pronounced absorbance at 5.7 microns is shown along the vertical dotted line; compare with Figure 9.6). A, Eocene angiosperm leaf cuticle; B, extant* (Acer) *Angiosperm leaf cuticle; C,* Psilophyton *cf.* forbesii; *D,* P. princeps; *E.* Sawdonia; *F,* Crenaticaulis; *G,* Gosslingia; *H,* Eohostimella; *I,* Archaeopteris; *J,* Tetraxylopteris; *K,* Oocampsa; *L,* Chaluria; *M,* Pertica; *N,* Renalia; *O,* Protosalvinia arnoldii; *P,* Orestovia; *Q,* Parka; *R,* Pachytheca; *S,* Spongiophyton; *T,* Prototaxites; *U,* Botryococcus; *V,* Nematothallus.

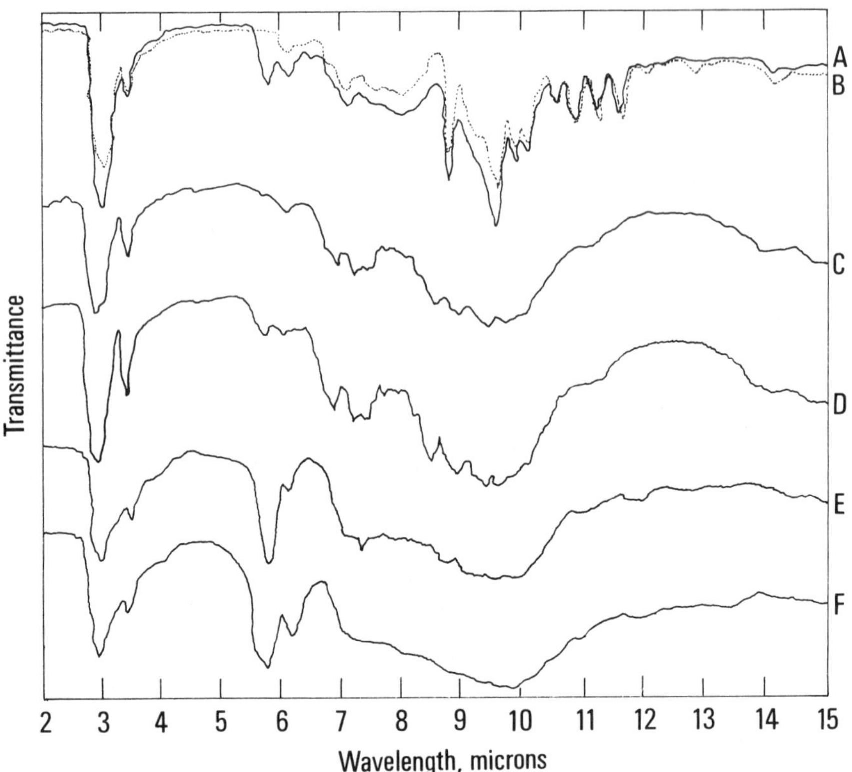

FIGURE 9.6 *Infrared spectra: A, tar from cotton cellulose; B, levoglucosan; C, cotton cellullose; D, residue from cotton cellullose (16% loss by volatilization); E, oxidized cellullose; and F, its residue after 20T loss by volatilization. B to D lack a strong absorption at 5.7 microns, while tars, residues, and oxidized cellulose show a similar spectrum to that characteristic of fossil plant residues.*

resemblance between these spectra and those of fossil plants. However, the spectra of various fossil debris (Figure 9.5) and of cellulose residues differ from that of cotton cellulose in that the former lack an absorption band at approximately 5.7 microns. This band may be attributed to one of the various types of carbonyl groups that absorb in this region. These data are not interpreted to suggest that "cuticles" of fossil plants are predominantly cellulosic residues. Numerous experiments have shown that cellulose is degraded more rapidly than other cell-wall constituents, particularly lignin and cutinic acids (Flaig, 1968, pp. 202-203). However, tars derived from oxidized cellulose, upon volatilization, yield chemical moieties identified from numerous coal tars, e.g., humic acids, melanoidins. Chemical extractions of cotton cellulose tars indicate that various phenolic constituents within the cell wall contribute to the characteristic infrared absorptions seen and are similar to various phenolic lignin-like residues isolated from fossil plants.

Elemental gravimetric analyses play an important and useful role in geochemical studies and have many economic applications, particularly as regards the formation of coal. Dormans et al. (1957) defined coalification tracks for different coal macerals based on

H/C:O/C atomic ratios, and the relative separation seen in these coalification "trajectories" supports the concept that various plant parts (e.g., spores, wood) generate different maceral types (e.g., exinite, vitrinite). The degree, or rank, of coalification is measured commonly by chemical parameters such as carbon content. Fossil plants can individually be coalified during compression and elemental analyses have been used to define broad taxonmic subgroupings (see Niklas and Chaloner, 1976, Fig. 3). Similarly, various diagenetic mechanisms such as dehydration, decarboxylation, and demethanation, can be quantitatively related. For example, analyses of the Precambrian architach *Chuaria* and the Devonian thalloid plants *Orestovia* and *Parka* indicate that dehydration sequences were predominant in their diagenesis, while decarboxylation and demethanation have defined the "coalification" of the Upper Devonian alga, *Protosalvinia* (see Niklas, 1976c).

Organic chemical extractions of specific plant taxa have yielded a large body of data on possible biochemical relationships. By analogy to the chemical taxonomy of extant species, ths new area of research has been called *paleochemotaxonomy*. The term is not meant to suggest a congruence between living and fossil plants, but rather a program of analyses attempting to determine chemical similarities and differences between fossil plant groups (see Niklas and Gensel, 1977). The use of chemical compositions to elucidate taxonomic affinity has value in cases where the structure of a fossil leaves its relationship to living or other fossil forms in doubt. Chemical criteria in defining taxonomic relationships, however, must be used with caution. Studies indicate that dependence upon a simple category of compounds for such a purpose is unreliable since various geologic factors cause the degradation, removal, or transformation of organic constituents. For example, Hohn and Meinschein (1977) have shown that no consistent interspecific differences could be detected in the chromatograms of nonaromatic hydrocarbon extracts made from two *Nyssa* isolated (Early Tertiary Brandon Lignite) from a lake. Similar studies by one of the authors (KJN) indicated that rapid transformation mechanisms effect major alterations in the fatty acid/n-paraffin profiles of even recent sediments. The relative abundances, isolated from the alga *Botryococcus*, of these two organic fractions with respect to the depth of burial in a lake bottom (San Cristobal, Galapagos), indicate a rapid disappearance of carbon-number preference, i.e., the relative distribution of normal fatty acid and of normal paraffins may be described by a carbon-preference index (CPI), which is defined as the mean of two ratios that are determined by dividing the sum of the even-carbon-numbered acids over a given range ($C_{24}C_{36}$). Large CPI values can be expected in most modern sediments due to the ubiquity of even-carbon-numbered acids and odd-carbon-numbered paraffins in biological systems. The rapid alterations seen in the Galapagos sediments may be the result of step-by-step decarboxylation of fatty acids forming alkyl radical intermediates that react to give normal paraffins with one less carbon atom. Kvenvolden (1966) has shown similar relationships between CPI values and age in some Lower Cretaceous sediments.

Statistical analyses of the total organic chemical profiles derived from fossil remains can be used to determine the various relationships between physical and chemical factors. Figure 9.7 shows a principal-component analysis of selected vascular and non-vascular plants where the data are weighted to emphasize the relative geologic age of each taxon (Table 9.1). Each plant fossil is represented by a

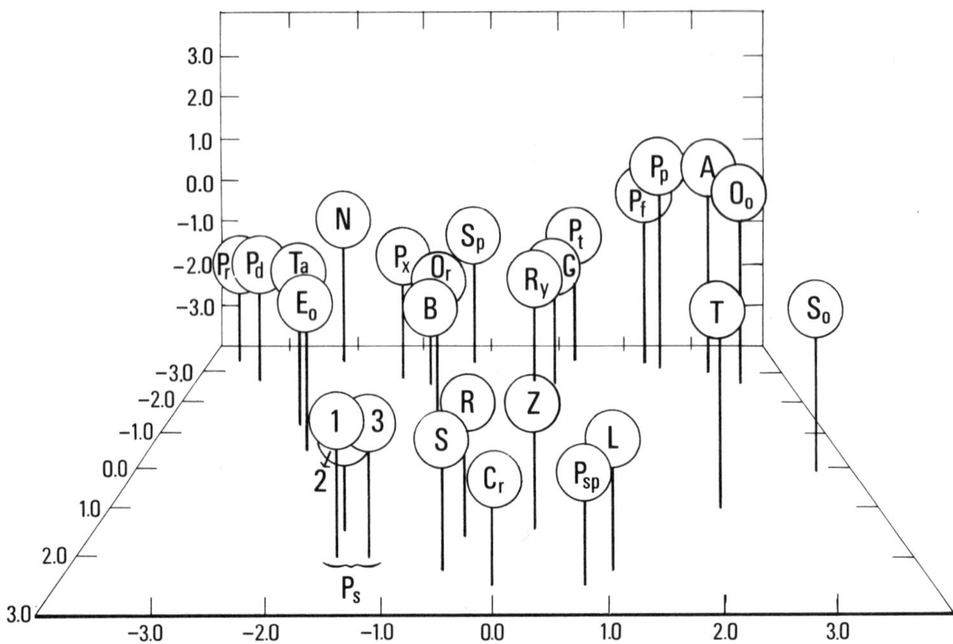

FIGURE 9.7 ***Principal component analysis of selected plant fossils showing the statistical relationship between respective geologic age and chemical composition.*** ***Pr,*** Parka; ***Pd,*** Pachytheca; ***Ta,*** Taeniocrada; ***Eo*** Eohostimella; *N,* Nematothallus; ***Ps,*** Protosalvinia arnoldii *(1),* ravenna *(2), and* furcata *(3);* ***Px,*** Prototaxites; *B,* Botryococcus; *Or,* Orestovia; *Sp,* Spongiophyton; *S.* Sawdonia; *R,* Renalia; *Cr,* Crenaticaulis; *Ry,* Rhynia; *G,* Gosslingia; ***Pt,*** Pertica; *Z,* Zosterophyllum, *Psp,* Pseudosporochnus; *L,* Leclerquia; *Pf,* Psilophyton *cf.* forbesii; *Pp,* P. princeps; *A,* Archaeopteris; *Oo,* Oocampsa; *T,* Tetraxylopteris; *So,* Solenites; ***see Table 9.1 for taxonomic descriptions, ages, and life forms.***

sphere positioned in a cartesian system, the axes of which are dimensionless numbers representing coefficients of similarity. The fossils that are chemically similar and have roughly the same age have similar ordinations, while those plant remains having divergent chemical compositions and/or ages are separated to varying degrees. This three-dimensional plot is a very crude attempt to assess the age relationships of fossil compositions, and there exists considerable "noise" in this plot. Various taxa known to be related are separated, while some plant fossils having presumed divergent taxonomic affinities show similar ordinations. Since analyses of these taxa have shown suprageneric taxonomic relationships consistent with major plant groupings (see Niklas and Gensel, 1977), the repositioning of these taxa due to weighting of geologic age has resulted in a "chemical stratigraphy." The utility of such analyses in stratigraphic studies, however, is seriously limited, since chemical compositions of fossil plants are defined more on a basis of heat and pressure gradients than on absolute age. In many respects, Figure 9.7 is only a more subtle representation of Figure 9.2, where the relative geologic lifespans of various biological markers are given. Statistical analyses, however, of

TABLE 9.1 *Plant taxa used in statistical analysis of age (geologic) and chemical interrelationships (cf. Figure 9.7).*

Genus	Symbol	Age	Affinity	Life Form	References to Chemical Composition
Botryococcus	B	Mississippian	Chlorophyta	colonial	Niklas and Chaloner (1976
Parka	Pr	Upper Silurian-Lower Devonian	Alga	thalloid	" "
Pachytheca	Pd	Upper Silurian-Lower Devonian	Alga	filamentous spheres	Niklas (1976a)
Nemathothallus	N	Upper Silurian	Alga ?	cuticular sheets of cells	"
Prototaxites	Px	Upper Silurian-Lower Devonian	Alga ?	intertwining tubes	Niklas (1976b)
Spongiophyton	Sp	Middle Devonian	Alga	thalloid	"
Orestovia	Or	Lower Devonian	Alga ?	dichotomous branches	"
Protosalvinia	Ps	Upper Devonian	Alga	thalloid	Niklas (1976c)
P. arnoldii	1	"	"	"	"
P. ravenna	2	"	"	"	"
P. furcata	3	"	"	"	"
Eohostimella	E	Lower Silurian	Land plant (? vascular)	plant axis, hollow cylinders	Niklas and Gensel (1977)
Taeniocrada	Ta	Lower Middle Devonian	Land plant (? vascular)	plant axis	" "
Rhynia	Ry	Lower Devonian	Rhyniophytina	" "	" "
Cooksonia	Co	Middle-upper Ludlovian	Rhyniophytina	" "	" "
Pertica	Pt	Late Lower Devonian	Trimerophytina	" "	" "
Sawdonia	S	Middle Devonian	Zosterophyllophytina	" "	" "
Renalia	R	Lower Devonian	Rhyniophyte-zosterophyllophyte intermediate	" "	" "
Crenaticaulis	Cr	Lower Devonian	Zosterophyllophytina	" "	" "
Zosterophyllum	Z	Lower Devonian	Zosterophyllophytina	" "	" "
Gosslingia	G	Gedinnian-Siegenian	Zosterophyllophytina	" "	" "
Leclerquia	L	Middle Givetian	Lycophyta	" "	" "
Pesudosporochnus	Psp	Lower Givetian	Cladoxylopsida	" "	" "
Psilophyton principes	Pp	Lower Devonian	Trimerophytina	plant axis	Niklas and Gensel (1977)
P. cf. *forbesii*	Pf	Lower Devonian	Trimerophytina	" "	" "
Chaleuria	C	Lower or Middle Devonian	Progymnosperm	" "	" "
Tetraxylopteris	T	Upper Givetian	Progymnosperm (Aneurophytales)	" "	" "
Oocampsa	Oo	Middle Devonian	? Trimerophyte-progymnosperm intermediate	" "	" "
Archaeopteris	A	Lower Frasnian	Progymnosperm (Archaeopteridales)	" "	" "
Solenites	So	Jurassic	Gymnosperm (Czekanowskiales)	leaf segments	Niklas and Chaloner (1976)

well-defined and relatively brief periods of geologic time may provide a useful adjunctive approach to stratigraphic studies.

CONCLUSIONS

This chapter has attempted to review the general methods by which organic chemical constituents can be extracted, separated, and identified from sediments as well as to make reference to the potential use of biochemical markers (chemical fossils) in stratigraphic studies. The potential utility of paleobiochemistry is limited, due to the multivariate effects of physical and chemical parameters (diagenesis) on the organic chemical profiles of fossil debris and related rock strata. However, when descriptions of the thermal history and porosity of strata yielding fossils are given, the biological markers may provide a useful body of data, particularly if the stratigraphic interval in question is relatively small.

Chemical changes in fossils during sedimentation (Figure 9.1) are shown to produce a rapid loss of many organic constituents of plant and animal taxa, resulting in the differential preservation of various biological markers (Figure 9.2). Quantitative and qualitative changes in sediment chemistry are primarily effected through diagenesis with burial (Figures 9.3-9.7), and are known to be the result of geothermal gradients. Several aspects of molecular structure contributing to the thermal stability of organic compounds are: (1) linearity of chains involving a paraffinic structure, (2) the introduction of double bonds, (3) the presence of benzene rings, (4) relatively high molecular weights, (5) cross-linking, and (6) the absence of oxygen in the molecular backbone. While these aspects of structure are useful guides for determining geothermal stability, exceptions are prevalent, e.g., paraffinic structures are weakened by tertiary or quaternary carbons in the chain, while double bonds introduce weakness in bonds that are in the β position to them.

Chemical alterations of the original biochemistry of plants results in the formation of residues resistant to organic solvent extraction techniques (kerogen). Kerogen is present in all types of sediments and its properties may depend on biological sources as well as chemical, physical, and microbial environments of dsposition. The structure of kerogen must be determined for each type of source material (algal or vascular plant derivation). In general, kerogen seems to consist structurally of polymethylene chains with some limited degree of cross-linking, a small amount of aromatic material, and unbranched material peripheral to the molecular "nucleus." In the case of algal derived kerogen, the kerogen nucleus appears to be derived from the condensation products of unsaturated oxygenated compounds that may or may not include the chlorophyll moiety. Kerogens associated with vascular plants, producing spores or pollen, contain chemical structures similar to sporopollenin.

Statistical or mathematical analyses of plant kerogens indicate that their chemical compositions may offer a basis for assessing taxonomic affinities. By analogy with the use of chemical criteria in the classification of extant plants, this may be referred to as "paleochemotaxonomy." Discrepencies in clustering analyses, (Figure 9.7) where the chemical data are weighted for the geologic age of each fossil, may be the result of methodology, as well as of the chemical convergence of diverse taxa due to diagenesis. Cluster techniques yield age associations that are superimposed upon chemical relationships that may not reflect *bona fide* taxonomic affinities. The date presented do, however, indicate that chemical profiles may be useful in determining stratigraphic relationships over short intervals.

ACKNOWLEDGEMENTS

The authors with to thank Mr. Ralph Rocklin (photography), Ms. Adrianne Fusco (for Figure 9.7), Mr. Larry Bilden (computer printouts), and Ms. Bernice Winkler (manuscript preparation). This study was supported in part by an N.S.F. grant (DEB 76-82573) to the junior author.

REFERENCES

Allen, J., and A. G. Douglas, 1974. Alkanes from the Pyrolytic Degradation of Bituminous Vitrinites and Sporinites. In *Advances in Organic Geochemistry, 1973*, B. Tissot and F. Bienner, eds. Paris: Editions Technip pp 203-206.

Barghoorn, E. S., W. G. Meinschein, and J. W. Schopf 1965. Paleobiology of a Precambrian Shale. *Science* 148:461-472.

van der Berg, M.L.J.; J. W. Leew; and P. A. Schenck 1974. Constitutional analysis of kerogens and lignite by ozone. In *Advances in Organic Geochemistry 1973*, B. Tissot and F. Bienner, eds. Paris: Editions Technip, pp. 163-178.

Boulter, D.; M. V. Laycock; J. Ramshaw; and E. W. Thompson 1970. Amino acid sequence studies of plant cytochrome c, with particular reference to Mung Bean cytochrome c. In *Phytochemical Phylogeny;* J. B. Harborne, ed. London and New York: Academic Press, pp. 179-186.

Boylan, D. B.; Y. I. Alturki; and G. Eglinton, 1968. Application of gas chromatography and mass spectrometry to porphyrin microanalysis. In *Advances in Organic Geochemistry 1968*, P.A. Schenck and I. Havenaar, eds. Oxford: Pergamon Press, pp. 227-240.

Brooks, J. 1970. Chemical Constituents of Various Plant Spore Walls, Ph.D. thesis, University of Bradford, April 1970.

Brooks J. 1977a. Organic Matter through the Precambrian and its bearing on the History of the Atmosphere and Hydrosphere. Proceedings "The Geological Evidence for the Evolution of the Atmosphere and the Hydrosphere." *Proc. Imperial College London*, 1979.

Brooks, J. 1977b. Diagenesis of Organic Matter: some microbiological, chemical and geochemical studies on sedimentary organic matter. *3rd Internat. Symp. Environment. Biogenochem., Proc.* Wolfenbuttle, West Germany, 1977, pp. 287–308.

Brooks, J. 1977c. Diagenesis: A critical survey of palynological and geochemical studies. *Proceedings of ICP Congress on Stratigraphic Palynology and Palynological Applications to Fossil Fuel Exploration*, Lucknow, India, 1977 (in press).

Brooks, J., and M. D. Muir 1977. Precambrian chemistry and the origin of life. *Proc. Sym. Acritarchs, 4th Internat. Palynological Conf.* Lucknow, India, 1977 (in press).

Brooks, J.; M. D. Muir; and G. Shaw, 1973. Chemistry and morphology of Precambrian microorganisms. *Nature* 244:15-18.

Brooks, J., and G. Shaw. 1969. Identity of sporopollenin with older kerogen and new evidence for the possible biological source of chemicals in sedimentary rocks. *Nature* 220-678-680.

Brooks, J. and G. Shaw 1972. The geochemistry of sporopollenin. *Chemical Geology* 10:69-75.

Brooks, J., and G. Shaw, 1973. *Origin and Development of Living Systems.* London and New York: Academic Press.

Brooks, J., and G. Shaw 1977. Ozone reactions with saturated aliphatic compounds. *3rd Internat. Conf. Ozone, Proc.*, Paris, 1977 (in press).

Burgess, J. D. 1974. Microscopic examination of kerogen (dispersed organic matter) in petroleum exploration. In "Carbonaceous Materials as Indicators of Metamorphism," R. E. Dutcher, ed. *Geol. Soc. Am. Spec. Paper 153*, pp. 19-30.

Burlingame, A. L., and H. K. Schnoes. 1969. Mass spectrometry in organic geochemistry. In *Organic Geochemistry*, G. Eglinton and M. T. J. Murphy, eds. Berlin: Springer-Verglag, pp. 89-160.

Burlingame, A. L.; P. A. Haug; H. K. Schnoes; and B. R. Simoneit. 1969. Fatty acids derived from the Green River Formation Oil Shale by extraction and oxidation-a review. In *Advances in Organic Geochemistry 1968* P.A. Schenck and I. Havenaar, eds. Oxford: Pergamon PRess, pp. 68-71.

Cane, R. F. 1976. The origin and formation of oil shale. In *Oil Shale*, T. F. Yen and G. V. Chilingarian, eds. Amsterdam: Elsevier Scientific, pp. 27-60.

Casagrande, D. J., and G. W. Hodgson 1974. Generation of homologous porphyrins under simulated geochemical conditions. *Geochim. Cosmochim. Act.* 38:1745-1758.

Casagrande, D., and K. Siefert 1977. Origins of sulfur in coal: Importance of the ester sulfate content of peat. *Science* 195:675-676.

Cummins, J. J., and W. E. Robinson 1972. Thermal degradation of Green River kerogen at 150° to 350°C. *U.S. Bur. Mines Rep. Invest., 7620.*

Darlingwater, J. 1973. Trilobite cuticle microstructure and composition. Palaeontology 16:827-838.

Degens, E. T. 1969. Biogeochemistry of stable carbon isotopes. In *Organic Geochemistry*, G. Eglinton and M. T. J. Murphy eds. Berlin: Springer-Verlag, pp, 304-329.

Djuricic, M. V.; R. C. Murphy; D. Vitoroviv; and K. Bieman. 1971. Organic acids obtained by alkaline permanganate oxidation of kerogen fromthe Green River Shale. *Geocheim, Cosmochim. Acta* 35:1201-1207.

Djuricic, M. V.; D. Vitorovic; and B. Ilic 1974. New structural information obtained by stepwise oxidation of kerogen from the Aleksinac (Yugoslavia) shale. In *Advances in Organic Geochemistry 1973*, B. Tissot and F. Bienner eds. Paris: Editions Technip, pp. 177-189.

Dormans, H. N. M.; F. J. Huntijens; and D. W. van Krevelen 1957. Chemical structure and properties of coal: 20. Composition of the individual macerals (vitrinites, fusinites, micrinites and exinites). *Fuel* (London) 36:321.

Dungworth, G.; A. McCormick; T. G. Powell; A.G. Douglas. 1971 Lipid components in fresh and fossil pollen and spores. In *Symposium on Sporopollenin*, J. Brooks et al eds, New York: Academic Press, pp. 512-544.

Durand, B., and J. Espitalié. 1976. Geochemical studies on the organic matter from the Donala Basin (Cameroon) — II. Evolution of kerogen. *Geochim. Cosmochim. Acta* 40:801-808.

Eck, R. V., and M. O. Dayhoff. 1969. Paleobiology. In *Organic Geochemistry*, G. Eglinton and M. J. Murphy eds., London: Longman Group pp. 196-212.

Flaig, W. 1968. Biochemical factors in coal formation. In *Coal and Coal Bearing Strata*, D. G. Murchison and T. S. Westoll eds. London: Oliver and Boyd, pp.. 197-232.

Gelpi, E.; H. Schneider; J. Mann; and J. Oro 1970. Hydrocarbons of geochemical significance in microscopic algae. *Geochim. Cosmochim. Acta.* 34:603-612.

Gransch, T. A., and E. Eisma 1968. Characterization of insoluble organic matter of sediments pyrolysis. In *Advances in Organic Geochemistry 1966.* G.D. Hobson and G.C. Speers, eds., London; Pergamon Press, pp. 407-426.

Han, J.; E. D. McCarthy; W. Van Hoeven; and M. Calvin 1968. Organic

Geochemical Studies, II. A preliminary report on the distribution of aliphatic hydrocarbons in algae, in bacteria and in a recent lake sediment. *Proc. Natl. Acad. Sci.* 59:29-33.

Harper, M. L. 1971. Approximate geothermal gradients in the North Sea basin. *Nature* 230(S291):235-236.

Hodgson, G. W., and E. Peake. 1961. Metal chlorin complexes in recent sediments as initial precursors to petroleum porphyrin pigments. *Nature* 191:766-767.

Hohn, M. E., and W. G. Meinschein. 1977. Fatty acids in fossil fruits. *Geochim. Cosmochim. Acta* 41: 189-193.

Kvenvolden, K. 1966. Molecular distributions of normal fatty acids and paraffins in Lower Cretaceous sediments. *Nature* 209:573-577.

Libert, P. 1974. Sur quelques proprietes physico chemiques du kerogen et la contribution possible de la sporopollenine a sa genese. Ph.D. thesis, University of Bordeaux, France, 1974.

Martin, J. T., and B. E. Juniper. 1970. *The Cuticles of Plants*. Leeds: Edward Arnold, 347p.

McIver, R. D. 1967. Composition of kerogen-clue to its role in the origin of petroleum. *7th World Petrol. Congr.* (Mexico 1967). London: Elsevier, pp. 25-36.

Muir, M.D., and P.L. Giles. 1977. Electron microscopy in studies of organic matter. In *Coalification and Carbonization of Sedimentary Organic Matter*, J. Brooks, ed. New York: Academic Press.

Nagy, B., and L. A. Nagy. 1968. Investigations of the Early Precambrian Onverwacht sedimentary rocks in South Africa. In *Advances in Organic Geochemistry* P.A. Schenck and I. Havenaar eds. Oxford: Pergamon Press, pp. 209-216.

Niklas, K. J. 1976a. Chemical examinations of some non-vascular plants. *Brittonia* 28:113-137.

Niklas, K.J. 1976b. Chemotaxonomy of *Prototaxites* and evidence for possible terrestrial adaptation. *Rev. Palaeobot. Palynol.* 22:1-17.

Niklas, K. J. 1976c. Organic chemistry of *Protosalvinia (=Foerstia)* from the Chattanooga and New Albany Shales. *Rev. Palaeobot. Palynol.* 22:265-279.

Niklas, K. J., and W. B. Chaloner. 1976. Chemotaxonomy of some problematic Palaeozoic plant fossils. *Rev. Palaeobot. Palynol.* 22:81-104.

Niklas, K. J., and P. G. Gensel. 1977. Chemotaxonomy of some Paleozoic vascular plants. Part II: Chemical characterization of major plant groups. *Brittonia* 29:100-111.

Niklas, K.J., and D.E. Giannasi. 1977a. Flavonoids and other chemical constituents of fossil Miocene *Zelkova* (Ulmaceae). *Science* 196:877-878.

Niklas, K. J., and D. E. Giannasi. 1977b. Geochemistry and thermolysis of flavonoids. *Science* 197:767-769.

Niklas, K. J., and E. K. Schofield. (In press). Vertical chemical analyses of *Botryococcus*-rich sediments from El Junco, San Cristobal, Galapagos.

Philip, R. P., and M. Clavin. 1976. Possible orgin for insoluble organic (kerogen) debris in sediments from insoluble cell-wall materials of algae and bacteria. *Nature* 262:134-136..

Robbins, E. I. 1977. Geothermal Gradients. In *Geochemical Studies on the COST No. B-2 Well, U.S. Mid-Atlantic Outer Continental Shelf Area*, P. A. Scholle, *Geological Survey Circular 750*. pp. 44-45.

Schmidt-Collerus, J. J., and C. H. Prien 1974. Investigations of the hydrocarbon structure of kerogen from oil shale of the Green River Formation. *Am. Chem. Soc., Div. Fuel Chem. Prepr.* 19:100-108.

Schopf, J. M. 1975. Modes of fossil preservation. *Rev. Palaeobot. Palynol.* 20:27-53.

Simoneit, B. R., and A. L. Burlingame. 1973. Carboxylic acids derived from Tasmanite (Tasmania) by extractions and kerogen oxidations. *Geochim. Cosmochim. Acta* 37:595-610.

Staplin, F. L. 1969. Sedimentary organic matter, organic metamorphism, and coal and gas occurrence. *Bull. Canadian Petroleum Geol.* 17(1):47-66.

Swain, F. M. 1969. Fossil carbohydrates. In *Organic Geochemistry*, G. Eglinton and M. T. J. Murphy, eds. London: Longman Group., pp. 374-400.

Swain, F. M.; J. M. Bratt; and S. Kirkwood. 1967. Carbohydrate components of some paleozoic plant fossils. *Jour. Paleontology* 41:1549-1554.

Teichmüller, M., and R. Teichmüller, 1968. Geological aspects of coal metamorphism. In *Coal and Coal Bearing Strata*, D. Murchison and T. S. Westoll eds. New York: Elsevier, pp. 233-268.

Tissot, B., and F. Bienner. 1974b. Ketones derived from the oxidative degradation of Green River Formation oil shale kerogen. In *Advances in Organic Geochemistry 1973*, B. Tissot and F. Bienner eds. Paris: Editions Technip, pp. 191-201.

Towe, K. M., and Urbanek. A. 1972. Collagen-like structures in graptolite periderm. *Nature* 237:443-445.

Triebs, A. 1935. Chlorophyll-and Haminderivate in bituminosen Gestreinen, Erdolen, Erdwachsen und Asphalten. *Ann. Chem.* (Warsaw) 510:42.

Triebs, A. 1936. Chlorophyll-und Haminderivate in organischen mineralstoffen. *Angew. Chem.* 49:682.

Urban, J. B. 1972. A re-examination of chitinozoa from the Cedar Valley Formation of Iowa with observations on their morphology and distribution. *Bull. Am. Paleontology* 63:5-13.

Watts, C. D., and J. R. Maxwell. 1977. Carotenoid diagenesis in a marine sediment. *Geochim. Cosmochim. Acta* 41:493-497.

Wehmiller, J. R.; P. E. Hare; and G. A Kujala. 1976. Amino acids in fossil corals: racemization (epimerization) reactions and their implications for diagenetic models and geochronological studies. *Geochim. Cosmochim. Acta* 40:763-776.

Welte, D. H. 1965. Relation between petroleum and source-rock. *Bull. Am. Assoc. Petrol. Geologists* 49:2246-2260.

Welte, D. H. 1972. Petroleum exploration and organic geochemistry. *Jour. Geochem. Explor.* 1:117-136.

Welte, D. H. 1974. Recent advances in organic geochemistry of humic substances and kerogen. A review. In *Advances in Organic Geochemistry 1973*. B. Tissot and F. Bienner eds. Paris: Editions Technip, pp. 3-13.

Yen. T. F. 1976. Structural aspects of organic components in oil shales. In *Oil Shale*, T. F. Yen and G. V. Chilingarian eds. Amsterdam: Elsevier Scientific, pp. 127-148.

INDEX

A

B

C

D

E

F

G

H

I

J

K

L

M

N

O

P

Q

R

S

T

U

V

W

X

Y

Z